Lecture Notes in Electrical Engineering

Volume 64

Subhas Chandra Mukhopadhyay · Henry Leung (Eds.)

Advances in Wireless Sensors and Sensor Networks

Springer

Subhas Chandra Mukhopadhyay
School of Engineering and Advanced Technology (SEAT)
Massey University (Turitea Campus)
Palmerston North
New Zealand
E-mail: S.C.Mukhopadhyay@massey.ac.nz

Henry Leung
Department of Electrical and Computer Engineering
2500 University Drive NW
University of Calgary
Calgary, Alberta T2N 1N4
Canada
E-mail: leungh@ucalgary.ca

ISBN 978-3-642-12706-9　　　　　　　e-ISBN 978-3-642-12707-6

DOI 10.1007/978-3-642-12707-6

Library of Congress Control Number: 2010925140

© 2010 Springer-Verlag Berlin Heidelberg

This work is subject to copyright. All rights are reserved, whether the whole or part of the material is concerned, specifically the rights of translation, reprinting, reuse of illustrations, recitation, broadcasting, reproduction on microfilm or in any other way, and storage in data banks. Duplication of this publication or parts thereof is permitted only under the provisions of the German Copyright Law of September 9, 1965, in its current version, and permission for use must always be obtained from Springer. Violations are liable to prosecution under the German Copyright Law.

The use of general descriptive names, registered names, trademarks, etc. in this publication does not imply, even in the absence of a specific statement, that such names are exempt from the relevant protective laws and regulations and therefore free for general use.

Typeset & Coverdesign: Scientific Publishing Services Pvt. Ltd., Chennai, India.

Printed on acid-free paper

9 8 7 6 5 4 3 2 1

springer.com

Guest Editorial

This Special Issue titled "Advances in Wireless Sensors and Sensor Networks" in the book series of "Lecture Notes in Electrical Engineering" contains the invited papers from renowned experts working in the field. A total of 15 chapters have described the advancement in the area of Wireless Sensors and Sensor Networks, Healthcare and diagnosis, Instrumentation architectures, Energy harvesting and scavenging, Signal processing, Wearable technologies, Measurement techniques, Design and prototyping in recent times.

In recent times wireless sensors and sensor networks have become a great interest to research, scientific and technological community. Though sensor networks have been in place for more than a few decades now, the wireless domain has opened up a whole new application space of sensors. Wireless sensors and sensor networks are different from traditional wireless networks as well computer networks and therefore pose more challenges to solve such as limited energy, restricted life time, etc.

This book intended to illustrate and to collect recent advances in wireless sensors and sensor networks, not as an encyclopedia but as clever support for scientists, students and researchers in other to stimulate exchange and discussions for further developments.

We do sincerely hope that the readers will find this special issue interesting and useful in their research as well as in practical engineering work in the area of biomedical sensing technology. We are very happy to be able to offer the readers such a diverse special issue, both in terms of its topical coverage and geographic representation.

Finally, we would like to whole-heartedly thank all the authors for their contribution to this special issue.

Subhas Chandra Mukhopadhyay, Guest Editor
School of Engineering and Advanced Technology (SEAT),
Massey University (Turitea Campus)
Palmerston North, New Zealand
S.C.Mukhopadhyay@massey.ac.nz

Henry Leung, Guest Editor
Department of Electrical and Computer Engineering
University of Calgary
2500 University Drive NW,
Calgary, Alberta T2N 1N4
leungh@ucalgary.ca

Dr. Subhas Chandra Mukhopadhyay graduated from the Department of Electrical Engineering, Jadavpur University, Calcutta, India in 1987 with a Gold medal and received the Master of Electrical Engineering degree from Indian Institute of Science, Bangalore, India in 1989. He obtained the PhD (Eng.) degree from Jadavpur University, India in 1994 and Doctor of Engineering degree from Kanazawa University, Japan in 2000.

He is working currently as an Associate professor with the School of Engineering and Advanced Technology, Massey University, Palmerston North, New Zealand. His fields of interest include Smart Sensors and Sensing Technology, Electromagnetics, control, electrical machines and numerical field calculation etc.

He has authored/co-authored over 200 papers in different international journals and conferences, edited eight conference proceedings. He has also edited six special issues of international journals and five books with Springer-Verlag as guest editor.

He is a Fellow of IET (UK), a senior member of IEEE (USA), an associate editor of IEEE Sensors journal and IEEE Transactions on Instrumentation and Measurements. He is in the editorial board of e-Journal on Non-Destructive Testing, Sensors and Transducers, Transactions on Systems, Signals and Devices (TSSD), Journal on the Patents on Electrical Engineering, Journal of Sensors. He is in the technical programme committee of IEEE Sensors conference, IEEE IMTC conference and IEEE DELTA conference. He was the Technical Programme Chair of ICARA 2004, ICARA 2006 and ICARA 2009. He was the General chair/co-chair of ICST 2005, ICST 2007, IEEE ROSE 2007, IEEE EPSA 2008, ICST 2008 and IEEE Sensors 2008. He has organized the IEEE Sensors conference 2009 at Christchurch, New Zealand as General Chair during October 25–28, 2009.

Dr. Henry Leung received his Ph.D. degree in electrical and computer engineering from the McMaster University, Canada. He is now a professor of the Department of Electrical and Computer Engineering, University of Calgary, Canada. Before that he was with the Defence Research Establishment Ottawa, Canada, where he was involved in the design of automated systems for air and maritime multi-sensor surveillance. His research interests include chaos, computational intelligence data mining, information fusion, nonlinear signal processing, multi-media, sensor networks and wireless communications.

Table of Contents

Security for Wireless Sensor Networks – Configuration Aid 1
   Thomas Newe, Victor Cionca, and David Boyle

Low Power Wireless Buoy Platform for Environmental
Monitoring 25
   Arne Sieber, Juergen Markert, Christian Woegerer,
   Michele Cocco, and Matthias F. Wagner

A Detail Performance Evaluation of the Novel Mechanisms
Ensuring Maximum Connectivity and Data Transmission between
Nodes, Based on the Heuristics Under 5-Color Clustered Response
Approach 43
   Mahdi Nasrullah Al-Ameen

Inventory Management in the Packaged Gas Industry Using
Wireless Sensor Networks 75
   A. Mason, A. Shaw, and A.I. Al-Shamma'a

An EM-IMM Method for Simultaneous Registration and Fusion of
Multiple Radars and ESM Sensors 101
   Dongliang Huang and Henry Leung

Locatable, Sensor-Enabled Multistandard RFID Tags 125
   D. Brenk, J. Essel, J. Heidrich, and R. Weigel

Optimal Sensor Network Configuration Based on Control Theory ... 151
   Takashi Takeda and Toru Namerikawa

Optimal Local Map Registration for Wireless Sensor Network
Localization Problems 177
   Yifeng Zhou and Louise Lamont

Wireless Sensor Network: Application to Vehicular Traffic 199
   Jatuporn Chinrungrueng, Saowaluck Kaewkamnerd,
   Ronachai Pongthornseri, Songphon Dumnin,
   Udomporn Sunantachaikul, Somphong Kittipiyakul,
   Supat Samphanyuth, Apichart Intarapanich,
   Sarot Charoenkul, and Phakphoom Boonyanant

Thermal Energy Harvesting for Wireless Sensor Nodes with Case
Studies 221
   C. Knight and J. Davidson

IEEE 1451.5 Standard-Based Wireless Sensor Networks 243
    Eugene Y. Song and Kang B. Lee

Fuzzy Based Optimized Routing Protocol for Wireless Sensor
Networks 273
    P. Manjunatha, A.K. Verma, and A. Srividya

Energy Aware Sensor Group Scheduling to Minimise Estimated
Error from Noisy Sensor Measurements 283
    Siddeswara Mayura Guru and Suhinthan Maheswararajah

Smart Home for Elderly Using Optimized Number of Wireless
Sensors 307
    A. Gaddam, S.C. Mukhopadhyay, and G. Sen Gupta

Estimation of Packet Error Rate at Wireless Link of VANET 329
    Hao Jiang, Yang Yang, Jun Xu, and Lin Wang

Author Index 361

Security for Wireless Sensor Networks – Configuration Aid

Thomas Newe, Victor Cionca, and David Boyle

University of Limerick,
Limerick, Rep. of Ireland
`thomas.newe@ul.ie`

Abstract. The range of application scenarios for which WSN technology is suitable implies a number of responsibilities. One of the most important responsibilities for researchers and designers has been the establishment of a means to secure both the network and the information sensed and disseminated within. In many cases the security of the application is vital to its successful deployment and usefulness. In order to understand what is necessary to properly secure a WSN implementation, all of the known threats to such a network must be identified and addressed. The stability and effectiveness of a WSN can be adversely affected through a number of attacks other than merely eavesdropping. There are also a number of various points at which an adversary can attempt to attack a network; such as at varying layers in the protocol stack. It is therefore necessary to consider all of the various attacks to which WSN's are susceptible. These attacks can be broadly classified as attacks against the privacy of network data, denial-of-service (DOS), impersonation or replication attacks, routing attacks and physical attacks. The prevention of such attacks depends largely on the correct implementation/configuration of security protocols within the network itself. This configuration is a non-trivial task which can involve intensive knowledge of both security and WSN's, which the user may not have. What the user does know however is how the network is used in the application and what data it carries. The decisions made to configure the necessary security can be constrained by the aforementioned user's knowledge of the application space. Therefore by identifying a set of controlling parameters it should be possible to automate the security configuration process through the use of a configuration tool.

1 Introduction

The work reported here describes a GUI tool that recommends a suite of security protocols able to give the best and most efficient protection against attacks for the application described by the user's parameters. The set of protocols that are used cover the most important security issues such as: key distribution, key establishment, data security and authentication. Through careful analysis of the sensor network design space, a set of parameters are defined that are needed to aid the selection of the protocols. Some of these parameters are defined by the user, and others are derived from the user's values, but all the decisions needed to identify the appropriate security suite are taken based on facts that should be known or are available to the user.

Currently the configuration tool can only be used as a configuration helper [1] as it only displays a list of protocols suitable for a given application. The next phase of this work is to expand the tool so that it assists throughout the entire network deployment operation: once the appropriate protocols are determined they can be connected to the application and installed on the motes. This installation on the motes directly will further remove the user from the complexity of implementing a secure wireless sensor network.

2   Motivation

One of the challenges identified by the International Telecommunication Union (ITU) in their report on *"The Internet of Things"* [2] is that of privacy and security. When we start being surrounded by these smart objects that are moving around gathering information concerning our lives, behaviour or habits there will be great concern regarding the security of that information. There are currently many protocols dealing with all the aspects of security. What is really needed is a way to adapt these protocols to the user's needs, to make them easily configurable, as some people might ignore privacy and some may be obsessed with it.

The European Parliament defines [3] network and information security as the ability of a network or an information system to resist, at a given level of confidence, accidental events or unlawful or malicious actions that compromise the availability, authenticity, integrity and confidentiality of stored or transmitted data and the related services offered by or accessible via these networks and systems. Most books and papers discussing the subject will give a definition along similar lines, but what is needed here is how to define security using the network and application parameters.

> What is the set of parameters needed to give a complete definition?
> How do we quantify different levels of security (low or high)?

The answers to these questions are needed in order to design and implement a configuration tool able to recommend (as a first step) a set of security protocols to use for a given wireless sensor network application. The tool presented here addresses these issues and offers a user the ability to select appropriate security protocols for a specific application without fully understanding the complex security issues involved.

3   Threats to Wireless Sensor Networks

In many cases the security of a WSN application is vital to its successful deployment and usefulness. In order to understand what is necessary to properly secure a

WSN implementation, all of the known threats to such a network must be identified and addressed. Otherwise, the successful operation of the WSN is at risk, i.e. in a military application if the enemy learns the contents of the messages sent, the application becomes redundant. Also, from a medical perspective; physiological patient data is legally required to be kept confidential and therefore, must be secured. The main threat in these cases, eavesdropping, (figure 1) is conquered through the encryption of the sensed and disseminated data. This is one of the foundations of secure communications, together with node authentication with respect to WSN's.

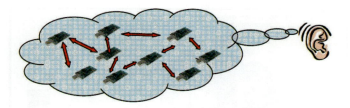

Fig. 1. Eavesdropping on a Wireless Sensor Network

The stability and effectiveness of a WSN can be adversely affected through a number of attacks other than merely eavesdropping. There are also a number of various points at which an adversary can attempt to attack a network; such as at varying layers in the protocol stack. It is therefore necessary to consider all of the various attacks to which WSN's are susceptible. These attacks can be broadly classified as attacks against the privacy of network data, denial-of-service (DOS), impersonation or replication attacks, routing attacks and physical attacks.

Table 1 presents an overview of the various attacks that can be applied to WSN's at various levels of the protocol stack.

Table 1. Attacks to WSN's, Applied to the Networking Stack

Layer	Attack
Application Layer	Overwhelming sensor stimuli DOS – Path Based Attack Physical Capture Attack – Reprogramming Monitoring, Eavesdropping – cryptanalysis
Transport Layer	DOS – Flooding, De-synchronization
Network Layer	DOS – Neglect & Greed, Homing, Misdirection (Spoofing), Black Holes, Flooding Sybil Worm Hole Attack
Data-link/MAC Layer	DOS – Collision, Exhaustion, Unfairness Interrogation Sybil – Data aggregation, Voting
Physical Layer	DOS – Jamming, Tampering Sybil

DOS (Denial-Of-Service) attacks are the most common threat to WSN security. They can be seen in many forms, and are defined as any attack that can undermine the network's capacity to perform expected functions. Specific not only to WSN's but to all wireless networks, "jamming" the channel with an interrupting signal is an effective attack, as are flooding or collision attacks. At the physical layer, jamming is a popular DOS attack. It can either be intermittent or constant jamming, both of which will have a detrimental impact upon the network. The number of nodes required to perform this attack may be as small as one (one malicious node may be enough to disrupt the intended operation; see Figure 2).

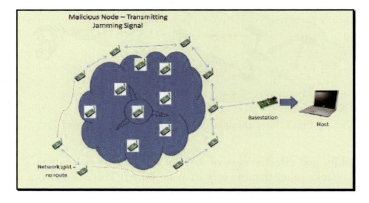

Fig. 2. Denial-of-Service Attack on a WSN

At the data link layer, collision attacks are a common DOS attack, whereby an adversary will intentionally violate the communications protocol; this may only mean changing a small part of the packet which would lead to an error in the checksum in an attempt to generate collisions. This would require the retransmission of any packet affected by the collision. This contributes to unnecessary consumption of the network nodes' resources. Misdirection is another attack at the network layer; whereby a malicious node could route a message to an incorrect node. In a similar fashion, a node may refuse to route messages at all, possibly disabling communication with some parts of the network. The transport layer is susceptible to DOS in the form of flooding. This attack is as simple as sending multiple connection requests to a network member node. Because resources must be allocated to handle the request, these resources will eventually diminish and become exhausted, rendering the node useless. De-synchronisation is another attack at the transport layer. A pair of nodes may be forced into the synchronisation recovery protocol by disrupting some of the packets being transmitted between these two nodes and maintaining correct timing [4]. This also leads to unnecessary resource consumption and exhaustion. The application layer is susceptible to a path based DOS attack, whereby the attacker may insert spurious or replayed packets into the network. As the packet is forwarded to the destination, energy and bandwidth are consumed by forwarding nodes. This attack can starve the network of authentic data transmission as resources along the path to the basestation are consumed.

The Sybil attack can be described as that of a malicious node taking on multiple identities. This node can then launch a number of attacks. This attack is most effective at the higher layers of the communication protocol [5]. At the physical layer, the Sybil attack is performed by compromising a legitimate node or by fabricating a new one. Identities are therefore acquired either through theft or the fabrication of new ones. The malicious node can then behave as if it was a number of nodes, participating at different points in the network. At the data link layer, the Sybil attack has two variations; negative reinforcement and stuffing of the ballot box. Data aggregation is employed in WSN's to reduce power consumption. If a node is continually providing large amounts of inaccurate data, then an aggregated data packet returned by the network would be corrupted. Similarly, a voting scheme may be corrupted via a Sybil attack, whereby the malicious node's various identities could mean that it could stuff the ballot box in its own favour. At the network layer, a Sybil attack could be used to compromise most multi-path routing schemes. If the information provided by the malicious node, and its multiple identities, is taken into account in the routing tables of nodes, then the decisions taken by the nodes to route messages to other network members would be at risk of failure.

Traffic analysis is a combination of monitoring and eavesdropping. Through monitoring the number of packets sent or received by specific nodes, it could lead to inferred information as to the role of that node in the network. The adversarial node could then behave like a normal network node, attracting packets and re-routing them incorrectly, for example. These packets could be sent to nodes performing analysis of the network data etc. [6].

There are numerous different kinds of "Hole" attacks. These include worm-hole attacks, black-hole attacks and sink-hole attacks. A worm-hole attack can disrupt routing by convincing nodes that are usually many hops away from the base station that they are only one or two hops away. There are no definitive ways to ensure security against a worm-hole; notwithstanding, some attempts have been made to better understand the problem [7]. In a wormhole attack, malicious nodes may re-route messages received in one particular part of a sensor network over a low latency link and replay them to a different part of the network. Resulting from the nature of wireless transmission, it is possible for an adversary to create a wormhole for packets that are addressed to other nodes (since it can overhear other wireless transmissions) and tunnel them to a colluding adversary in another location. This tunnel can be created using a variety of means, including: an out of band hidden channel (possibly a wired link), packet encapsulation or high powered transmission. The tunnel creates the illusion that the endpoints are very close to one another by ensuring that packets appear to arrive sooner, and via less hops, compared with packets sent over regular routes. This permits the adversary to undermine the correct operation of the routing protocol by controlling various routes in the network [8]. Black-hole and sink-hole attacks are similar in that they advertise zero routes to the base station via a specific malicious node. Some routing protocols will then elect to route a number of packets via this node (as it appears to be the lowest cost, and this is usually sought), and a large amount of network data could be lost. The neighbours of the malicious node select this route and compete for bandwidth, and through this process, energy is wasted and resources consumed. When a hole or partition is created in the network, it is considered to be a black hole.

Many of the applications of WSN's require that the sensor nodes are deployed in harsh, even hostile, environmental conditions. As a result of their small size, unattended operation and distributed sensor field deployments, they are susceptible to a number of attacks against their physical integrity. If a network node is destroyed, there may be limited damage done to the integrity of the network itself. If, however, the node can be physically tampered with, it may be possible for an adversary to extract sensitive information from the device. Cryptographic secrets may be compromised, the circuitry may be altered or the node may be reprogrammed. It has been proven that the Mica2 motes can be compromised in less than one minute [9], and a warning is provided by the manufacturer. Tamper resistant hardware protection is a plausible solution, given the intended deployment environment, coupled with the security level required by an application. Such attacks may be preventable through the incorporation of encrypted micro-controller cores and encrypted program memory into WSN devices. This translates to an increase in the cost of hardware, which may be prohibitive to potential application designers, and could be considered counterintuitive with respect to the development of WSN's using low-cost devices. Another method of physically attacking a WSN is to overload a node with sensor stimuli (additional heat, motion etc.). This would cause the node to propagate inaccurate information throughout the network; rendering the deployed application near useless.

There are a number of methods available to thwart the attacks mentioned here. These primarily involve the use of cryptographic protocols to ensure the confidentiality and authenticity of sensor data and network participants. Although these are the primary defences against attack, there exist other mechanisms that are useful for hindering some of the attacks, such as: Tuning of sensors to defend against overwhelming sensor stimuli and anti-replay and redundancy mechanisms to ensure defence against replay attacks through the use of a freshness counter. These mechanisms are outside the scope of the work presented here which concentrates on the use of basic cryptographic/security primitives for node and network protection.

4 Security Primitives

There are many facets and goals to security, with some being dependent on others. Although cryptographic operations are generally considered computationally secure these days, they need to be properly used and combined in order for the entire system to be considered secure. Often communication protocols that simply rely on strong ciphers exhibit weaknesses which are not obvious, but sooner or later will be discovered by cryptanalysts. The path to securing a system is portrayed in figure 3. Each step/primitive of the process depicted in figure 3 is discussed below and section 5 looks at some protocols suitable for use with wireless sensor networks, for each of these primitives.

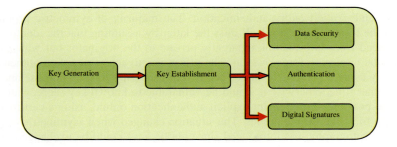

Fig. 3. Primitives needed to secure a system

Key Generation: To achieve secrecy of communication cryptographic keys are required, be they symmetric or asymmetric. Although cryptography also relies on algorithms and it is still common for new algorithms to be kept secret, one of the principles of cryptography, as stated by Auguste Kerckhoffs in his "La Cryptographie Militaire" [10] is that the strength should not lie in the algorithm but in the key. As history has taught, algorithms can always fall into the enemy's hands. One of the requirements of secure encryption is that the key be totally random: if we have a key of n bits any key must have the probability of being used of $1/2^n$. There should not be any connections between keys or between the individual bits in a key; otherwise an attacker may be able to reduce the size of the keyspace, thus making possible a brute force attack. So when choosing a key it is best to generate it using a random source, either a true random physical source or a software pseudo-random number generator which can be seeded with a unique value, like local time expressed with millisecond or even microsecond precision.

Key Establishment (KE): Before the second half of the 20th century all secured communication was performed through symmetric means, meaning the same secret being shared by the originator of the communication and the receiver. This raises the problem of transmitting the secret (*key distribution*) in a secure way and at first it was done using trusted couriers or code-books, supplying keys for up to several months, assuming the use of one key per day. In [11], Shannon explained that the probability of identifying a key used to encrypt a set of messages, when the size of the set increases, only stays constant in theoretical secrecy systems. In practical, non-perfect, systems this probability will increase and ultimately, the cryptanalyst is able to ascertain the key. As communication technology progressed, the number of messages transmitted per day increased rapidly and it became evident that a fast, secure and reliable mechanism to exchange keys was needed. As far as general modern cryptography is concerned, the battle between code makers and code breakers has been won by the former, as the majority of attacks still possible will somehow involve brute force. Key establishment protocols are still in a somewhat precarious position, mostly because they involve communication: there are many attacks against them and it is difficult to create a protocol that is completely secure. The primary concern of key establishment is to allow access to the key only to the right parties and to make sure that the parties are who they say they are. These are two forms of entity authentication, namely implicit and explicit entity authentication. The majority of symmetric protocols exhibit the former, but many will miss the latter by assuming that the identity can be deduced from information apparently private to the originator but which is in fact public or easy to forge or steal. The safest way to achieve explicit authentication is by using a previously shared secret, but used only for this purpose and not at any other time, otherwise it would be vulnerable to known key secrecy attacks. It is also possible to implement explicit authentication online, using a certification authority to vouch for the parties. The situation changed when asymmetric or public-key cryptography was introduced since public keys could be exchanged in the open and there is was no need for a secure exchange to take place.

A new key establishment technique was introduced by Diffie and Hellman [12], where the parties obtain the same secret not by explicitly sending it but by computing it using their own private and public key pair plus the public key of the peer. This mechanism is known as *key agreement* and it usually relies on the difficulty of the discrete logarithm problem or large number factoring. The most renowned key agreement protocol (asymmetric) is the Diffie and Hellman protocol. There are several protocols that perform key agreement using only symmetric operations, but they require priori secrets, whereas public key/asymmetric systems generally don't. In many modern systems where power consumption is of concern a mix of asymmetric and symmetric systems are employed in order to reduce computational overheads thereby reducing power consumption (discussed later). The two possible methods of key establishment (*agreement* or *distribution*) for both symmetric and asymmetric cryptographic systems are summarised in figure 4 below. This figure indicates that for key agreement the key is established through some computation while for key distribution some communications means is necessary.

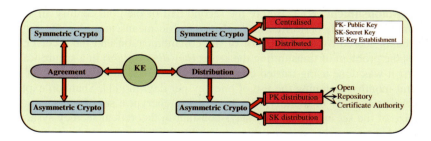

Fig. 4. Methods used for Key Establishment (KE)

In addition to key distribution there exists another key establishment method called key predistribution. Key predistribution can happen two ways: either at network deployment, by installing the key in the networked device, or later on, by distributing the key using a secure channel. In [13] Stajano and Anderson propose several innovative key predistribution techniques, involving physical contact or close proximity between the keying and keyed device. The most common key predistribution techniques in use are: single network-wide (master key), server shared, pairwise, lambda secure and probabilistic.

Data Security: Data security generally translates as data confidentiality and can be achieved using strong encryption algorithms. Depending on the type of cipher used (symmetric or asymmetric) there may or may not be the need to establish keys between the communicating parties. As many popular encryption algorithms are considered safe (3DES, AES, RSA, RC5 etc.) the security provided by the cipher relies mostly on the key and how it is managed or communicated. Weak keys or weak establishment methods can create favourable conditions for an attacker and a skilled cryptanalyst can extract the original plaintext information from the ciphertext.

Authentication: On a list of priorities for a secure system, authentication ranks highly, since without authentication there can be no secrecy. Authentication accounts for many different things: data integrity which is important even when there are no adversaries; entity authentication or identification which helps prove that a peer in communication is who he says he is; data authentication to bind data to its origin; data freshness and timeliness, supposed to preserve the order of messages and prevent the same message being retransmitted by an adversary. For key establishment, authentication is needed in the first step, where the communicating parties will authenticate each other before commencing a session. Authentication, as previously mentioned, can be used to thwart many of the security threats to wireless sensor networks.

5 Security Protocols

Wireless sensor network literature is abundant with security protocols that propose a wide variety of approaches to solving the problems associated with wireless networks. However only a small number are actually implemented and even fewer are commonly used. A brief description of protocols covering the major security primitives of key predistribution, key establishment, data security and authentication is presented here.

Key predistribution: Due to the large sizes considered for sensor networks and the limitation in mote memory, pairwise schemes where nodes share a key with each other node, seems inappropriate. To this end probabilistic, random or λ-secure key predistribution schemes were proposed. The paper that first proposed probabilistic predistribution of pairwise keys was by Eschenauer and Gligor [14]: keys are distributed to nodes from an initial pool, so that any two nodes can share a key with a certain probability. Nodes that don't share keys will have to initiate a key establishment process through a secure path. The main problem with probabilistic predistribution is that the attacker may have all the keys in the network by compromising a small set of nodes. The protocols that followed Eschenauer and Gligor's approach were attempts to improve the slope of probability of network compromise as a function of number of nodes compromised. Chan et al proposed [15] that in the initial scheme for two nodes to be securely connected they should share q keys instead of two, which increases the attacker's effort. This is known as q composite predistribution. Their paper also proposes multipath key update to improve key establishment post predistribution, and random predistribution. Random predistribution is a more resilient scheme than pairwise where nodes are randomly paired together and one key will be shared by only a pair of nodes. This method is as resilient as pairwise keying, the difference being that nodes are connected at random. Another class of key predistribution protocols are the ones based on Blom's [16] scheme which achieves the same security as pairwise keying as long as no more than λ nodes are compromised. Du et al [17] combined the random scheme with Blom's scheme, in that an initial pool of m key spaces is defined and every node is assigned $t < m$ spaces at random. Nodes sharing a key space can then determine their shared key using Blom's scheme. The motivation is that while Blom's scheme guarantees any two nodes will be able to share a key, this is not always needed; therefore nodes will be randomly paired. A similar approach is presented by Liu and Ning [18] only they use bivariate polynomials instead of MDS (Maximum Distance Separated [16]) codes, still achieving the same results. Their protocol was implemented as TinyKeyMan and is available for TinyOS. While these protocols present interesting approaches to key distribution the fact that they rely on probabilities and do not guarantee 100% connectivity could be a problem for sensor networks. There are also protocols that implement traditional techniques, such as LEAP [19], which pre-distributes pairwise keys for a node to share with neighbours, and another key that it will share with the base station. These keys are used later for cluster and group key establishment. Chan and Perrig propose PIKE [20] where nodes are represented as a two or three dimensional matrix and nodes on the same row or column will share a key. Two nodes that do not share a key will need to find an intermediate node, one that shares keys with both, to exchange a pairwise key.

Key establishment: Research into key management for sensor networks has focused largely on key predistribution due to its inherent difficulties. There are however protocols for key establishment, as in distribution or agreement, but the majority of them are approached using public key cryptography, with digital signatures and certificates. Malan's EccM [21], implements Diffie-Hellman key agreement using elliptic curve cryptography. In [22] Abi-Char et al propose an authenticated key agreement protocol between a server and a client where a challenge is sent by the server and the client identifies itself by digitally signing the challenge response. The key is derived from the hashed value of the user's response and the user's private key. Since public key cryptography is time and power consuming for motes, there are some approaches which try to shift any computational burden to more powerful nodes such as basestations. They are called hybrid schemes and they mix symmetric and asymmetric cryptography, to allow a session key to be generated or distributed between parties. The best approaches are by Huang et al in [23] and Wong and Chan in [24]. The latter requires the client node to perform only one public key encryption, leaving the rest for the more powerful server, and it is resilient against repudiation and interleaving session attacks. There is one special approach worth mentioning, by Anderson et al [25] who, by analysing the attacker model, argue that in most cases there is little probability for an attacker to be eavesdropping at the very moment of key deployment, since the exact timing would be unknown. Therefore the authors propose an open key distribution which can later be updated through multiple paths, which would make it harder for attackers. Their approach is different because the rest of the protocols usually start with the assumptions that the attacker is all powerful and can monitor all the communication in the network at any time. This model is possibly too strict as it can be said that underestimating an adversary can result in a security breech.

Data security: Data security mostly deals with data confidentiality, so the choices for sensor networks are not that vast. The SNEP protocol from SPINS [26] and TinySec [27] are two of the earliest approaches to data security in WSN's. SNEP uses RC5 in counter mode (CTR – figure 5), with the counter not transmitted but stored as an internal state. The counter is also used to provide weak data freshness, which prevents replay attacks; strong freshness can be achieved if messages are sent with a challenge. If the counters go out of sync, a synchronisation protocol can be performed. The type of keys used in the protocol is not clearly stated in the paper but they seem to be node-base station, so the security can be specified as end-to-end.

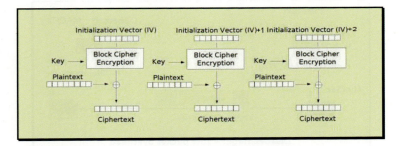

Fig. 5. Counter Mode Encryption (CTR)

TinySec uses Skipjack [28] in CBC (Cipher Block Chaining-see figure 6) mode and the main focus is integration with the operating system to achieve minimum overhead in time and energy consumption. The structure of the TinyOS packet is modified to remove some unneeded fields like CRC, which is replaced by the message authentication code. A good design choice of TinySec made is that when sending a packet, the cryptographic operations of authentication and encryption will be performed while the packet preamble is being sent. Therefore, there is no time overhead when sending a message.

A more recent approach is MiniSec by Luk et al [29] which uses Skipjack in offset codebook (OCB) mode (figure 7). The advantage of using OCB is that authentication and encryption can be performed at the same time, as opposed to CBC, which needs CBC-MAC for authentication as CBC alone only produces ciphertext. MiniSec also uses a counter to provide data freshness but while the counter is 64 bits, only the lower three bits are included in the packet. This allows 2^3 packets to be lost without the counters needing to be synchronised. The focus of

MiniSec is on replay protection, especially in broadcast communication. MiniSec achieves good energy consumption, with the authors claiming a reduction to a third of the values obtained by TinySec. One disadvantage however is the increased packed size needed.

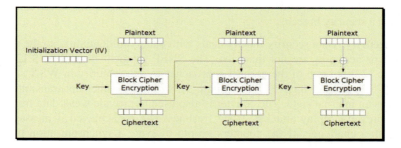

Fig. 6. Cipher Block Chaining (CBC)

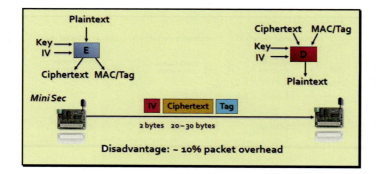

Fig. 7. MiniSec operation

Authentication: The same protocols that provide data security for sensor networks also provide authentication. TinySec and MiniSec use CBC-MAC and OCB respectively, with an overhead for the MAC of five and three bytes respectively. The SPINS protocol suite uses CBC-MAC for authentication within SNEP, with an overhead of eight bytes but also proposes μTESLA [30] for broadcast authentication. μTESLA uses symmetric key authentication with delayed key disclosure. The keys are part of a one way key chain, with the last key being distributed to all nodes in advance so that they are able to check the authenticity of received keys.

6 Configuring Security

Technology that is unusable is either art or dead technology. There are many great technical ideas in the past that went unnoticed just because they were badly designed. Sensor networks have lots of potential but they need to be more user friendly. A lot of research work is still being done on WSN, not only for protocol design but also to define the behaviour, functionality and general concept of WSN's. Their usage is still mostly limited to experienced personnel for research projects. Ultimately the complexity of configuring a WSN depends on how the network will be used and for what application. Some users will settle for a "full package" comprising a deployed network with protocol suite plus data analysis applications for static or mobile clients etc. Other users may want more freedom to configure the network and use it for different application types. More so, the network configuration may change even with non-complex operations, such as adding or removing nodes, or modifying the application details. In all cases the network has to be as accessible and as comprehensible to the user as possible. The interface to the network should be through tools that display and monitor the network status, that gather data and present it in an understandable format. The configuring of the network must take place through easy to use and easy to follow GUI based tools. For all this the following requirements can be stated:

- Assume the user has very poor knowledge about configuring WSN's.
- All the internal tool decisions should be made based on parameters that the user understands and is able to modify.
- Parameter values should be objective and known to the user.

Security is a fundamental part of any technology that involves information and/or communication. If sensor networks are to be used by more than academia they need to be secured so that the attacks presented previously can be defended against. The main issue is that most users of technology, WSN technology included, will not know what cipher they should use, how strong the cipher should be (besides "the strongest"), how keys should be distributed, etc. Even within the research community itself security comes last when wireless sensor network technology is developed. This is evident by the fact that the TinyOS operating system has only one security protocol included, TinySec. Tinysec however is not compatible with the newer generations of motes and also only provides data security as there is no included support for key establishment and other security primitives. However, given this there are many other protocols designed to provide these primitives. The main issue is how to combine these protocols and make them work together so that all the security issues exposed in the previous section are covered. Every application can have specific security requirements; for example, untrusted individuals or enemies are unlikely to steal nodes from a home WSN, but they are expected to do so in a military WSN deployed on enemy territory. These requirements must be met with the choice of ciphers, key management and so forth otherwise the end result might be an unsecured network or an overly secure one that unnecessarily consumes power and has a short life.

One can now see that configuring security on a WSN is a non-trivial task often involving intensive knowledge of security requirements, the application area and the WSN configuration. Generally a user will not have knowledge of all these areas and may see knowledge of the application itself to be enough. For a secure application this knowledge is not sufficient. However the decisions made to configure security can be constrained by the application space: it should be possible to determine a set of parameters that controls both the application and its security suite; these parameters would allow automatic security configuration. In the next sections a tool that recommends a suite of security protocols able to give the best and most efficient protection against attacks for the application domain described by the user's parameters is outlined. All the decisions needed to identify the appropriate security suite(s) are taken based on facts that will be known or available to a user.

6.1 Protocols Supported

With the functional peculiarities of WSN's comes a new set of security issues that have previously been discussed, physical attacks, jamming attacks, network holes etc. In this work the traditional primitives of: key predistribution, key establishment, data security and authentication are considered, as it can be argued that the basic primitives can prevent the majority of the aforementioned security attacks, if those primitives are correctly used and configured. Routing attacks can be avoided with link layer security; jamming attacks can be thwarted with frequency hopping while physical attacks are still an open problem with the best solution so far being hardware self-destruction. It can be argued that if network formation and control can be assured in relation to sending/forwarding of data or querying of the network then the network can be considered safe against the majority of attacks. To this end the following protocols (summarised in table 2) have been selected for use with the configuration tool.

For key predistribution only general schemes are used as they theoretically involve no message exchanges on the wireless network itself so there are no protocols to consider. For key establishment, classic protocols like Otway-Rees, Station-to-Station and MQV were used, in addition to others [31]. The vulnerabilities and attacks against these protocols have already been analysed and are well documented in the research literature, therefore their use and the security they offer is well documented [31]. The following were selected for the configuration tool:

- *Centralised key transport with symmetric ciphers*: Centralised Needham – Schroeder (N-H KDS), Otway Rees (O-R), Wide Mouthed Frog (WMF);
- *Distributed key transport with symmetric ciphers*: Key update (KU), Shamir's protocol (Shamir);
- *Key transport with asymmetric ciphers*: Public Needham-Schroeder (N-H PK), X509;
- *Key agreement with symmetric ciphers*: ZigBee Symmetric Key Key Establishment (SKKE);
- *Key agreement with asymmetric ciphers*: Diffie-Hellman (D-H), Matsumoto-Takashima-Imai (MTI), Menezes-Qu-Vanstone (MQV), Station-to-Station (STS).

A comparison of the protocols is presented in table 2. The table indicates the computational requirements in number of symmetric (sym) and asymmetric (asymm) operations, the communication is specified in number of messages (msg) per node, whether the nodes must be synchronised or not (if they use timestamps) and the attacks the protocols are vulnerable to.

The following key is used to denote attacks [32] in table 2:

TF - Type Flaw, **KF** - Key Freshness, **Rep** - Replay, **Imp** - Impersonation, **MiM** - Man in the Middle, **FImp** - Forward Impersonation, **MS** - Multiple Session, **SFImp** - Impersonation by Signature Forgery, **UKS** - Unknown Key Share, **SG** - Small Group, **FS** - Forward Secrecy, **KC** – no Key Confirmation.

Table 2. Comparison of key establishment protocols

Classification	Protocol	Computation	Comm	Synchro	Attacks
Centralised Key Transport with Symmetric Ciphers	N-H KDS	3 sym ops	5 msg	Yes	KF, Rep
	O-R	2 sym ops	4 msg	No	KC, TF
	WMF	1 sym op	2 msg	Yes	KC, MS
Distributed Key Transport with Symmetric Ciphers	KU	1 sym op	1 msg	Yes	FS, KC
	Shamir	2 mod exp	3 msg	No	Imp, KC
Key Transport with Asymmetric Ciphers	N-H PK	3 asymm ops	3 msg	No	KF, FImp
	X509	2 asymm + 1sign + 1 verify	2 msg	Yes	SFImp
Key Agreement with Symmetric Ciphers	SKKE	2 MAC	3 msg	No	FS, UKS
Key Agreement with Asymmetric Ciphers	D-H	1 asymm op	2 msg	No	MiM, Imp, FS
	MTI	3 asymm ops	2 msg	No	UKS, SG
	MQV	1 asymm op	2 msg	No	UKS
	STS	4 asymm + 2 symm ops	3 msg	No	UKS

For data security and authentication WSN specific protocols are used, where the decisions made in the design process reflect the hardware constraints of sensor nodes. These include TinySec and MiniSec which are already implemented, the ZigBee security stack [33] because it is a standard and implementations for TinyOS are currently being delivered, and SPINS since it is the only protocol providing broadcast authentication with μTESLA.

6.2 Application Based Parameters

It has been stated that sensor networks must be configured to fit an application. To this end a set of parameters that are determined by the application space and in turn determine which security protocols are to be used to secure that application need to be found. Generally a user knows the details of the WSN application and therefore can loosely specify the security requirements of the application: military and hospital applications require "high security", whereas environment monitoring can settle with "lower security". It is difficult to transpose this kind of subjective values, numerical or string scales, onto technical matters. After analysing the application space and security requirements for sensor networks the following list of parameters will be used in the configuration tool: As mentioned before, the user-controlled parameters are not sufficient to determine the protocols: the internal parameters need to be derived from them. To this extent, direct parameters determine protocols, while indirect parameters determine the values of other parameters.

These parameters are controlled by the user:

- Application type
- Number of nodes
- Network topology
- Network dynamics
- Communication pattern
- Performance
- Timing constraints
- Available power
- Communication frequency (sampling rate)
- Network lifetime

The following list is of internal or derived parameters:

- Power consumption per packet
- Environment security level
- Data security level required
- Ciphers
- Semantic security
- Attacks

Now all of the parameters will be discussed and the role of each one defined.

Application type is an indirect parameter and it determines the data and environment security level. The applications considered were *military*, which requires high data security and presumes a hostile environment. *Medical*, has high data security requirement and operates in a public environment. *Environment* monitoring has a low data security requirement and public environment. Finally *home* applications have a low security requirement and operate in trusted environments. It is possible to extend the application type to determine the communication frequency, network lifetime and the possible attacks, however for the moment this parameter is just used to enable the security level to be specified as low, medium or high to better suit the user.

The **number of nodes** in the network is a direct parameter. It is influential in the selection of key predistribution protocols, since it determines the number of keys that must be stored on a node, and this is limited by the available memory space. Therefore, pairwise predistribution is suitable for smaller networks; using a master key seems to work for larger networks, but the larger the network the higher the possibility of node compromise which would lead to total security failure. So master key should also be used for smaller networks. Server shared keys can be used independently of network size, same as probabilistic schemes.

Network topology is a direct parameter, which could be automatically selected based on application type and network size. Here however, it is provided by the user and it determines the selection of key predistribution schemes and key establishment protocols. Networks with centralised topology make use of a server shared key and of centralised key distribution protocols using symmetric cryptography. Clustered topologies can use server shared keys but they can also use group or master keys within the clusters themselves; the same key establishment protocols are required. Mesh networks demand the use of pairwise or probabilistic key predistribution but a master key can also be used. Since communication is not strictly vertical or horizontal we can have distributed key establishment or key agreement, if the device has sufficient computation power.

Network dynamics state if a network is static or dynamic because nodes are mobile or they can be added or removed from the network. It is a direct parameter supplied by the user and again determines key predistribution: dynamic networks cannot use pairwise keys as for every change the entire network must be updated.

The **communication pattern** shows how data is moved around the network: in clusters, for data aggregation nodes can communicate randomly to compare measurements; outside clusters the data can move vertically, or hierarchically, since reports are sent to the base station which can send back further commands. It is a direct parameter and determines key management: random communications uses pairwise, probabilistic or master keys, otherwise the server would be a bottleneck. Hierarchical communication patterns are suitable for master and server shared keys and centralised key distribution.

Performance together with **timing constraints** define the ciphers (cryptographic primitives) used. A good review of the computational abilities of different motes is presented in TinyECC [34] for the execution times of public cryptography algorithms, and the results show that there are roughly two classes of performance: those that can perform the algorithm in acceptable time (under one second) (called high end algorithms) and those that can't (called low end algorithms). The timing constraints indicate the amount of time that can elapse from the moment data has been sensed to when it is reported on the central server. If the sensed data represents critical events like a fissure in a bridge cable or a fire, the timing constraints are tight; otherwise they are relaxed. Coming back to cipher selection, the tool defines four cipher types which are commonly used in sensor networks: symmetric ciphers Skipjack, RC5 and AES which can be performed on low end devices as determined in [35] and ECC for asymmetric cryptosystem, used for key establishment protocols and which should only be performed on high end devices. The timing constraints define the same classification for tight and relaxed constraints. Timing constraints are also used as a direct parameter to select key establishment protocols based on the number of exchanged messages.

The **available power** supply, with the **communication frequency** and the **network lifetime** are indirect parameters which define the maximum power consumption per packet using the formula:

$$\text{Max consumption per packet} = \frac{\text{power supply}}{\text{comm\_freq} * \text{net\_lifetime}}$$

This value is then compared to the measured or reported values for the data security and authentication protocols to determine if a protocol can be used in the defined circumstances without exhausting the power supply ahead of time.

Environment security determines if the area where the network is deployed is likely to be open to attackers who can steal, corrupt or physically damage the nodes. Depending on the parameter's values the potential damage that an attacker can do can be limited by choosing the appropriate key predistribution scheme: in hostile environments, attacks will be intentional so pairwise keys or server shared keys should be used; in public environments, nodes can be damaged or stolen by accident so probabilistic schemes can also be used; finally master keys should only be used in trusted environments where there is no or limited risk of node compromise, typically home environments.

The **data security level** is the only parameter with subjective values, high or low security, but its value is determined by the application type. The security level is used in the selection of ciphers and cryptographic algorithms, for example

Skipjack and RC5 offer relatively low security while AES and public algorithms offer high security.

The **cryptographic primitives** determine which data authentication and security as well as key establishment protocols will be used, since all define their preferred ciphers. Included in the configuration tool are Skipjack for TinySec and MiniSec, RC5 for SPINS, AES for ZigBee and public key (ECC) for key establishment or digital signatures.

Semantic security is a general term but it is used as an indication of the number of packets that can be secured with a cipher and one key, until the initialisation vector (IV) overflows thus weakening the security. It is an intermediate parameter computed as the maximum number of packets sent in the planned lifetime (comm_freq * net_lifetime), and it is compared to the maximum number of IVs a protocol provides. If there are more packets than IVs and the protocol does not account for key update, it will be considered unsafe and not selected.

In the GUI tool the **attacks** are used to compare two security protocols of the same primitive. A list of attacks (table 2 above) and vulnerabilities has been compiled [32] and ordered depending on the effort they put on the attacker. The list in descending order is: type flaw, key freshness, replay, denial of service, impersonation, man in the middle, impersonation by forwarding, multiple session, impersonation by signature forgery, unknown key share, small group, forward secrecy, with type flaw having requiring an effort of one. Most of the attacks target key establishment, while replay and denial of service are also a threat to data security and authentication. Figure 8 shows the attacks and what key establishment protocol is affected by each attack. The numbers represent the danger level of the attack, 1 being the lowest level and 12 the highest.

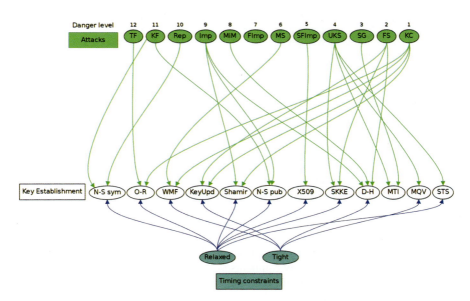

Fig. 8. Attacks and Key Establishment protocols

6.3 Configuration Scenario

A scenario will better explain the way security is configured based on application parameters. Consider a hospital where every room has a WSN of less than one hundred nodes, deployed on patients to monitor their vitals. Data should be reported to a central computer from each sensor every second, and the maximum time a patient is considered to stay in the hospital under monitoring is one month. The nodes are cheap and the hardware is low end, they are powered by two AA batteries giving roughly 3000 mAH each. The nodes should be mobile to allow patients to move around, and common sense dictates that a clustered topology should be used, with each patient being a cluster of sensors.

Since this is a medical application in a public environment a high level of security is required (although the environment could also be considered trusted under certain conditions, not considered here). The public environment eliminates master keys, and the clustered topology, hierarchical communication and node mobility restrict the choices to using server shared keys. Key establishment protocols are restricted to centralised distribution, and the safest protocol is found to be the Wide Mouthed Frog (got from figure 8 where the attack values are summed and the lowest summation equals the safest protocol), vulnerable only to multisession attacks.

Next, for data security, the ciphers are determined from the performance, timing constraints and security level: public algorithms are eliminated as medical applications require a high level of security, therefore the AES algorithm (see figure 9 for selection chart) is left. The ZigBee security stack is chosen, as the others use either RC5 or Skipjack. ZigBee data authentication can also be used, but for broadcast authentication the current selection of protocols does not provide one that uses AES. Therefore, the tool presents all the broadcast authentication protocols and the user can select one they deem to be most appropriate.

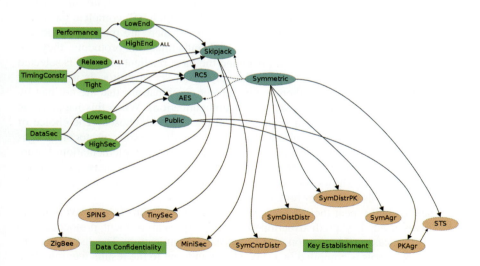

Fig. 9. Security Protocol Selection Chart

While this is not ideal it is planned to extend the tool in the future to offer a greater choice of algorithm and protocol.

What was presented here indicates how the tool makes its decisions based on user input and predetermined relationships between algorithms, protocols and applications.

7 Tool Design

The previous sections demonstrated through the use of tables and charts how a complete security system/protocol can be selected from basic application parameters. The aim of the project discussed in this work was to automate as much as possible this process through the use of a GUI based tool. In this section the tool is described through some screen shots of the user interface in the hope that the reader will see how a configuration aid like this would greatly simplify the inclusion of security in a WSN implementation.

The design and implementation of the security configuration tool is centred on the user perspective and the set of parameters and security primitives described previously. To create a good user interface, a logical program flow and identify the main objects involved, the steps a user might take if asked to describe a network application was analysed. The first step is to identify the application type and the type of device used. These would reveal the desired level of security for data and for the environment and would indicate what types of ciphers can be used. Then he or she would select the number of nodes, the topology, the way data is transferred around in the network, and whether the nodes will be mobile. These parameters enable the selection of the key predistribution scheme and a class of key establishment protocols. Lastly, when expected network lifetime and communication frequency are specified the tool can compute the maximum number of packets and the maximum consumption per packet which will complete the selection of data security and authentication protocols. If there are more than one key establishment protocols or data security and authentication protocols, they are ordered based on the total attacker effort and the most resilient protocol is selected for every category. Because only a relatively small set of protocols is currently included for every category, there are parameter combinations that are not met by any protocol (especially data security and authentication). When this happens the user is provided with a list of protocols and one must be chosen. The choice of protocols available in the tool was based on usability (exotic protocol designs were avoided), thorough analysis and most of all, ease of implementation. Figure 10 shows the GUI interface the user sees when the application/network is specified and figure 11 shows the resulting interface which recommends a series of security protocols to the user for each security primitive.

The tool will repeat the previous step (protocol selection) for the five security primitives used: key predistribution schemes, key establishment protocols, data security, unicast authentication and broadcast authentication. The list of recommended protocols is compiled in a state object and sent to the results window. If in a category more than one protocol was selected (internal to the tool) they are

Security for Wireless Sensor Networks – Configuration Aid

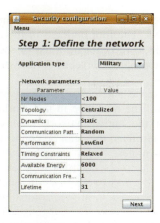

Fig. 10. Define the Network

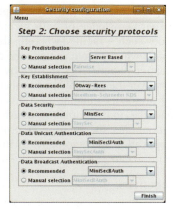

Fig. 11. Security Protocol selection

ordered by attacker effort and the most secure is presented to the user. As stated already, some combinations of user parameters will yield no protocols in which case the user is presented with a list of all the protocols in that category and can select as he pleases.

In its current form the automated security configuration tool has only illustrative purposes as it does not affect the network deployment. This will change as future versions of this tool will be integrated with a security configuration framework that will allow a security enabled WSN to be automatically deployed.

8 Conclusion

The difficulties of sensor network protocol design apply in particular to security protocols. For starters, security is an adversarial problem where the opponent's moves are not known, which makes it a very complex task. WSNs provide many new security problems due to the high number of nodes, the unattended functionality or the particular traffic flows. Most of the present protocols have come to rely on public cryptography which is too computational to be used on the lower powered motes. In addition every application from the wide range that WSNs are used for has its own specific security requirements. One cannot design a "general purpose" security architecture since in some cases it will lead to low security and in others, to unnecessary or wasted computation or power. A comparative example would be a network deployed in one's home to monitor temperature and a military network used for battle field surveillance. The former does not need a lot of security whereas for the latter security is critical. If an architecture is designed for a home deployment and it is then deployed to the battle field it would be completely unsecure, while the reverse would be wasteful. The solution is to make security configurable for every application. The protocols and ciphers are there, all that is needed is to connect them in such a way as to achieve the ideal security level for the application. In addition, it should be easy for regular users or network deplorers' to configure security without having extensive knowledge of both security and WSNs. The tool outlined in this work provides such a facility whereby the tool is able to recommend a list of security protocols which would provide the desired and appropriate level of security for the application parameters specified by the user.

However the implemented tool can only be used as a helper as it only displays a list of protocols suitable for a given application. The next step in this work is to expand the tool so that it assists throughout the entire network deployment operation: once the appropriate protocols are determined they can be connected to the application and automatically installed on the motes.

References

[1] Cionca, V., Newe, T., Dadarlat, V.: A tool for the Security Configuration of Sensor Networks. In: Sensors and Their Applications XV, Edinburgh, UK (October 2009) (to appear)

[2] The internet of Things. Technical report. International Telecommunication Union (2005) ISBN: 92-61-11291-9

[3] Regulation (EC) No 460/2004 of the European Parliament and of the Council of 10, establishing the European Network and Information Security Agency (March 2004)

[4] Sarma, H., Kar, A.: Security Threats in Wireless Sensor Networks. In: 2006 International Carnahan Conference on Security Technology, Kentucky, USA, October 16-20 (2006)

[5] Newsome, J., Shi, E., Song, D., Perrig, A.: The Sybil Attack in Sensor Networks: Analysis and Defenses. In: Proceedings of the Third international Symposium on Information Processing in Sensor Networks, Berkeley, California, USA, April 26-27 (2004)

[6] Walters, J.P., Liang, Z., Shi, W., Chaudhary, V.: Wireless Sensor Net-works Security: A Survey. In: Xiao, Y. (ed.) Security in Distributed, Grid and Perva-sive Computing, Aurbach Publications, ch. 17, CRC Press, Boca Raton (2006)

[7] Buttyan, L., Dora, L., Vajda: Statistical Wormhole Detection in Sensor Networks. In: Molva, R., Tsudik, G., Westho, D. (eds.) Security and Privacy in Ad-hoc and Sensor Networks. LNCS, pp. 128–141. Springer, Heidelberg (2005)

[8] Tun, Z., Maw, A.H.: Wormhole Attack Detection in Wireless Sensor Networks. In: Proceedings of World Academy of Science, Engineering and Technology, December 2008, vol. 36, pp. 549–554 (2008)

[9] Hartung, C., Balasalle, J., Han, R.: Node Compromise in Sensor Networks: The Need for Secure Systems, Department of Computer Science, University of Colorado, Boulder, Technical Report CU-CS-988-04. Technical Report (2004)

[10] Kerckhoffs, A.: La Cryptographie Militaire. Journal des sciences militaires IX (1883)

[11] Shannon, C.E.: Communication theory of secrecy systems. Bell Systems Technical Journal 28, 656–715 (1949)

[12] Diffie, W., Hellman, M.: New directions in cryptography. IEEE Transactions on Information Theory 22(6), 644–654 (1976)

[13] Stajano, F., Anderson, R.J.: The resurrecting duckling: Security issues for ad-hoc wireless networks. In: Proceedings of the 7th International Workshop on Security Protocols, London, UK, pp. 172–194. Springer, Heidelberg (2000)

[14] Eschenauer, L., Gligor, V.D.: A key-management scheme for distributed sensor networks. In: CCS 2002: Proceedings of the 9th ACM conference on Computer and Communications Security, pp. 41–47. ACM, New York (2002)
[15] Chan, H., Perrig, A., Song, D.: Random key predistribution schemes for sensor networks. In: SP 2003: Proceedings of the 2003 IEEE Symposium on Security and Privacy, Washington, DC, USA, p. 197. IEEE Computer Society, Los Alamitos (2003)
[16] Blom, R.: An optimal class of symmetric key generation systems. In: Proc. of the EUROCRYPT 84 workshop on Advances in cryptology: theory and application of cryptographic techniques, pp. 335–338. Springer, New York, Inc. (1985)
[17] Du, W., Deng, J., Han, Y.S., Varshney, P.K.: A pairwise key predistribution scheme for wireless sensor networks. In: CCS 2003: Proceedings of the 10th ACM conference on Computer and communications security, pp. 42–51. ACM, New York (2003)
[18] Liu, D., Ning, P.: Establishing pairwise keys in distributed sensor networks. In: CCS 2003: Proceedings of the 10th ACM conference on Computer and communications security, pp. 52–61. ACM, New York (2003)
[19] Zhu, S., Setia, S., Jajodia, S.: Leap: efficient security mechanisms for large-scale distributed sensor networks. In: CCS 2003: Proceedings of the 10th ACM conference on Computer and communications security, New York, NY, USA, pp. 62–72. ACM, New York (2003)
[20] Chan, H., Perrig, A.: Pike: Peer Intermediaries for Key Establishment in sensor networks. In: Proceedings IEEE INFOCOM 2005. 24th Annual Joint Conference of the IEEE Computer and Communications Societies, March 13-17, vol. 1, pp. 524–535 (2005)
[21] Malan, D., Welsh, M., Smith, M.: A public-key infrastructure for Key Distribution in TinyOS based on Elliptic Curve Cryptography. In: First IEEE International Conference on Sensor and Ad Hoc Communications and Networks (SECON), Santa Clara, CA, USA (October 2004),
[22] Abi-Char, P.E., Mhamed, A., El-Hassan, B.: A fast and secure elliptic curve based authenticated key agreement protocol for low power mobile communications. In: The 2007 International Conference on Next Generation Mobile Applications, Services and Technologies, 2007. NGMAST 2007, 12-14 September, pp. 235–240 (2007)
[23] Huang, Q., Cukier, J., Kobayashi, H., Liu, B., Zhang, J.: Fast authenticated key establishment protocols for self-organizing sensor networks. In: WSNA 2003: Proceedings of the 2nd ACM international conference on Wireless sensor networks and applications, pp. 141–150. ACM, New York (2003)
[24] Wong, D.S., Chan, A.H.: Efficient and mutually authenticated key exchange for low power computing devices. In: ASIACRYPT 2001: Proceedings of the 7th International Conference on the Theory and Application of Cryptology and Information Security, London, UK, pp. 272–289. Springer, Heidelberg (2001)
[25] Anderson, R., Chan, H., Perrig, A.: Key Infection: smart trust for smart dust. In: Proceedings of the 12th IEEE International Conference on Network Protocols ICNP 2004, October 2004, pp. 206–215 (2004)
[26] Perrig, A., Szewczyk, R., Tygar, J.D., Wen, V., Culler, D.E.: Spins: Security Protocols for Sensor Networks. Wireless Networks 8(5), 521–534 (2002)
[27] Karlof, C., Sastry, N., Wagner, D.: Tinysec: A Link Layer Security Architecture for Wireless Sensor Networks. In: SenSys 2004: Proceedings of the 2nd International Conference on Embedded Networked Sensor Systems, pp. 162–175. ACM, New York (2004)

[28] SKIPJACK and KEA algorithm specifications. Technical report, U.S. National Security Agency (May 1998), http://csrc.nist.gov/groups/ST/toolkit/documents/skipjack/skipjack.pdf
[29] Luk, M., Mezzour, G., Perrig, A., Gligor, V.: Minisec: A Secure Sensor Network Communication Architecture. In: IPSN 2007: Proceedings of the 6th International Conference on Information Processing in Sensor Networks, pp. 479–488. ACM, New York (2007)
[30] Perrig, A., Canetti, R., Tygar, J.D., Song, D.: Efficient Authentication and Signing of Multicast Streams over Lossy Channels. In: IEEE Symposium on Security and Privacy 2000, pp. 56–73 (2000)
[31] Clark, J., Jacob, J.: A survey of authentication protocol literature (1997), http://cs.york.ac.uk/jac/papers/drareview.ps.gz
[32] Panti, M., Spalazzi, L., Tacconi, S.: Attacks on cryptographic protocols: A survey. Technical report, Istituto di Informatica, University of Ancona, Ancona, Italy (October 2002)
[33] ZigBee Alliance. Zigbee Specification (2007), http://www.zigbee.org/
[34] Liu, A., Ning, P.: TinyECC: A configurable library for elliptic curve cryptography in wireless sensor networks. In: International Conference on Information Processing in Sensor Networks IPSN 2008, April 2008, pp. 245–256 (2008)
[35] Law, Y.W., Doumen, J., Hartel, P.: Survey and benchmark of block ciphers for wireless sensor networks. ACM Transactions on Sensor Networks 2(1), 65–93 (2006)

Low Power Wireless Buoy Platform for Environmental Monitoring

Arne Sieber[1,2], Juergen Markert[3], Christian Woegerer[4],
Michele Cocco[5], and Matthias F. Wagner[3]

[1] IMEGO, Gothenburg, Sweden
[2] SP Marine AB, Gothenburg, Sweden
[3] University of Applied Sciences, Frankfurt, Germany
[4] Profactor GmbH, Seibersdorf, Austria
[5] Tuscan Archipelago National Park, Isle of Elba, Italy

Abstract. A novel sensor buoy platform was developed for detailed environmental monitoring of coastal areas. Using many buoys in a mesh network on low power ZigBee basis allows dramatic reduction of the overall power consumption, enabling an autonomous operation on a single battery set over 3y. Each buoy acts as sensor node but also as router forwarding messages from neighbor buoys to a centralized server. Thus special attention had to be directed to network synchronization and the handling of processor sleep states. One –wire temperature sensors were used for temperate profiling. Moreover the platform is equipped with several analog and digital inputs allowing the connection of additional sensor and sensor systems. The fist buoy network was designed based on 2.5 GHz RF modules (Atmels ATmega1281 together with the AVR Z-Link AT86RF230). According the specifications of the manufacturer under perfect conditions the maximum transmission range is about 4km. A maximum buoy to buoy distance of approximately 400m could be achieved. Instead of 2.5 GHz technology the final version of the buoys was then equipped with 868 MHz transceivers. A buoy to buoy distance of 3km was possible.

1 Introduction

The global climate changes are expected to influence the behaviour of marine organisms. In the past several events of Mucilage massive blooming were observed, which led in many cases to the death of the Red Gorgonia. A novel wireless buoy platform was developed, enabling monitoring of environmental parameters especially temperature profiling with high spatial resolution. This novel tool will be of importance for marine environmental preservation, as it allows obtaining a reliable picture of the present scenario along with a prediction of future changes. Moreover it will lead to the better understanding of inter-dependencies of Mucilage blooming, water temperature, currents [1] i.e. the field of tolerance of a species to the principal factors of its environment) and of mucilage impact on gorgonians in the Tyrrhenian sea (Italy).

Mucilage's massive blooming is a periodical phenomenon in Tyrrhenian Sea (Italy) that is occurring with a seasonal frequency. Also the type and the extent of the impact of mucilages depend on the season [2]: three species of algae (Nematochrysopsis marina, Chrysonephos lewisii and Acinetospora crinita) constitute the principal components of the mucilaginous aggregates. In general, the first two species occur during the spring season, down to 20 m, while A. crinita occurs at greater depths. In July, when the mucilages reach their maximum development, C. lewisii is the predominant species. This species mainly affects E. cavolinii and E. singularis while A. crinita mainly affects P. clavata, which colonizes greater depths.

(a) (b)

Fig. 1. a). P. clavata wood (Formiche della Zanca, Isle of Elba, Tuscany Archipelago) **b).** Mucilage affection of P. clavata colonies (Pianosa Island, Tuscany Archipelago, 2007)

Empirical considerations often connect the abnormal appearance of mucilage to a warmer water temperature during spring months as well as to the lack of early summer gales. For the better understanding of the relation between abnormal mucilage blooming and the water temperature gradient, a system is required to enable temperature profiling with high spatial resolution. Buoys, equipped with temperature sensors along the mooring can therefore be used. The high spatial resolution requires then a large amount of such buoys.

Buoys can also be equipped with additional sensors, that then allow measurement of chemical and physical parameters. Next to standard sensors like wind velocity and direction, currents and ions also sensors for pesticides and heavy metals are of interest to determine environmental pollution.

Research on and monitoring of marine mammals with the help of Passive Acoustic Localization (short PALS) is another application where small mesh size buoy networks are useful. Conservation of marine habitats and marine species presents an enormous challenge. EU member and candidate states are also facing another challenge - a network of protected nature conservation sites, called Natura 2000, has to be designed and properly managed. The latter is not possible without reliable monitoring tools and protocols to provide accurate and complete scientific data.

A fundamental prerequisite for efficient marine management is knowledge of biological and physical diversity of habitats and of species that inhabit them. In the Mediterranean, very little is known on the distribution, population dynamics, feeding and reproduction strategies of the cetacean species migrating or inhabiting those waters, thus making it impossible to use the information from stranded animals, and interpret, for example, high levels of pollutants in terms of their potential harm on the rest of the population. A proper conservation and management plan for marine mammal species must be based on data received through long term scientific research on population size and trend, distribution, migration, habitat utilization, health and reproductive status and population viability. This data will improve our knowledge base on the ecology of the species and improve our understanding on the relation of human activities and marine mammals, enabling us to effectively implement European conservation policies.

Dolphins can be tracked and localised with the help of PALS: For example the Bottlenose dolphins vocal repertoire consists of three classes – the broadband short-duration (sonar) clicks, broadband pulsed sounds, called burst pulses and narrowband frequency-modulated whistles. These classes have already been studied, analyzed and modelled [3]. The sonar clicks are emitted by the dolphin in highly directional sound beams with peak energies at frequencies ranging from 40 to 130 kHz with source levels of up to 220 dB. Sometimes also frequency parts in the audio band are present. Burst pulse sounds are a general classification given to such. Barks, mews, chirps, and pops made by dolphins only under emotional duress, when they are angry, frightened, upset, or frustrated, are classified generally as burst pulse sounds. Whistles are primarily used for communication between dolphins and show frequencies between 2 and 21 kHz with peak energies at a frequency around 9 kHz and durations between 250 ms and 4s.

The dolphin whistle is characteristic for each individual and maintains its uniqueness for the dolphins life. It should be therefore feasible to identify individual dolphins by their whistle. Hydrophones allow a recording of underwater sounds like the dolphin clicks and whistles. Sounds travel in saltwater at the speed of approximately 1590 m/s, so depending on the distance between the animal and the hydrophone, a sound emitted from some point in water is received at a remote location after some predictable time delay. A buoy equipped with a hydrophone and appropriate sound processing hardware can be used to monitor dolphin activity. Using several of these buoys in a network with a defined mesh size, allows on the one hand a survey of a larger area and on the other hand a precise the localization of the dolphin with the help of triangulation of dolphin sounds. This feature also enables dolphin tracking.

Several studies have been made on the localisation and tracking of marine mammals especially the bottlenose dolphin with the help of PALS in pools and restricted, relatively small areas in coastal waters or channels. The amount of hydrophone-equipped buoys is restricted to a small number like three or four. Each buoy is equipped with a hydrophone, a preamplifier and a radio transmitter (each buoy is transmitting on its unique frequency) [4]. The buoy transmits all recorded data to a remote computer. Data reception, processing and analysis are then performed with a main computer. Due to the amount of data needed to be processed by the main computer real-time processing is in most cases not applicable.

To be able to survey large areas, the mesh size should be approximately 2 km (so for example 36 buoys to cover an area of 100 km²). A mesh network of PALS buoys allows then monitoring of and further research on:

- population size
- population trend
- distribution, migration and habitat utilization
- effects of environmental parameters
- human influences (as example: boat traffic, oil industry....)

The present paper details the design of a first experimental buoy network, that was designed for high spatial monitoring of temperature profiles. The first approach was based on Long Range ZigBee modules. The 2.5 GHz technology did not turn out to be sufficient reliable and also the maximum transmission range was quite short. A second generation of buoys based on low power 868 MHz technology was developed.

2 WSN for Habitat Monitoring

Increasingly Wireless Sensor Networks (WSN) are being used for habitat monitoring. Projects in biology and life sciences have special requirements for which WSN are especially suited [6]:

- Miniaturization
- Adaptive sensor design
- Localization
- Tracking
- Data aggregation
- Time synchronization
- Energy efficiency
- Interfaces to global networks

A possible network architecture is shown in figure 2. WSNs are deployed in the region of interest. The WSN is coupled via gateways to a cellular network, satellite networks or directly to the Internet. Depending on the project goals additional requirements, like reliability, real-time response etc. have to be designed into the system.

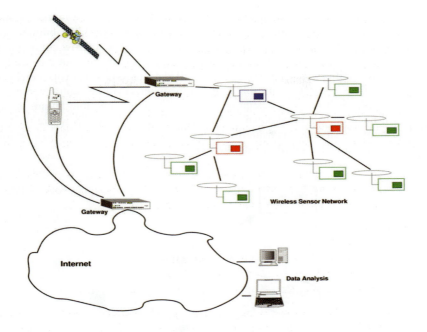

Fig. 2. Wireless Sensor Network Architecture

3 Buoys for Environmental Monitoring

Buoys for monitoring of environmental parameters are normally custom designed according to the requirements. Therefore different buoys especially designed for a specific task are available on the market. One drawback of these custom designed buoys is their high price starting from around 10.000€ per buoy. Additional to this, actually available buoys use satellite transmission technologies like Geos, Inmarsat, ARGOS. ORBCOM or alternatives like VHF or GSM. All these systems have in common relatively high power consumption, presenting the need of a large power supply, diesel generators or solar panels. These needs then define the overall dimension of the buoy and with that, also the overall weight.

The costs for a proper mooring plus deployment and recovery are rising with size and weight of a buoy, thus a miniaturization enabled by a low power design is one of the main goals pursued in this proposal. The present platform was designed as an open architecture with a variety of analogue and digital interfaces, thus allowing an easy attachment of additional (already existing) sensor modules or hardware, that were not specifically developed for this buoy. Furthermore this allows also an easy integration into existing monitoring facilities. This approach leads to a universally applicable buoy.

The low price of the overall buoy allows a deployment of a large amount of buoys thus enabling a monitoring of the coastal areas with high spatial resolution.

The buoys form a telemetry network (figure 3). Different from traditional telemetry usually applied on buoys – like satellite supported or VHF that both have a high power consumption - long range RF network design will allow communication and data transmission between the buoys and a server, without the need of powerful transmitters. Each buoy can forward data received from its neighbor, thus a buoy does not need to be able to reach the receiver of the central computer. Moreover the mesh network design is self healing, thus a possible failure of one buoy will not lead to a breakdown of the whole network. In cases where the network of buoys is situated far from the central server, one of the buoys can optionally be equipped with a GSM or satellite modem.

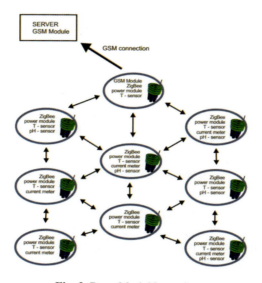

Fig. 3. Buoy Mesh Network

The main specifications of the present platform are:

- easy adaptable modular design
- easily configurable
- thus: buoy can easily be adapted to specific requirements.
- ultra low power design
- ultra low power design allowing power supply via batteries
- long autonomous operation without maintenance
- small size
- lightweight
- easy and cheap mooring (due to the small size and the low weight)
- easy deployment and recovery
- robust design

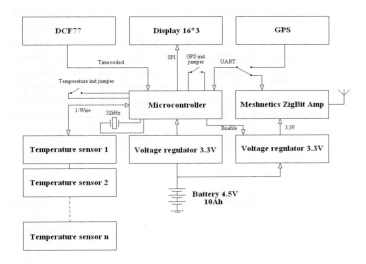

Fig. 4. System Design

4 System Design

Figure 4 details the overall system design of the buoy platform. Core component is a 8 Bit RISC AtMega644p microprocessor [Atmel] operated at 1 MHz. Its main specifications are:

- 64-Kbyte self-programming Flash Program Memory
- 4-Kbyte SRAM, 2-KByte EEPROM, 8 Channel 10-bit A/D-converter
- 1.8 - 5.5 Volt Operation
- Two 8-bit Timer/Counters with separate prescalers and compare modes (one with a 32kHz crystal input)
- One 16-bit Timer/Counter with separate prescaler
- Real time counter with separate oscillator
- Six PWM channels
- 8-channel, 10-bit ADC
- Differential mode with selectable gain at 1x, 10x or 200x
- Byte-oriented Two-wire Serial Interface
- Two programmable serial USART
- Master/Slave SPI serial interface
- Programmable watchdog timer with separate on-chip oscillator
- On-chip analog comparator

One low drop linear voltage regulator is used to supply the microprocessors main board. The Meshnetics ZigBit Amp module is interfaced via UART. For power management purposes the ZigBit Amp module is powered via a own voltage regulator. During "sleep periods", this regulator is disabled thus the ZigBee module is not supplied anymore. Alternatively a GPS can be connected to the microprocessors UART.

A 16x3 characters LCD is used for data visualization and for initialization. A DCF77 module is connected to one digital I/O. The temperature sensors are connected to another I/O pin. A separate FET to pull up the line is not required, as alternatively the pin can be set as output pin. Moreover the 1-wire bus line is pulled up with a 1,2 k resistor. A 32 kHz crystal is connected to the internal 8 bit timer 2. With a prescaler of 128, an overflow occurs exactly every second. This is the basis for the real time clock.

5 Networking

ZigBee is an open standard for short range wireless networks based on the Physical Layer and the Media Access Control from IEEE 802.15.4, focusing on minimizing the overall power consumption and at the same time maximizing network reliability.

IEEE 802.15.4 and ZigBee are offering the developer three kind of devices to form a so called personal area network (PAN).

- End-devices, which in general periodically collect data and transmit them. Furthermore for minimizing power consumption in such end-devices advanced sleep modes are implemented.
- Routers play an important role in a network, they collect data from end-devices and forward them to the destination (like another router or to the final coordinator). For the correct function of a network a router always has to be active. Thus sleeping modes are not possible and therefore are also not implemented.
- Coordinator: One routers in a PAN is usually configured as coordinator. It's main function are parameterization and management of the PAN and the collection of the networks data.
- The principal structure of a ZigBee network is shown in figure 5.

Typical ZigBee modules use the 2.4 GHz Band (16 channels) with transmission powers between 0.5mW to 10mW. Additionally in the United States 10 channels in the 915 MHz Band, and in Europe one Channel at 868 MHz can be used for ZigBee applications. The typical transmission range is 100m in free line of sight with a maximum data transmission rate of 250kBit/s. This short transmission range is not sufficient for dater transmission in a buoy network where we expect distances of 1-2 km.

A detailed market research pointed out 3 manufacturers that produce ZigBee modules suitable for larger communication ranges. The XBee 802.15.4 OEM RF module [Digi International] is an IEEE 802.15.4 compliant solution supporting mesh networking (series 2) and has with a maximum output of 60mW (+18dBm) a maximum transmission range of 1.6 km. Meshnetics Germany is a company providing a complete and user friendly solution for wireless communication based on ZigBee. Actually they deliver two module types: ZigBit, and the enhanced version

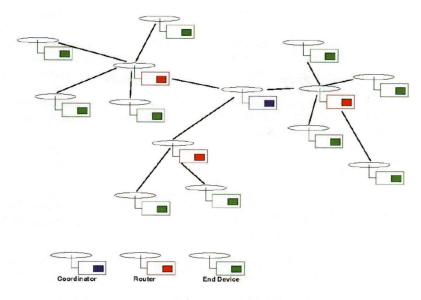

Fig. 5. Typical ZigBee Network

ZigBit Amp, software development kits and demo kits to minimize the overall development time. Both modules are available with a variety of antennas or antenna-connectors. The Meshnetics ZigBee modules are based on the Atmel ZigBee bundle consisting of the 8 Bit Risc ATmega 1281v micro controller and the AVR Z-Link AT86RF230 transceiver. Additional onboard amplifiers on the ZigBit-Amp module increase the transmission power by +20dBm and improve the receiving sensitivity to -104dBm enabling according to Meshnetics wireless communication up to 4 km.

6 Time and Network Synchronization

Clock synchronization is a well known topic in the computer area and has already been widely investigated. It is the main way to assure that networking systems have a common notion of time required for time scheduled operations. As already stated in the introduction, a temperature measurement every 6h should be sufficient. To maximize the autonomous maintenance free operation, the power consumption during these measurements has to be reduced to a minimum. During this period, data transmission is not necessary, thus the ZigBee receiver can be switched off and the microprocessor enters a sleep mode. The modules have to be time synchronized in order to assure a synchronized "waking up" which is necessary to establish again a correctly working ZigBee network.

Clock synchronization mechanisms addressing especially wireless networks are discussed in [6]. The approaches there have in common, that they are based on interactive communication, which increases the overall data volume transferred on the network.

An alternative to a synchronized time basis can be using a wireless radio time signal like the DCF77 from Frankfurt, Germany, or the HBG from Prangins, Switzerland. Unfortunately it turned out, that the signal could not be reliable received in the test area on Elba, Italy.

One alternative that can be acceptable for the present platform is using the internal timer 2 as real time clock. With a prescaler a 32 kHz crystal and a prescaler of 128, an overflow occurs every 1s. This overflow generates an interrupt. A time variable is increased by one. If this time variable equals the predefined "wake up time", the microprocessor exits "sleep mode". For clock synchronization, the system time from the Host computer, which is connected to the ZigBee Network coordinator is broadcasted. Such a broadcast is forwarded to all connected nodes in the network with a maximum delay of a few seconds. This delay can be neglected in a application where data have to be transmitted only approximately every 6h.

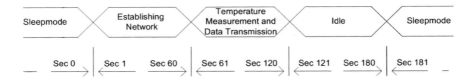

Fig. 6. Time schedule

Figure 6 details the time schedule. After a sleep period, 1 minute is dedicated for establishing the network connection to assure a complete and correct network operation before data are sent. One minute should be enough, even if the buoys clocks are desynchronized by some seconds. After receiving a data package with valid temperature values, the coordinator sends an acknowledge message. Each buoy maintains the network operation for in total 3 minutes in order to assure that all nodes had enough time to send their data packages.

7 Power Supply

The major overall power consumption of a buoy depends on:

- sampling rate (how many measurements per day)
- working/standby power consumption
- main computer board and data storage
- telemetry

Typical buoys use solar panels, wind turbines or diesel engines to provide power for long autonomous operation of battery powered solutions for shorter operation periods.

The present approach takes advantage of today's state of the art low power electronics with enhanced sleeping modes, where during no operation cycles the power consumption is reduced to a minimum, low power sensor electronics and ZigBee based low power telemetry. This dramatically reduced power consumption allows powering the whole buoy for a long period of time from a battery pack and avoiding necessities for large buoy constructions due to the need of solar panels. Another important benefit from this low power design is the minimized overall size and weight of the buoy. This allows easy deployment and mooring thus minimizing the overall operation costs for the buoy.

A pack of 3 Mono type alkaline battery cells (~10Ah) was selected as power supply.

8 Temperature Profiling

Many techniques are known for temperature measurements. Analog sensors are often based on resistors with a temperature dependent conductivity (NTC or PTC). For temperature monitoring with high spatial resolution it is necessary, to mount a larger amount of temperature sensors (n sensors) on the mooring line. Using analog sensors this would result in at minimum n+1 analog lines. Especially in underwater applications, where electrical components have to be carefully sealed against the surrounding water, is is necessary, to reduce the overall number of electrical lines to a minimum. When using analog sensing elements, also the resistance of the sensor cables may be the cause for errors.

To avoid errors resulting from long analog lines and to reduce the overall necessary lines digital sensors can be used on a bus. Several digital sensors are available today equipped with a variety of sensor interfaces. As stated above, it is important to reduce the interfacing lines to a minimum to achieve a reliable mechanical design resulting in low cable costs, low assembly time and efficient sealing. Maxim offers digital temperature sensors, where several can be used on two lines. One line is ground, the other line is used for power supply and for communication.

Using Maxims 1-Wire the theoretical maximum of sensors on one bus is 248 (64 bit serial code consisting of 8 bit family code, 48 bit part number and 8 bit checksum). This (low speed) bus again allows to interface the sensors with 2 lines – again one for ground and one for power and data communication. The Maxim DS18B20P is such a parasite powered temperature sensor. Especially its specifications listed in the following are interesting for the present work.

- Device can be powered from data line (parasite powered)
- Power supply range is 3.0V to 5.5V
- Measures temperatures from -55°C to +125°C
- ±0.5°C accuracy from -10°C to +85°C
- Thermometer resolution is user selectable from 9 to 12 Bits

One drawback when using one wire devices is that the physical location on the bus is not given. Even though each sensor has a unique serial code, mounted on the bus just from the serial number it is not possible to read out its physical position, unless the sensors are not carefully preselected and ordered before the assembly. For the present project this means that we can read out all the temperatures from the sensors on the bus, but as its location is not known, it is also not known at what depth each value was measured.

Mounting all the DS18B20P series in parallel on the bus allows fast assembly and cheap sealing. For the first prototype, the sensors were directly soldered on a 2 leads cable (figure 7a) and the contacts were sealed with 2 components Epoxy resin. Over the cable and the device a special heat shrink tube with a water resistant layer of glue is placed for mechanical protection and sealing purposes (figure 7b).

To regain the location information, a special "sensor identification procedure" is necessary. In this procedure a user is requested to warm up one sensor after the other – according to its physical location on the bus. Based on the detection of this temperature changes the physical order of the sensors is obtained and then stored in the EEPROM.

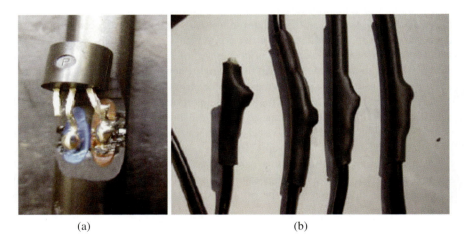

(a)   (b)

Fig. 7. a). sensor soldered directly on the cable. **b).** heat shrink tube covering the sensor on the cable.

9   Transmission Range Tests

First range tests were carried out on land using Meshnetics ZigBit AMP development kits together with their high gain antennas. Distance measurements were carried out with a Garmin Etrex GPS receiver. Although 4km are advertised by Meshnetics, a maximum transmission distance of 1km was achieved.

Several buoys were deployed in coastal waters near Pisa, Italy. The antennas on the buoys were mounted approximately 1.5m above the water surface. The maximum transmission range that allowed reliable networking was approximately 600m. This got worse in the case of waves and rain.

This short and unreliable transmission is by far not acceptable for a buoy network. A second generation of buoys was designed incorporating the XBee-PRO 868 RF module from Digi International. Its specifications are summarized in the following:

- 868 MHz short range device (SRD) G3 band for Europe
- Software selectable transmit power
- 40 km RF LOS w/ dipole antennas
- 80 km RF LOS w/ high gain antennas (TX Power reduced)
- Simple to use peer-to-peer/point-to-multipoint topology (see Figure 8 for possible network topologies)
- 128-bit AES encryption

Buoys equipped with the XBee-Pro 868 RF were tested in coastal regions near Göteborg, Sweden. Transmission was successfully tested up to 3km, but then the test had to be aborted due to bad weather conditions.

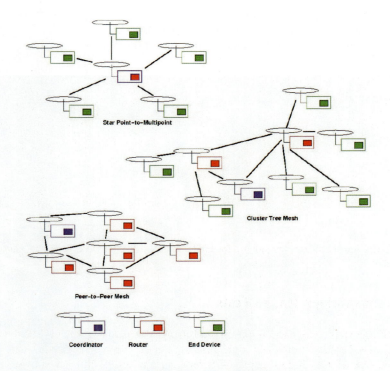

Fig. 8. Peer-to-peer/point-to-multipoint topology

10 Buoy Deployment

Once a proper mooring place for a buoys if found, first of all an anchoring point is deployed. After mooring the buoy to the anchor, the buoy electronics is configured by attaching the deployment vessels GPS via NMEA0813 and powering up the buoy. In this initialization phase the actual coordinates are written into the EEPROM of the AtMega644p.

Alternatively a GPS module can be optionally installed on a buoy, enabling the buoy to determine its position itself. Considering that moorings can break leading to the loss of buoys this makes sense, as with such an installed GPS the buoy is able to detect its position and an eventual position change. If the mooring broke, the buoy may send its coordinates to the server enabling an easier recovery.

11 Results and Conclusion

Several Buoy prototypes were built up. Their specifications are:

- height: 1.5m (over water)
- weight: 3,5 kg
- positive buoyancy: 5 kg
- power supply: 3 x 1,5V alkaline D cell, (10Ah)
- power consumption during transmission: 60 mA

Fig. 9. First field tests near Pisa, Italy

- power consumption in sleep mode: ~60 µA
- autonomous maintenance free operation: >3 y

Several digital and analog inputs allow easy interfacing of additional sensors. Low weight, low cost and small size and long maintenance free operation are the key for a deployment of a larger amount of buoys. Initially ZigBit AMP modules from Meshnetics were used, but their transmission range was not sufficient.

For first evaluation of the transmission range and the network reliability, the buoys were deployed in coastal waters near Pisa, Italy. Unfortunately the maximum mesh size for a reliable network operation turned out to be about 200 – 500m dependent on the sea and weather conditions.

A second generation of buoys was build up where the ZigBit AMP 2.5 GHz module was substituted with a Digi International 868 MHz modem. Data transmission was successfully tested successfully up to 3km buoy to buoy distance.

12 Ongoing and Future Work

Actual work includes now on the one hand testing of the 868 MHz data transmission technology. Digi Internationals 868 MHz modules perform well, but only support star networks. The group at the University of Applied Science is now working on an implementation that also will support mesh networks.

For what concerns further hardware development a modular approach will be followed. Figure 10 shows the new design of the buoy platform, where several

Fig. 10. Universal buoy platform design

modules can be stacked. 2 modules that will always be present are the power supply and the core microprocessor board. Standard optional modules are the networking RF modem, the GPS and the GSM board.

Envisaged optional modules are:

- AD converter module, offering several analogue I/O with a programmable amplification and resolution
  - Module for amperometric sensors
  - Module for potentiometric sensors
- Digital I/O module, for control and readout of external components
- Mechatronic module, for the automatisation of pumps and electromagnetic valves; One output of WP2 will be a novel measurement chamber, where the sensors can be integrated. This chamber will be situated inside the buoy. A pump allows filling the chamber with sea water and also washing the electrodes with a conditioner
- GPS Module, equipped with a SIRF III receiver allowing accurate position detection
- GSM/GPRS/UMTS modem, used for data uplink
- Satellite Uplink board
- Additional battery packs if, for example, a sensor system has a higher power consumption or a longer autonomous operation is required
- Solar panel controller, in cases where sensor systems require higher currents and long maintenance free autonomous operation is required, an alternative power supply can be solar panels that are used to recharge the batteries of the buoy.

This modular design allows equipping a buoy according to the actual needs – thus not every buoy must be equipped with the same sensors, while maybe each buoy is equipped with a temperature sensor only every second one may have a current meter and a pH sensor included. If the server is situated far from the network, one of the buoys may be equipped with a GSM module for data upload via a normal cellular network, or, in the case of no network coverage, a satellite uplink modem can be integrated. When using the buoy as a drifter, an additional GPS module may be a requirement.

Additional sensors that may be mounted:

- temperature
- several temperature sensors along the mooring line for profiling
- humidity
- wind velocity and direction
- currents and current direction
- UV radiation

While vertical profiling of, for example, the water temperature can be performed easily by mounting various cheap digital temperature sensors along the mooring line, other environmental parameters require more sophisticated sensor systems. For vertical profiling, recording measurement parameters over the depth, state of the art tools are water samplers that are carried by an ROV or AOV, custom

Fig. 11. Water sampler docked to the buoy

solutions like the Pontoon-Mounted Vertical Profiling System form YSI, or, for example, the ARGO floats, that can dive autonomously to a maximum depth of 2000m, record depth, temperature and salinity, and after a dive return to the surface to transmit the recorded data. ROV/AOV based solutions require a manual analysis of the collected samples. The above mentioned Profiling system and the ARGO floats can only measure limited amounts of parameters. Some sensors cannot operate under high pressure, thus a deployment on a measurement device at a specific depth is not possible. A versatile buoy is now under development (figure 11), that can be equipped with a variety of sensors according to the actual monitoring needs.

A novel Autonomous Water Sampler that can be optionally deployed on a buoy, takes advantage of the measurement facilities of the buoys. Its main functions will be:

- sliding down the mooring line to a predefined depth
- collecting a water sample
- "crawling" back to the surface along the line
- dock on the surface buoy and feed the collected sample into the measuring chamber of the buoy

To perform these tasks over a long period, also in the Autonomous Water Sampler lowest power components will be deployed. Its main components will include:

- Electronic Hardware:
- Microcontroller board
- Telemetry module
- Power supply
- Depth sensor

- Motor controller
- Valve controller
- Housing
- Motor unit (in oil) to climb drive the unit up and down along the mooring line
- Inner water reservoir
- Water inlet and water outlet (with valves)

Thus not only surface measurements can be performed, the same list of parameters can be recorded from every requested depth. Detailed profiling is possible.

References

[1] Hutchinson, G.E.: Concluding remarks, Cold Spring Harbor Symposium. Quant. Biol. 22, 415 (1957)
[2] Giuliani, S., Lamberti, V.C., Sonni, C., Pellegrini, D.: Mucilage impact on gorgonians in the Tyrrhenian sea. Science of The Total Environment 353, 340–349 (2005)
[3] Greco, M., Gini, F., Verrazzani, L.: Analysis and modeling of the acoustic signals emitted by the Mediterranean bottlenose dolphins. In: Proceedings of the 3rd IEEE International Symposium on "Signal Processing and Information Technology", ISSPIT 2003 (2003)
[4] Janik, V.M., Van Parijs, S., Thompson, P.M.: A two-dimensional acoustic localization system for marine mammals. Marine Mammal Science 16, 437–447 (2000)
[5] Polastre, J., Szewczyk, R., Mainwaring, A., Culler, D., Anderson, J.: Analysis of Wireless Sensor Networks for Habitat Monitoring. Wireless Sensor Networks. Springer, Heidelberg (2004)
[6] Sundararaman, B., Buy, U., Kshemkalyani, A.D.: Clock Synchronization for Wireless Sensor Networks: A Survey, Department of Computer Science. University of Illinois, Chicago March 22 (2005)

A Detail Performance Evaluation of the Novel Mechanisms Ensuring Maximum Connectivity and Data Transmission between Nodes, Based on the Heuristics Under 5-Color Clustered Response Approach

Mahdi Nasrullah Al-Ameen

Bangladesh University of Engineering and Technology
Dhaka, Bangladesh

Abstract. To efficiently manage the sensor networks the topology of the entire network has to be discovered by the monitoring node. In this paper, a novel topology discovery algorithm for sensor networks is proposed. The algorithm finds a set of distinguished nodes, using whose neighborhood information the approximate topology of the network is constructed. Only these distinguished nodes reply back to the topology discovery probes. These nodes logically organize the network in the form of clusters comprising nodes in their neighborhood. Topology discovery algorithms form a tree of clusters rooted at the monitoring node, which initiates the topology discovery process. This organization is used for efficient data dissemination and aggregation, duty cycle assignment and fault tolerance of the network system. The unpredictable behaviors of sensor networks have made it a vital point that how the operational nodes will be managed when a node in the network fails. In this paper, novel fault tolerance mechanism for sensor networks is proposed based on clustered response approach on considering different scenarios that may come to consideration when a node fails; thus ensuring maximum connectivity among operational nodes after the failure of a node. The mechanism explains how the information packets transmitted to the faulty node can be cached by an operational node. After being repaired the faulty node is reinstalled to operational state and the mechanism of getting the repaired node connected to the network is proposed in this paper. Reverse traverse mechanism has been proposed in this paper as a part of fault tolerance mechanisms, which ensures that the number of clusters is not increased when a faulty node is repaired and re-connected to the network. A novel mechanism for duty cycle assignment based on clustered response approach has been proposed in this paper. The proposed mechanism clearly defines how a packet of information is transmitted between a pair of clusters. In this case, a set of nodes from a cluster is selected by the cluster head for communication with another cluster. So, each cluster has a specific set of selected nodes to communicate with a certain cluster and this node-selection mechanism is discussed in detail in this paper. Priority factor of nodes and also their reliability and energy factors have been considered to constitute this selection mechanism. Distance between nodes and nodes within the communication region are other parameters for the selection mechanism which are evaluated using fuzzy evaluation. The mechanisms proposed in this paper are distributed and highly scalable.

1 Introduction

Rapid technological advances in the electronic industry in recent years have facilitated the development of wireless technology that allows for the deployment of a large number of small sensor devices in a physical area of interest. The purpose of the sensors is to gather report and analyze data collected from their environment where certain phenomenon, such as movement or temperature change, is being monitored [4].

Equipped with a portable radio transceiver, each sensor has the ability to communicate with neighboring sensors that fall within its transmission radius. However the devices are low-powered with finite resources and this imposes restrictions on their data processing operations and communication activities. Issues to consider due to the limited energy resources of the sensors include the use of energy-efficient routing protocols, the management and diagnosis of dynamically changing network topologies, methods of data aggregation, collision detection and avoidance. All of the above contribute to the ultimate goal of ad hoc sensor networks, which is to maintain network integrity and ensure its lifespan is maximized [6].

In order to be able to perform topology control operations for the purpose of network management, it is essential that a sensor network self-organizes into a desirable topology structure that provides communication and sensing coverage under stringent energy constraints. For data-gathering applications of sensor networks where the typical traffic is many-to-one, the topology of the network is modeled as a type of connected graph. Such graphs guarantee a connected communication topology that simplify the dissemination, aggregation and routing of information in a network. In particular, the extracted topology of a sensor network is often represented as specific types of graphs: spanning trees, dominating sets, and relative neighborhood graphs all have the desired connectivity property. Sensor networks therefore use these types of graphs to self-configure by generating an efficient topology over which information can be transmitted [19], [3], [5].

The topology of the network is susceptible to frequent change. In mobile sensor networks this is due to the dynamic nature of the devices. Although not affected by movement, the topology of static networks must also be adaptable to change because sensors can become faulty or exhaust their resources. In such circumstances, neighboring sensors need to locally reorganize themselves to maintain overall network connectivity [1].

Topology management and control is important because it helps achieve the global objective of extending the network lifetime. It does this by providing information about the network state in terms of the number of active/alive nodes present and the connectivity (reach ability) map of the system. Topology is also a key attribute because it aids in network performance analysis. Accurate knowledge of network topology is a prerequisite to critical network management tasks such as proactive and reactive resource management and utilization [13].

There has been a lot of research into the design of control algorithms used in sensor networks for the purposes of broadcasting, routing and scheduling of transmissions. The complexity of the solutions is influenced by the topology of the network and they can be solved once the topology is defined [12].

In this paper 5-color clustered response mechanism has been proposed to discover the topology of the network and this mechanism is designed in a way so that the nodes in sensor networks get and store the required information to simply and efficiently implement fault tolerance and duty cycle assignment mechanisms, proposed in this paper. Reverse traverse mechanism has been proposed in this paper as a part of the fault tolerance mechanisms that ensuring the number of clusters is not increased when the operational nodes are reconnected at the failure of a node.

Duty cycle assignment mechanism ensures the maximum connectivity between each pair of clusters. Each cluster has a minimal number of nodes, which are active to transfer packets between a parent-child cluster pair. The duty cycle assignment mechanism, proposed in this paper ensures a high percentage of successful data transmission between clusters. The selection mechanism of nodes for the set to communicate with certain cluster has been developed in such a way that efficiently handles lower reliability and priority factors of nodes and ensures a high rate of success in the event of transmitting packet to the end node.

Through simulation the performances of the proposed mechanisms have been analyzed and found providing satisfactory outputs that stand for the high efficiency of the mechanisms; proposed in this paper.

2 Related Work on All Layers of Protocol Stack

Many protocols that take the unique characteristics of wireless ad hoc networks have been developed. Among them energy efficiency, routing and MAC layer protocols have attracted most attention [23]. Self-configuration and self-organizing mechanisms are needed in sensor networks because of the requirement of unattended operation in uncertain, dynamic environments. Some attentions have been given to developing localized, distributed, self-configuration mechanisms in sensor networks, and studying conditions under which they are feasible.

Sensor networks are characterized by severe energy constraints because the nodes will often operate with finite battery resources and limited recharging. The energy concerns can be addressed by engineering design at all layers. Some of the energy concerns are being addressed at the hardware and architecture level. At the physical layer, there is now a significant body of work on minimizing energy costs by adjusting the transmit powers of nodes while achieving global network properties such as connectivity. At the link layer, some of the works have focused on energy-efficient medium access schemes suitable for sensor networks. At the networking layer, meta naming of data and data-aggregation during routing has been proposed and analyzed as a significant means for energy savings. At the application layer, it has been recognized that energy savings can be obtained by pushing computation within the network in the form of localized and distributed algorithms [23], [1].

One of the main advantages of the distributed computing paradigm is that it adds a new dimension of robustness and reliability to computing. Computations done by clusters of independent processors need not be sensitive to the failure of a small portion of the network. Wireless sensor networks are an example of large scale distributed computing systems where fault-tolerance is important. For large-scale sensor networks to be economically feasible, the individual nodes necessarily have to be low-end inexpensive devices. Such devices are likely to exhibit unreliable behavior. It is important to guarantee that faulty behavior of individual components does not affect the overall system behavior [21].

3 Literature Review of Topology Discovery Mechanisms

Several topology discovery mechanisms have been proposed in the past. [8] uses a clustering scheme to discover the network topology. The cluster heads are dynamically chosen based on geographic location or network connectivity, and the Management Information Bases (MIBs) at cluster heads are used to gather topology information.

For mobile ad hoc networks, [27] proposes a mobile agent based framework to distribute topology information. The agents migrate to least visited nodes and update topology caches in visited nodes. Getting the network topology at the initiating node would entail either waiting for an agent to visit it or querying every node for their topology caches. [14], [17] introduce the concept of virtual backbones (which essentially is a dominating set) for wireless networks. [9], [10] describe routing in ad hoc networks using minimal connected dominating sets.

In [18], authors give bounds on approximation algorithms for finding Connected Dominating Sets (CDS). Using each of the above methods, a simple topology discovery algorithm can be designed to query the dominating nodes, which provide their neighborhood lists. Work in [11] expands on previous work done in [12]. By selecting a subset of nodes, approximate topology is created by merging their neighborhood lists. Each node is aware of its neighbors since it can listen to the shared communication channel. The resolution of the topology depends on the cardinality and structure of the chosen set of nodes. TopDisc algorithm in [30] includes request propagation with three colors and then with four colors to discover the topology of sensor networks and is based on the simple greedy log (n)-approximation algorithm for finding the set cover.

The mentioned schemes utilize a demand-driven based approach, where the topology of the sensor network can be extracted at any time when necessary from the base station. My approach differs since I suggest an event-driven method where individual nodes are responsible for sending update messages to the base station when they detect a topological change in their own descendant sub-tree. The next section describes in detail my proposed topology discovery algorithm and the underlying concepts of its design.

4 5-Color Clustered Response Mechanism

In clustered response approach the network is divided into set of clusters where each cluster is represented by one node, called the cluster head. A sensor network is considered to be an undirected graph G {V, E} with vertices V and Edges E. Consider H as a set of cluster heads, and N_i as the neighborhood list of node i, where i∈ H. In this case following conditions hold.

1. $N = \cup N_i$.
2. $\forall\, x \in N_i$, edge $(x, i) \in E$.

Communication overhead for clustered response approach is dependent on the number of clusters that are formed and the path length connecting the clusters. So for minimum communication overhead following problems are needed to be solved.

- Find minimum cardinality set of cluster heads which have to reply back.
- Form a minimal tree with the set of the cluster heads.

In fact these are combinatorial optimization problems. Moreover, for optimal solution one needs global information about the network whereas the nodes only have local information. So in this paper heuristics have been given that provide approximate solutions to the problems.

In the mechanism to be discussed, 5 different colors represent 5 possible states of a node. Topology request contains the id and location information of the node from which it is broadcasted.

- Initially all nodes are white. The node, which has not received any topology request, is white. The topology request is begun to be broadcasted from a root node which is also termed as monitoring node. Monitoring node becomes yellow, broadcasts topology request to nodes within its communication region and starts a timer. (figure 1)

Fig. 1.

- A white node becomes yellow on getting a topology request and sends an acknowledgement to the parent node with its location information. (figure 2)

Fig. 2.

- As the timer is stopped, the color of a yellow node becomes red if it receives acknowledgement at least from one node. Red node establishes a priority queue based on the location information of its children yellow nodes. Distant node gets higher priority. The red node sends 'broadcast trigger' message to the node with the highest priority. (figure 3)

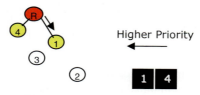

Fig. 3.

- On getting the message a yellow node broadcasts topology request with its location information to nodes within its communication region and starts a timer. (figure 4)

Fig. 4.

- According to the mentioned mechanism node: 2 broadcasts topology request and starts a timer after getting a 'broadcast trigger' message from node: 1. As there is no white node within the communication region of node: 2 it gets no acknowledgement when the timer is on. Its color becomes gray as timer spans. Of course a gray node has no child. A gray node sends a message with its updated state to its parent red node. (figure 5)

Fig. 5.

- Broadcast trigger message is sent to the next node in the priority queue. (figure 6)

Fig. 6.

A Detail Performance Evaluation of the Novel Mechanisms

- Node: 3 becomes gray and informs its parent about the updated state. A red node becomes black on receiving 'updated-state' message from all the nodes in its priority queue and sends its topology information aggregated with the topology information of its children nodes to its parent red node. (figure 7)

Fig. 7.

- Node: R sends 'broadcast trigger' message to node: 4 after getting topology information from its child black node: 1. According to the described mechanism the monitoring node gets the complete topology of the network and becomes black. A black node is always a cluster head and a gray node is the part of a cluster of its parent black node. (figure 8)

Fig. 8.

In this mechanism a non-white node does not respond to a topology request but stores the node-id and location information of the topology request sender node.

4.1 Heuristics Behind 5-Color Clustered Response Mechanism

The heuristic behind giving the farthest node the highest priority in the priority queue is explained as follows: The coverage region of each node is the circular area centered at the node with radius equal to its communication range. Then the number of nodes covered by a single node could be proportional to its coverage area times the local node density. The number of new nodes covered by a forwarding node is proportional to its coverage area minus the already covered area. This is illustrated in figure 9.

Coverage region of node: A is colored blue. New area (outside the coverage are of node: A) covered by node: B is colored light green and the new area (outside the coverage are of node: A) covered by node: C is colored green.

In figure 9, the distance of node: B from node: A is greater than the distance of node: C from node: A. Thus node: B covers larger new area than the new area covered by node: C.

Fig. 9. Illustration for distance heuristic of 5-color clustered response mechanism

5   A Novel Duty Cycle Assignment Mechanism

In this paper to propose the duty cycle assignment mechanism, black nodes are considered to be cluster heads and other nodes that are not cluster heads are considered to be gray nodes. According to the proposed mechanism in this paper, a set of gray nodes is selected by the cluster head to communicate with another cluster. In this case, a gray node may be in multiple sets of nodes. Through the collaboration between the parent and child black nodes, the sets of gray nodes are selected for each cluster pair. To describe the mechanism, a number of phases have been considered in this paper.

5.1   Phase 1 (Determining Recommended Number of Gray Nodes)

Consider, node: bc is a child black node of node: bp and node: bp has to determine the recommended number of gray nodes to be in set: s1 for communication with the cluster of node: bc.

$$G_a = F_a \cdot G_t \tag{1}$$

Where, G_t = Total number of gray nodes in the cluster of node: bp.
         G_a = Recommended number of gray nodes for set: s1.
         F_a = Allocation factor.

The value of F_a is determined dynamically when the set: s1 is formed by node: bp. Factors that are considered to determine the value of F_a are as follows.

5.1.1   Number of Children Black Nodes

On considering the energy constraints of a node, a gray node can be assigned to a predefined maximum number of sets for communication with other clusters. Thus with increase in the number of children black nodes of node: bp, the number of gray nodes, to be assigned to set: s1 decreases linearly up to a threshold (Th1) and then becomes constant. The following membership function defines the affect of number of children black nodes on F_a.

$$\mu_1 = \begin{cases} \dfrac{1}{n}, & \text{for } n_{min} \leq n < n_{max} \\ \dfrac{1}{n_{max}}, & \text{for } n \geq n_{max} \end{cases} \tag{2}$$

n = number of children black nodes.
n_{min} = minimum possible number of children black nodes.
n_{max} = Number of children black nodes defining Th1.

5.1.2 Priority Factor of the Cluster of a Child Black Node

Each node in a sensor network is assigned a priority by the monitoring node and each cluster head calculates a priority factor (P_f) by adding the individual priorities of the nodes in its cluster with its own priority. Node: bp is acknowledged about P_f of the cluster of node: bc in the response of 'send priority factor' message.

As P_f increases for the cluster of node: bc, more gray nodes are likely to be assigned to set: s1. In this case, F_a is linearly dependent on P_f.

$$\mu_3 = \frac{P_f - a_{min}}{a_{max} - a_{min}} \quad (3)$$

This equation defines the effect of P_f.
In equation (3),
a_{min} = minimum possible priority of any cluster.
a_{max} = maximum possible priority of any cluster.

5.1.3 Reliability Factor of Gray Nodes

Node: bp calculates the reliability factor (P_{rt}) by adding the reliability of its children gray nodes. An initial reliability is assigned to each node by the monitoring node and later on it is updated for a node each time it is reinstalled to operation after the failure. P_r is defined by following function.

$$P_r = \frac{MTBF}{MTBF + MTTR} \quad (4)$$

MTBF = Mean time between failure.
MTTR = Mean time to repair.

P_{rt} is calculated by node: bp by adding the P_r of each child gray node. As P_r decreases, more gray nodes are likely to be assigned to set: s1.

5.1.4 Energy Factor

Energy of the gray nodes in set: s1 is used in a distributed fashion for communication with the cluster of node: bc. Thus, with the increase in energy factor, more nodes are likely to be assigned to set: s1. Energy factor is measured from the following equation.

$$\mu_4 = \frac{P_p - b_{min}}{b_{max} - b_{min}} \quad (5)$$

P_p = Recommended energy, should be available to the gray nodes of set: s1 to communicate with the cluster of node: bc. It is defined by the monitoring node based on the number of nodes and priority of the cluster of node: bc.

b_{min} = Minimum possible recommended energy.
b_{max} = Maximum possible recommended energy.

5.1.5 Calculating F_a Using Fuzzy Evaluation

Four criteria have been described for the evaluation of F_a and each criterion is assigned a numerical evaluation by fuzzy membership function. Thus F_a can be calculated by aggregation of these four criteria .So,

$$F_a = w1*u1 + w2*u3 + w3*(1/P_{rt}) + w4*u4 \tag{6}$$

Where w_i ($w_i \geq 0$) denotes the corresponding weights for criterion: i (i = 1... 4) such that

$$\sum_{i=1}^{4} w_i = 1 \tag{7}$$

To note, a dominant criterion should have a higher value as its weight.

5.2 Phase 2 (Node Selection Mechanism for a Parent Node)

Node: bp selects the gray nodes for set: s1 and a priority queue is formed in this case whose maximum length is equal to G_a. The factors that are considered to set the priority of a gray node (let us say this node: gn) to be placed in this queue (let us say this queue: pq) are as follows.

5.2.1 Number of Nodes (Belong to the Cluster of Node: bc) within the Communication Region of Node: gn

As the number of nodes (belong to the cluster of node: bc) within the communication region of node: gn increases, it likely stands with the higher priority for communication with the cluster of node: bc.

This factor can be calculated by using an equation similar to equation (5) where P_p can be replaced by N_p (number of nodes within the communication region of node gn, belonging to the cluster of node: bc), b_{min} and b_{max} can be defined respectively as minimum and maximum possible number of nodes within the communication region of a node in this case.

5.2.2 Distance between Node: gn and Node: bc

Information can be directly sent by node: gn to a node in the cluster of node: bc, which is within the communication region of node: gn. But when the receiving node in the cluster of node: bc is out of the coverage region of node: gn, the information packet is transmitted through node: bc (the cluster head) to the receiving node. Thus with the decrease in the distance between node: gn and node: bc, the priority of node: gn is likely to be increased in queue: pq.

A Detail Performance Evaluation of the Novel Mechanisms 53

This factor can be calculated by using an equation similar to equation (5) where P_p can be replaced by D_p (distance between node: gn and node: bc), b_{min} and b_{max} can be defined respectively as minimum and maximum possible distance in this case.

5.2.3 Reliability and Available Energy Factor of Node: gn

The priority of node: gn is likely to be increased in queue: pq with the increase in its reliability and the available energy for the transmission of information packet.

Reliability can be measured from equation (4) and available energy factor can be measured by using an equation similar to equation (5) where P_p can be replaced by R_p (available energy of node: gn for the transmission of information packet), b_{min} and b_{max} can be defined respectively as minimum and maximum possible available energy in this case.

By using fuzzy evaluation (similar to equation 6) for the mentioned factors, the selection factor, S_f is determined for node: gn and the node with maximum S_f gains the first position in queue: pq.

5.3 Phase 3 (Node Selection Mechanism for a Child Black Node)

5.3.1 Step 1

Node: bc determines the recommended number of gray nodes that can be assigned to set: s2 (a set of gray nodes from the cluster of node: bc to communicate with the cluster of node: bp), which depends on 4 factors. (Detail mechanism to measure the factors has been described in phase 1).

The factors have been mentioned below:

- Number of children black nodes of node: bc.
- Priority factor of the cluster of node: bp.
- Reliability factor of the cluster of node: bc.
- Energy factor of the cluster of node: bc.

5.3.2 Step2

The list of nodes in queue: pq is sent to node: bc and node: bc selects the gray nodes for set: s2 to communicate with the cluster of node: bp. The nodes in queue: pq are considered by node: bc one after another, according to their positions in the queue.

Consider node: gn as the first node in queue: pq. Node: bc selects a gray node from its cluster, which is within the communication region of node: gn. In this case, node: bc selects a gray node (let us say this node: gs) by considering the factors that have been described as follows. (Detail mechanisms to measure the factors have been described in phase 2).

- Number of nodes (belong to the cluster of node: bp) within the communication region of node: gs.
- Distance between node: gs and node: bp.
- Reliability and available energy factor of node: gs.

A selection factor (S_{f2}) is determined after imposing corresponding weights on these factors. The selections of node: gs means it has gained greater S_{f2} than any other nodes in this case.

Node: bc then considers the second node in queue: pq and according to the mechanism; already described, selects a gray node (node: gs is not considered in this case) to be assigned to set: s2. In this way, a gray node is selected for each node (which has at least one gray node that is within its communication region and belongs to the cluster of node: bc) in queue: pq and assigned to set: s2.

5.3.3 Step 3

After the first iteration through the queue: pq, if the selected number of nodes in set: s2 is less than the recommended number of nodes for set: s2, node: gn is considered again and another gray node within the communication region of node: gn, is assigned to set: s2 according to the mentioned mechanism. This iteration continues for a pre-specified number of times but the process is stopped when the number of nodes in set: s2 is equal to the recommended number of nodes that can be assigned to set: s2.

5.3.4 Step 4

As node: bc completes selecting the nodes for set: s2, it informs node: bp with an acknowledgement message that nodes have been assigned to the respective sets to ensure communication between this pair of clusters.

5.4 Transmission Mechanisms of Information Packets between a Pair of Clusters

- A black node selects a gray node from each set to be in active mode and other nodes of a set will be in sleep mode for this cluster pair but they may be in active mode for any other cluster pair. The 'mode set' message is sent from the parent node to the gray children nodes.
- A gray node may give up its active mode for a cluster pair when it has spent a certain amount of energy. It sends a signal to the parent black node and goes to sleep mode. The parent black node then selects another gray node from this set to be in active mode.
- The black node listens to all packets while these are being forwarded. The black node forwards a packet if the sending node is out of range of the active forwarding node.

6 Overview of the Scenarios When a Node Fails

- Watchdog timer in every node tracks "alive-signal" from its neighbors to ensure connectivity.
- Suppose, a neighbor of node: f does not receive this signal from node: f for a specific period of time.

- The neighbor identifies node: f as a faulty node and sends a "failure message" to the monitoring node with the node-id of node: f.
- Node: f goes into repair mode.
- Topology of the network is updated to maintain regular connectivity among operational nodes.
- Neighbors of node: f cache packets transmitted to node: f by source.
- Node: f is reinstalled to operation and "reinstalled message" is sent by node: f to its neighbors.
- Node: f is re-connected to the network and the topology of the network is updated again.
- Cached packets are transmitted to node: f.

7 Fault Tolerance Mechanism

In this paper, fault tolerance mechanism is discussed depending upon two different states of a node.

- Faulty State: When a node fails.
- Operational State: When the faulty node is repaired and reinstalled to operation.

7.1 Discussion of Fault Tolerance Mechanism at Faulty State of a Node

A node that has failed either may be a gray node or a black node. The faulty node (node: F) is detached from its parent. Two cases are considered in this respect.

- Node: F is the only child of node: P. When Node: F gets disconnected Node: P becomes gray and its updated topology information is sent to the upper level of network. (figure 10)

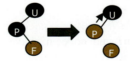

Fig. 10.

- Node: F is not the only child of node: P. When Node: F gets disconnected, Node: P sends its updated topology information to the upper level of network. (figure 11)

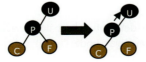

Fig. 11.

In the case of failure of a black node (node: F) its children (node:s1, s2a1,s2a2, s3a, s3b) become orphan which will be re-connected to the network by selecting new parent node.

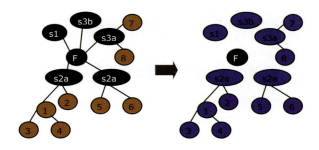

Fig. 12.

When a node identifies itself as orphan, its color changes to blue but it also stores its previous state (black/gray). If its previous color was black It broadcasts a "faulty state" message to the nodes of the tree rooted at itself and on getting this message the color of a node (i.e. node 7) is changed to blue. (figure 12).

Each child (i.e. node s1) of node: F broadcasts "parent find" message and starts a timer. Blue Nodes ignore the message and a non-blue node replies to the message with its current state (black/gray). On the basis of replies to the "parent find" message one of the three scenarios may come to consideration:

- Scenario 1: Node: s1 has received reply from at least one black node.
- Scenario 2: Node: s1 has received reply from no black node but from at least one gray node.
- Scenario 3: Node: s1 has received reply from no node.

7.1.1 Scenario 1

Considering figure 13, node: n1 and node: n2 have replied to the "parent find" message from node: s1. Node: n1 is closer to node: s1 than node: n2. So node: n1 has been selected as the new parent of node: s1.

Node: s1 sends "parent selection" message to node: n1 and node: n1 confirms with acknowledgement. Node: s1 then gets connected to the cluster of node: n1 and its color is reset to gray (its previous color). Afterwards Node: n1 sends updated topology information to the upper level of network. The scenario is illustrated in figure 14.

7.1.2 Scenario 2

This scenario comes to consideration when node: s2a1 receives reply from no black node but from at least one gray node.

A Detail Performance Evaluation of the Novel Mechanisms

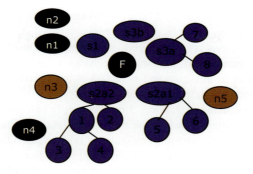

Fig. 13.

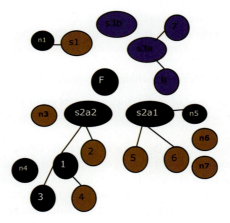

Fig. 14. Updated topology of the network after applying reverse traverse mechanism

a) Approach 1: According to figure 13, node: n5 is the nearest gray node to node: s2a1 and has been selected as its new parent. Color of node: s2a1 is changed to black (previous color).

Node: s2a1 broadcasts "parent found" message to the nodes of the tree rooted at itself and on receiving the message the color of a node is changed to its previous color (i.e. Color of node:5 is reset to gray). Node: s2a1 is connected to node: n5. Color of node: n5 is changed to black and its updated topology information is sent to the upper level of network. Figure 14 illustrates the scenario.

The number of clusters is increased in this approach but it is simple and minimum message-transmission is required to update the topology of the network in this case.

b) Approach 2: Following figure 13, node: s2a2 broadcasts the "search parent" message to each node of the tree rooted at itself and starts a timer. On receiving this message each node (node: 1, 2, 3 & 4) broadcasts the "parent find" message and starts a timer.

At this point two cases are considered. According to consideration 1, no black node responds to "parent find" message. Mechanism described in scenario 2 approach 1 is followed in this case and node: n3 is the nearest gray node to node: s2a2 in this case with which it will be connected.

According to consideration 2, node: 3 receives reply (to "parent find" message) from a black node (node: n4) when the timer is on. As the timer elapses node: 3 sends the "black parent discovered" message to node: s2a2. On getting this message the timer of the node: s2a2 is interrupted and immediately stopped. In case of multiple black nodes that have replied, black node with minimum distance is selected by node: 3.

Node: s2a2 sends an acknowledgement to node: 3. On getting the acknowledgement node: 3 sends "parent selection" message to node: n4 and gets connected to its cluster. The tree rooted at node: s2a2 is now connected to the rest of the network through node: n4. So the topology of the nodes of this tree has to be updated. In this case, reverse traverse mechanism has been described to update the topology of the network.

According to reverse traverse mechanism (consider fig 13):

- If the previous color of node: 3 was gray its color is changed from blue to black and it sends a "topology change" message to its current parent (node: 1).
- On getting this message the color of any node changes to gray if it has no other node connected to it except the node that has sent the message. But the color of a node changes to black if it has at least one other node connected to it. So node: 1 becomes black.
- If a node receives the message from its child, parent-child relationship is swapped between the two nodes and as a result the receiving node becomes the child of the transmitting node. But parent-child relationship remains unchanged when a child node receives the message from its parent. So in this case node: 3 is the parent of node: 1(parent-child relationship is swapped).
- Each node forwards the message to the nodes, connected to it after its color is changed to gray/black.

Figure 14 shows the updated topology of the network after applying reverse traverse mechanism.

In fact the increase/decrease of the number of clusters can be defined by 4 conditions. Conditions are described considering figure 15. In figure 15 the tree (rooted at node: 3b) is connected to the rest of the network (not shown in this figure) through node: A. Arrowhead represents the direction of the transmission of 'topology change' message. In figure 15, node-id stands for the sequence of receiving this message.

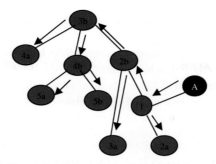

Fig. 15. Illustration for analyzing reverse traverse mechanism

- Number of clusters decreases if the previous color of node: 1 was black and node: 3b has no other child except the one (here it is node: 2b) from which it has received the 'topology change' message.
- Number of clusters remains unchanged if the previous color of node: 1 was black and node: 3b has at least one child except node: 2b.
- Number of clusters remains unchanged if the previous color of node: 1 was gray and node: 3b has no child except node: 2b.
- Number of clusters is increased if the previous color of node: 1 was gray and node: 3b has at least one child except node: 2b.

Here under 3 conditions (out of 4) the number of clusters is not increased and under 1 condition the number of clusters is decreased. So a higher probability exists of not increasing the number of clusters on applying the reverse traverse mechanism.

7.1.3 Scenario 3

In this scenario three cases are considered. Considering case: 1 following figure 14, node: s3b receives reply (to its "parent find" message) from no node and node: s3b has no child. So it is just disconnected from the rest of the network.

Case: 2 is considered for node: s3a which has children nodes. So node: s3a broadcasts "search parent" message to the nodes of the tree rooted at its own and starts a timer. The method of processing "black parent discovered" message has already been described in scenario 2 approach 2.

Consider no black node has replied to the "parent find" message when the timer remains on. Node: n6 is the nearest gray node (to node: 7) that has replied to parent find message from node: 7. Node: 7 sends "gray parent discovered" message to node: s3a with the distance information with node: n6. Node: n7 is the nearest gray node (to node: 8) that has replied to parent find message from node: 8. Node: 8 sends "gray parent discovered" message to node: s3a with the distance information with node: n7.

When the timer elapses node: s3a selects the gray node with minimum distance information (node: n6 in this case). Node: s3a sends an acknowledgement to node:

7. On getting the acknowledgement node: 7 sends "parent selection" message to node: n6. Color of node: n6 is changed to black and sends a confirmation message to node: 7. Node: 7 gets connected to node: n6. Topology is updated applying reverse traverse mechanism that has been already described. Node: n6 sends the updated topology information to the upper level of network.

Case: 3 is considered when node: s3a receives no "black parent discovered" message or "gray parent discovered" message when the timer remains on and thus the tree rooted at node: s3a is disconnected from the rest of the network.

7.2 Discussion of Fault Tolerance Mechanism at the Operational State of a Node

When a faulty node (node: f) is repaired and reinstalled to operation it has to be connected to the network. The topology of the network may have been changed after the failure of node: f. When node: f is reinstalled to operation it sends "get state" message to the nodes within its communication region and starts a timer. On getting the message a node replies with its current state (gray/black). At this point following approaches are followed for node: f to be connected to the network.

- If the current state of the previous (before node: f was faulty) parent node (let us say this node: p3) of node: f is black, node: f is connected to it as a gray node. Node: f sends a "parent selection after repair" message to node: p3. Node: p3 sends a confirmation message to node: f and transmits updated topology information to the upper level of the network. The number of clusters remains unchanged in this case.
- If the current state of node: p3 is gray, node: f selects the black node with minimum distance as its parent and sends it the "parent selection after repair" message to get connected. In this case number of clusters is not increased.
- If node: f gets response from no black node when its timer remains on, it is connected to node: p3 and the color of node: p3 is changed from gray to black. The number of clusters is increased in this case.

7.3 Method of Resetting the States of the Nodes

If node: f is re-connected to node: p3, it is possible to get the nodes to their previous (before node: f became faulty) states. When the topology of the network is updated after a node becomes faulty, each node whose topology is updated, stores its previous state. When node: f is connected to node: p3, node: f broadcasts "reset state" message and on getting the message each node gets back to its previous state and forwards the message to other nodes. In this case number of clusters remains unchanged at the end of the fault tolerance mechanism at the cost of increased message transmission.

A Detail Performance Evaluation of the Novel Mechanisms

7.4 Fault Tolerance Mechanism When a Node Fails Multiple Times

Each node stores its updated topology information after a node fails (let us say this f2) and if node: f2 fails for the second time, "restore state" message is broadcasted (with the node-id of the node: f2) by the node that has identified node: f2 as a faulty node and on getting this message a node restores its state to that state that has been stored when node: f2 became faulty for the first time. A considerable amount of message transmission is required for the fault tolerance mechanisms when a node becomes faulty for the first time. But it is reduced when a node fails multiple times.

7.5 Mechanism of Caching Packets Transmitted to a Faulty Node

When a node (node: f) fails, it cannot receive the information packet transmitted to it. The node (node: h) that has to forward the packet directly to node: f, caches the information packet for a specific period of time. Node: h starts a timer when it caches the packet.

When node: f comes to operational state it informs each node within its communication region by sending "reinstalled message" and if node: h receives this message within the period of time when its timer remains on, the timer is interrupted and stopped and it sends the cached packet to node: f. Packet is dropped if node: f is not reinstalled to operation within the period the timer remains on.

8 Performance Evaluation

Performances of the proposed mechanisms have been analyzed by using C/C++ compiler and MatLab simulator. Sensors are randomly deployed in a field of 100m x 100m. Number and communication range of the sensors have been varied according to the experiment requirements. The result represents average of these values for 100 different runs. Experiments 1 to 3 prove the efficiency of 5-color clustered response mechanism and the efficiency of the fault tolerance mechanism has been shown through the experiments 4 to 9. The efficiency of the proposed duty cycle assignment mechanism is shown through the experiments 10 to 13.

8.1 Experiment 1

The first experiment is designed to find the number of black nodes formed during topology discovery with the increase in total number of nodes. Figure 16 shows the number of black nodes formed at the end of 5-color clustered response mechanism and the corresponding values of centralized log (n)-approximate solution provided by the greedy algorithm for set cover. From the observation it is clear that the proposed mechanism works very nearly as well as the centralized algorithm.

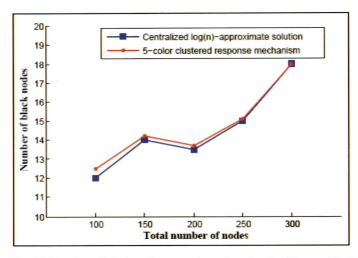

Fig. 16. Number of black nodes vs total number of nodes [Range: 20m]

8.2 Experiment 2

The second experiment illustrates the effect of communication region on the number of black nodes. The results are again compared with the centralized greedy set cover algorithm, shown in figure 17. Performance of the proposed mechanism is almost similar in this case and for higher communication range (greater than 33.7m) it is better indeed.

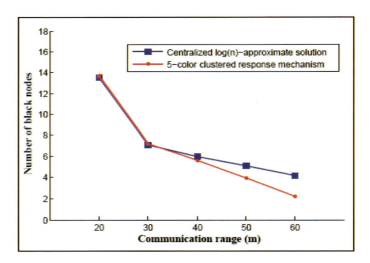

Fig. 17. Number of black nodes vs communication region [Total nodes: 200]

8.3 Experiment 3

5-color clustered response mechanism gives a reachability map based on the tree of clusters, comprising of black nodes and intermediate nodes between two black nodes, need to be active for reaching any part of the network. The average path length with the proposed mechanism is compared to the shortest path routing scheme. Figure 18 and 19 show that the proposed mechanism works very close to the shortest path routing.

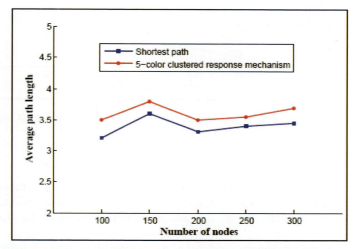

Fig. 18. Average path length vs total number of nodes [Range: 20m]

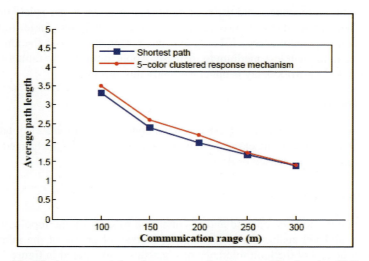

Fig. 19. Average path length vs communication range [Total nodes: 200]

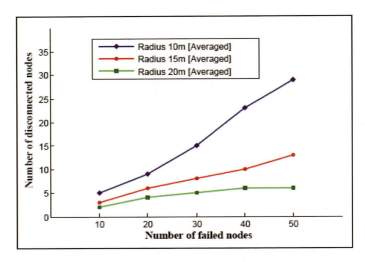

Fig. 20. Number of disconnected nodes vs number of failed nodes [Total nodes: 300]

8.4 Experiment 4

The experiment, represented by figure 20, has been designed to compare the number of disconnected nodes to the number of failed nodes. Disconnected nodes occur when a node fails and its children are unable to reconnect to another parent node.

The experiment demonstrates that as the density of the network is increased, the number of nodes that are able to reconnect is increased as the number of one-hop neighbor increases. For a radius of 20m, it has been observed that majority of nodes are able to reconnect and the total number of inactive nodes solely depends on the nodes that are dying. However, it is important to note that larger transmission radius means more energy is being expended. This also causes the amount of interference in the wireless medium to increase. Applications where fault tolerance is a priority and network delay is not can employ this approach.

8.5 Experiment 5

The topology of the network is updated when a node fails. In figure 21, it has been observed that in most of the cases the number of clusters remain unchanged and in very few cases it is increased or decreased.

8.6 Experiment 6

When the faulty node is repaired and re-connected to network the topology of the network is updated again. The percentage of increase/decrease of the number of clusters at the end of the fault tolerance mechanisms is illustrated in figure 22.

A Detail Performance Evaluation of the Novel Mechanisms 65

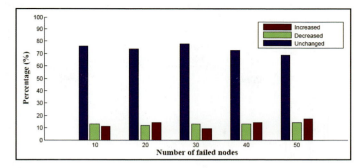

Fig. 21. Percentage of increase/decrease of clusters vs number of failed nodes when a failed node is disconnected from the network [Total nodes: 300, Communication range 20m]

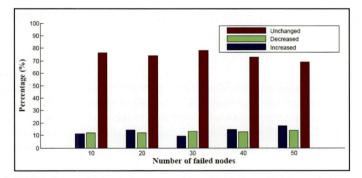

Fig. 22. Percentage of increase/decrease of clusters vs number of failed nodes when a failed node is re-connected to network after repair [Total nodes: 300, Communication range 20m]

8.7 Experiment 7

One of the important mottos of the fault tolerance mechanisms is not to increase the number of clusters which has been achieved to a satisfactory extent through the mechanism proposed in this paper. The performance of the caching mechanism has been represented by figure 23. The specific time (period of timer) to cache the packet is determined based on MTTR (mean-time-to-repair) and a predefined ratio. MTTR is a statistical data, which is updated regularly and each node is acknowledged about the MTTR of its neighbors. A ratio of period of timer to MTTR is initially defined and with the change in MTTR, period of timer is also changed to maintain this ratio. Caching is termed successful when the cached packet is transmitted to the faulty node after it is reinstalled to operation. With the increase in the period of timer and decrease in MTTR the rate of successful caching is increased.

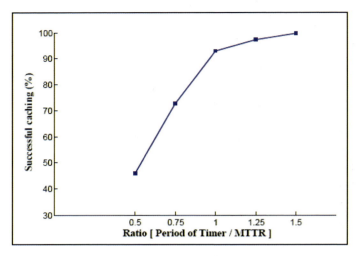

Fig. 23. Percentage of successful caching vs ratio [period of timer/ MTTR][Total nodes: 300, Communication range 20m]

8.8 Experiment 8

When a node fails multiple times, required message transmission is reduced compared to the required message transmission when a node fails for the first time. Reduction in message transmission is compared to the total number of nodes, illustrated in figure 24. It is observed that with the increase in the total number of nodes percentage of increase in message transmission is reduced to a greater extent.

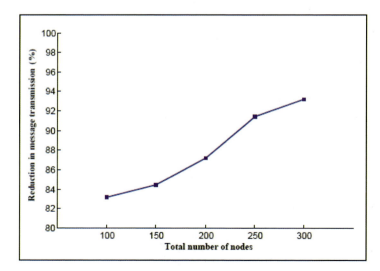

Fig. 24. Reduction in message transmission when a node fails multiple times vs total number of nodes [Communication range: 20m]

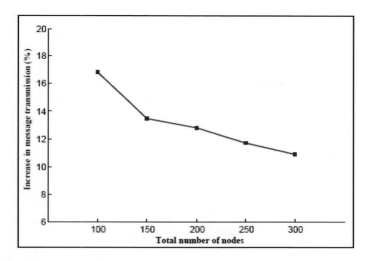

Fig. 25. Increase in message transmission to reset the topology of the network to the state before the failure of a node vs total number of nodes [Communication range: 20m]

8.9 Experiment 9

Fault tolerance mechanism proposed in this paper describes a method to reset the topology of the network to the state before the failure of a node when this node is repaired and re-connected to the network. But in this method message transmission is increased. This increase in message transmission is represented in figure 25 as the percentage of total message transmission required if the topology of the network is not reset to the state before the failure of the node, where it is compared with total number of nodes.

8.10 Experiment 10

The result of this experiment is represented by figure 26. It shows the efficiency of the duty cycle assignment mechanism proposed in this paper. The experiment shows that even with the increase of the total number of nodes the percentage of successful transmission remains at a good level. In fact the total number of nodes does not have any significant effect on the percentage of successful transmission and here is the success of the proposed mechanism that irrespective of the total number of nodes ensures a good percentage of successful transmission.

8.11 Experiment 11

This experiment is designed to analyze the relation between the percentage of successful transmission and the communication range of nodes. The result of the experiment is shown in figure 27 and it has been found that with the increase of the communication range of nodes, the percentage of successful transmission is increased.

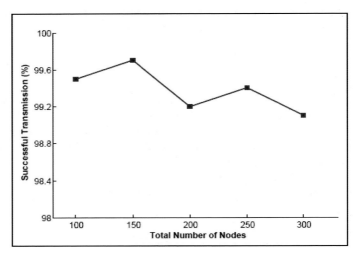

Fig. 26. Percentage of successful transmission vs total number of nodes [Communication range: 20m, Average priority factor: .7, Average reliablity factor: .7]

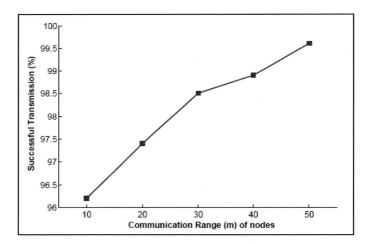

Fig. 27. Percentage of successful transmission vs communication range of nodes [Total nodes: 300, Average priority factor: .7, Average reliablity factor: .7]

According to the duty cycle assignment mechanism, proposed in this paper when a parent black node (node: b1) selects a gray node (node: g1) to be in the set to communicate with the cluster of a child black node (node: c1), it considers the number of gray nodes that are in the set formed by node: c1 to communicate with the cluster of node: b1 and within the communication range of node: g1. So with the increase of the communication range of nodes the probability of having a good number of gray nodes in the set to communicate with another cluster is increased that gives rise to the increase in the percentage of successful transmission consequently.

8.12 Experiment 12

Figure 28 represents the result of this experiment. According to the duty cycle assignment mechanism proposed in this paper, a parent black node (node: b2) considers the priority factor of the cluster of a child black node (node: Y) to determine the number of gray nodes to be in the set (set: s2) to communicate with the cluster of node: Y and the greater priority of the cluster of node: Y demands more number of gray nodes to be in set: s2. So percentage of successful transmission to the cluster of node: Y is increased with the increase in priority factor of this cluster.

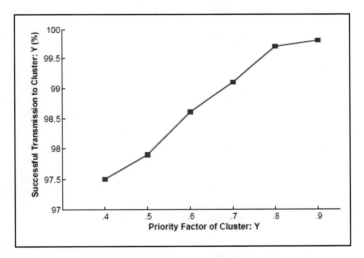

Fig. 28. Percentage of successful transmission to a cluster vs priority factor of this cluster [Communication range: 20m, Total nodes: 300, Average reliablity factor: .7]

8.13 Experiment 13

Reliability factor is of important consideration in this paper for duty cycle assignment mechanism. The proposed mechanism handles the reliability factor in a practical and efficient way that ensures not to decrease the percentage of successful transmission considerably even if the average reliability factor of nodes is decreased to a considerable extent.

Figure 29 shows the result of this experiment and it has been found that at a lower average reliability factor of nodes the percentage of successful transmission is quite good and it gets better with the increase in average reliability factor of nodes.

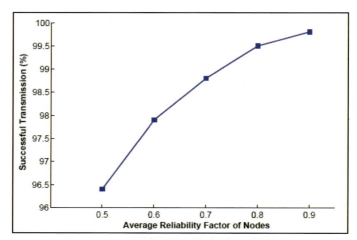

Fig. 29. Percentage of successful transmission vs average reliability factor of nodes [Communication range: 20m, Total nodes: 300, Average priority factor: .7]

The performance of a clustered response mechanism is termed to be better with the decrease in the number black nodes and average path length. The better performance of fault tolerance mechanism demands the decrease in the number of disconnected nodes and not to increase the number of clusters when the topology of the network is updated after the failure of a node. Rate of successful caching should be increased for better performance. Minimizing required message transmission is an important consideration to improve the performance of the fault tolerance mechanism. The efficiency of duty cycle assignment mechanism is proclaimed with the satisfactory rate of successful transmission. The experiments show that these requirements for satisfactory performances have been achieved to a good extent through the mechanisms proposed in this paper.

9 Conclusion and Future Work

In this paper I have described a distributed topology discovery algorithm for wireless sensor networks. I have focused on the specific task of topology generation and retrieval, as it has an important role in the management and control of energy constricted sensor networks. My work implements a simple event-driven approach, which reduces communication overhead and prolongs the lifetime of the network by utilizing a node reattachment method in the case of broken links due to faulty nodes. My work is based on previous research in the area of graph theory for the purpose of developing good topologies with desirable properties in terms of connectivity, throughput, interference, power consumption and delay.

My experiments provide results that indicate the algorithm I have developed can be used as a basic high level monitoring mechanism. It allows the base station node to maintain an up-to-date view of the topology tree of the network. The base is also able to observe the relative location of faults and problems as update messages return node ID, their hierarchical level and number of descendants. This is useful information since the base node could act proactively in the case where it observes an area of low coverage due to many faults. My simple algorithm can aid network administrators in obtaining an accurate topological view and, if necessary, more potent maintenance operations can be performed when there is a need for strong corrective action.

A large amount of future work in the area of topology control and management is indeed impending as wireless sensor devices become cheaper yet more capable of performing a wide range of surveillance and sensing tasks. My work can be extended upon in several directions. The potential for a more robust and adaptive re-attachment mechanism for faulty nodes can be investigated. Timing can be incorporated into my simulations and latency of data movement can be analyzed. Experiments can be performed to compare power consumption of nodes under different topology structures, and scheduling can be implemented in order to monitor the effects of interference on throughput. Improvement of tree building mechanisms can be explored for the benefit of data aggregation and reduction of node degree.

References

[1] Akyildiz, I.F., Su, W., Sankarasubramaniam, Y., Cairci, E.: Wireless sensor networks: a survey. Computer Networks 38, 393–422 (2002)
[2] Bao, L., Garcia-Luna-Aceves, J.J.: Topology Management in Ad Hoc Networks. In: Proceedings of the Fourth ACM International Symposium on Mobile Ad Hoc Networking and Computing (ACM MobiHoc 2003), Annapolis, Maryland, June 1-3 (2003)
[3] Borbash, S.A., Jennings, E.H.: Distributed Topology Control Algorithm for Multihop Wireless Networks. In: Proc. 2002 Int'l Joint Conf. Neural Networks, IJCNN, vol. 1, pp. 355–360 (2002)
[4] Bulusu, N., Estrin, D., Girod, L., Heideman, J.: Scalable coordination for wireless sensor networks: self-configuring localization systems. In: Proceedings of the Sixth International Symposium on Communication Theory and Applications, ISCTA 2001 (2001)
[5] Carle, J., Simplot-Ryl, D.: Energy-efficient area monitoring for sensor networks. Computer 37(2), 40–46 (2004)
[6] Chandra, R., et al.: Adaptive Topology in Hybrid Wireless Networks, http://www.cs.cornell.edu/people/ranveer/topology.pdf
[7] Chlamtac, I., Conti, M., Liu, J.J.-N.: Mobile ad hoc networking: imperatives and challenges. Ad-hoc Networks Journal 1(1), 13–64 (2003)
[8] Chen, W., Jain, N., Singh, S.: ANMP: Ad hoc network management protocol. IEEE Journal on Selected Areas in Communications 17(8), 1506–1531 (1999)
[9] Das, B., Bhargavan, V.: Routing in Ad Hoc Networks Using Minimum connected dominating Sets. In: IEEE International Conference on Communications ICC 1997 (June 1997)

[10] Das, B., Sivakumar, R., Bharghavan, V.: Routing in ad hoc networks using a virtual backbone. In: Proc. IEEE, IC3N 1997 (1997)

[11] Deb, B., Bhatnagar, S., Nath, B.: STREAM: Sensor Topology Retrieval at Multiple Resolutions, Computer Science Dept. Rutgers University

[12] Deb, B., Bhatnagar, S., Nath, B.: A topology discovery algorithm for sensor networks with applications to network management. In: Proceedings of the IEEE CAS Workshop on Wireless Communications and Networking, Pasadena, USA (September 2002)

[13] Deb, B., Bhatnagar, S., Nath, B.: Multi resolution state retrieval in sensor networks. In: Proceedings of International Workshop on Sensor Network Protocols and Applications, SNPA (2003)

[14] Ephremides, A., Wieselthier, J.E., Baker, D.J.: A design concept for trliable mobile radio networks with frequency hopping signaling. Proceeding of IEEE 75(1), 56–73 (1987)

[15] Ganesan, D., Krishnamachari, B., Woo, A., Culler, D., Estrin, D., Wicker, S.: Large-scale Network Discovery: Design Tradeoffs in Wireless Sensor Systems, Student Poster Presented at the 18th ACM Symposium on Operating System Principles, Banff, Canada (October 2001)

[16] Gerharz, M., Waal, C.D., Martini, P.: A Cooperative Nearest Neighbors Topology Control Algorithm for Wireless Ad Hoc Networks. In: IEEE ICCCN 2003, Dallas, TX (2003)

[17] Gerla, M., Tsai, J.T.C.: Multi cluster, mobile, multimedia radio networks. ACM J. Wireless Networks 1(3), 255–265 (1995)

[18] Guha, S., Khuller, S.: Approximation Algorithms for Connected Dominating Sets. In: European Symposium on Algorithm, pp. 179–193 (1996)

[19] Haas, Z.J., Deng, J., Liang, B., Papadimitratos, P., Sajama, S.: Wireless Ad Hoc Networks. In: Proakis, J. (ed.) Encyclopedia of Telecommunications, John Wiley, Chichester (2002)

[20] Jennings, E., Okino, C.: Topology control for efficient information dissemination in ad-hoc networks. In: 2002 Symposium on Performance Evaluation of Computer and Telecommunication Systems (SPECTS 2002), San Diego, CA, July 14-18 (2002)

[21] Krishnamachari, B., Sitharama Iyengar, S.: Efficient and Fault-tolerant Feature Extraction in Sensor Networks. In: Zhao, F., Guibas, L.J. (eds.) IPSN 2003. LNCS, vol. 2634, pp. 488–501. Springer, Heidelberg (2003)

[22] Li, N., Hou, J.C.: Topology Control in Heterogeneous Wireless Networks: Problems and Solutions. In: IEEE INFOCOM 2004, Hong Kong (March 2004)

[23] Li, X.Y., Wan, P.-J., Wang, Y., Yi, C.W.: Fault Tolerant Deployment and Topology Control in Wireless Networks. In: Proc. 4th ACM Intl. Symp. on Mobile Ad Hoc Networking and Computing (MobiHoc), Annapolis, MD (2003)

[24] McGlynn, M., et al.: Birthday protocols for Low Energy Deployment and Flexible Neighbor Discovery in Ad Hoc Networks. In: MobiHoc, LongBeach, CA (2001)

[25] Ni, S., Tseng, Y., Chen, Y., Sheu, J.: The Broadcast Storm Problem in a Mobile Ad Hoc Network. In: ACM MOBICOM 1999 (August 1999)

[26] Jianping Pan, Y., Hou, T., Cai, L., Shi, Y., Shen, X.: Topology control for wireless sensor networks. In: Proceedings of the 9th annual international conference on Mobile computing and networking, San Diego, CA, USA, pp. 286–299 (2003)

[27] R. Choudhury, R., Bandyopadhyay, S., Paul, K.: A Distributed Mechanism for Topology Discovery in Ad hoc Wireless Networks using Mobile Agents. In: Proc. of Workshop On Mobile Ad Hoc Networking & Computing, MOBIHOC 2000 (2000)
[28] Tim, N., Johann, H.: Local, Distributed Topology Control for Large-Scale Wireless Ad-Hoc Networks. In: International workshop and wirelesses hoc networks, University of Twente, Netherlands (2004),
http://www.cwc.oulu.fi/iwwan2004/slides/LPA_137.pdf
[29] Zhou, C., Krishnamachari, B.: Localized Topology Generation Mechanisms for Self-Configuring Sensor Networks. In: IEEE Globecom, San Francisco (December 2003)
[30] A Topology Discovery for Sensor Networks with Application to Network Management,
http://www.cs.rutgers.edu/dataman/papers/TopDisc.pdf

Inventory Management in the Packaged Gas Industry Using Wireless Sensor Networks

A. Mason, A. Shaw, and A.I. Al-Shamma'a

Liverpool John Moores University, Liverpool, United Kingdom

Abstract. Gas cylinders are used in many different situations, such as in research, industry, healthcare, and even the home. Due to demand in such a wide variety of situations, there is the inevitable ambition of gas suppliers to improve the efficiency of their business. To this end, a prototype Wireless Sensor Network (WSN) based inventory management has been implemented by the authors in order to provide such improved efficiency whilst also integrating sensors in order to monitor the condition of gas cylinders. This chapter discusses why such system is needed and what advantages it gives over existing technologies for this purpose. In addition the system is talked about in terms of its components, how it operates, its key features and also its preliminary experimentation results. Finally, conclusions are drawn on the current work, and how to proceed in the future is alluded to.

1 Asset Tracking and the Packaged Gas Industry

An asset is defined by the Oxford English Dictionary [1] as a "useful or valuable person or thing". In the context of this work, asset tracking refers to the knowledge an organisation possesses with regards to the location of their property as a result of its movement. One should not confuse an asset with the term inventory, since, although the two are commonly interchangeable, inventory is formally defined as "a list of items held in stock" [2]. Therefore in the packaged gas industry a gas cylinder is an asset, whilst a listing of all of the gas cylinders would be the inventory.

Such a distinction is important, since asset tracking is not a standalone procedure but more a component of a larger inventory management process. Inventory management is crucial to organisations for a number of reasons, and is ultimately the catalyst for this work. An inventory management system allows an organisation to know what assets it has available. However, this knowledge alone is often not particularly effective because there is the possibility of assets being stolen, misplaced or lost, all of which add cost. Therefore the addition of extra tools, such as asset tracking, into the inventory management process allows organisations to reduce such costs through maintaining greater asset visibility. Furthermore, concepts such as just-in-time (JIT) operation [3] can be implemented more effectively, thus allowing organisations, where applicable, to reduce their total inventory.

Within the supply chain of a packaged gas organisation, a gas cylinder may move to many different locations for many different reasons; this movement forms part of the life cycle of a gas cylinder, which is illustrated a flow diagram in Figure 1. An empty cylinder must first be filled with an appropriate gas. It will then be marked for delivery to a regional storage centre based upon demand for that gas. Once delivered to a storage centre, the gas will be stored for an unknown period of time. This may be days, months or even years in some cases – Figure 2 serves as an example of how gas cylinders may be found in a storage yard.

When a bottle is allocated for delivery to a customer it is loaded onto a delivery truck (an example of which is shown in Figure 3) and delivered. The customer will then consume the gas for their purposes – once it is completely consumed they may contact the gas company and request the empty cylinder be picked up. This empty cylinder will then be returned for refilling, but first it is inspected for any damage. If it is not serviceable, then it will either be repaired or disposed of.

Until quite recently, many within the packaged gas industry did not apply any method of tracking to their gas cylinders. This led to it being impossible to trace individual cylinders, meaning that an organisation would not know whether a particular cylinder was in stock, or with a customer. Storing information about individual cylinders was impossible, which is unacceptable when one considers the importance of cylinder maintenance. Cylinders have to be regularly tested, typically via hydrostatic methods [4], to see if they are corroded and if they are able to withstand their rated pressure. These tests are vital in order to ensure that cylinders are safe for use; if a cylinder cannot be identified, how does one prove that such tests were carried out?

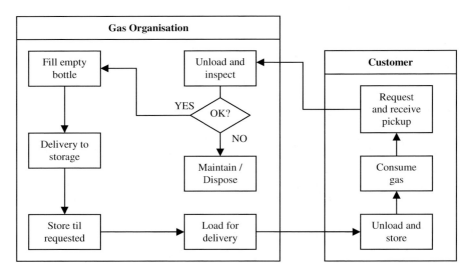

Fig. 1. The life cycle of a gas cylinder as it is stored, used and maintained

Inventory Management in the Packaged Gas Industry Using WSN

Fig. 2. Gas cylinders stored in a typical storage facility, grouped in metal racks

Fig. 3. Gas cylinders loaded onto a delivery truck ready to be delivered to customers

In terms of inventory management, many resorted to a best effort method by simply recording stock levels via manual counting. This afforded the ability to estimate whether or not a particular type of gas cylinder might be in stock. However, this system was open to errors in both the counting of stock levels and their communication. In addition, malpractice by customers was difficult to trace and so theft of gas was (and still is) common.

As a result, larger suppliers of gas cylinders, such as Air Products [5], have started using bar code systems. In addition, numerous third party companies, Pollution Control Industries [6] for example, have similar technologies which they will assist organisations in deploying. However these systems are largely manually operated, with personnel being equipped with hand held scanners (similar to that in Figure 5) in order to maintain inventory. Therefore, whilst it is now possible to track a cylinder as it moves through the organisation, is sent to customers, and is maintained, there is still the potential for error and theft since there is little way to automate the detection of cylinder movement within, or out of, a storage facility.

The problems faced as a result of poor tracking have been recognised by the packaged gas industry, but there appears to be little motivation to move to currently available alternative technologies in order to realise automated tracking. These alternatives exist in the form of large fixed barcode reader units, as well as in the form of RFID (Radio Frequency Identification) tags. Examples of both are shown in Figure 5.

Barcodes [8] are machine readable representations of information that rely upon optical equipment to detect (typically) the width and gaps of vertical black bars. Arguably they achieved commercial success in the 1980's [9], despite being first patented in 1952. Regardless it is difficult to find a purchasable item which does not have some form of barcode identifier and this is largely due to two reasons; barcodes are incredibly cheap, as well as having well recognised standards [10-11] such as UPC (Universal Product Code) and EAN (European Article Numbering). Further work in recent years by the bodies overseeing these standards has ensured that barcode readers are compatible with both.

Fig. 4. A Symbol MC3000 [7] handheld barcode scanner unit similar to those in use by some parts of the packaged gas industry

Inventory Management in the Packaged Gas Industry Using WSN   79

Fig. 5. (left) An industrial barcode scanner and (right) an example of an RFID tag

Barcodes have a distinct disadvantage however since they rely on optical line of sight for reading barcode information. Whilst industrial scanners (such as that shown in Figure 5) have large read ranges when compared to hand held devices (5m compared to 0.5m respectively), they cannot penetrate opaque objects. This means that automation of barcode tracking systems can be tricky if items are prone to movement or rotation, or if they are densely packed. Radio wave based technologies such as RFID have the upper hand here since radio waves can penetrate opaque objects and are therefore not subject to the same disadvantage. In terms of a practical example, if a large palette of items enters a warehouse then a barcode scanner would not be able to identify every item on the palette since it is likely that many items in the centre would be obscured. Therefore, a barcode on the palette itself may be read to identify it, and then this identity can be linked to a predefined list of contents, which are not necessarily guaranteed to be present. In theory however, RFID would not have such a problem, and so the entire contents of the palette could be listed in real-time as the palette enters the warehouse, without the staff even needing to be aware that any inventory mechanism is in place.

Companies such as Dell, Proctor and Gamble and Federal Express have all begun to make use of RFID in their supply chains since it able to easily provide high levels of visibility to assets as they move [12-13]. Wal-Mart is a key figure also in the RFID world since they are well known to have demanded their suppliers to attach barcodes to every box and palette shipped to them. The RFID tag itself, shown in Figure 5, is made up of an antenna for reception and transmission, some sort of housing material to allow it to maintain its structure (and often also so it can easily be attached to assets) and an integrated circuit (IC) which contains a serial number allowing the associated asset to be identified. It is also possible to include further information, but this is not often utilised in the supply chain as it increases the cost of the tags, which may be anywhere in the region of 10c - $10, depending on the particular application requirements. In addition to a tag, a reader device is required which interrogates the tags and causes them to respond; communication is directly between the tag and reader.

RFID uptake however is stunted by three major factors which are summarised below:

- *Environmental impact* – RFID tags are affected by their surrounding environment since radio waves display different propagation characteristics in different materials. Metals with a thickness of just a few µm may be able to block radio waves completely [14], which makes tracking assets such as gas cylinders difficult. Water also attenuates radio waves significantly and so can prove to be a problem.

- *Standardisation* – There are a number of available RFID standards, and whilst there are some efforts to provide global standardisation (e.g. GS1 [14], who oversee barcode standardisation), but there are still a number of non-compatible standards. In addition, as the technology is improved this problem worsens and the fear for many is that early adopters will be left with expensive proprietary tracking systems.

- *Cost* – A major factor owing to the fact that an RFID tag may cost many times more than a barcode. However, for many there is the possibility of saving vast quantities of money through implementation of JIT services, such that huge inventories are not a necessity. Therefore cost is an issue in some circumstances, but the greater asset visibility afforded by more intelligent tracking can be shown to actually provide savings in the long term [13].

2 Using Wireless Sensor Networks for Gas Cylinder Tracking

The work discussed in the remainder of this chapter is related primarily to tackling the issue of RFID tags becoming compromised in certain environments. One feature of asset tracking that becomes clear very quickly is that space is costly, and therefore at a premium in many cases. As a result, assets are usually packed very closely together – this is shown to some extent in Figures 2 and 3 where gas cylinders are stored in metal crates in close proximity. This work utilises this fact by allowing individual tags to communicate with each other over short distances, as opposed to the typical RFID model where tags have to communicate over long distances to a reader or interrogator device. In the RFID model, the long range communications can be attenuated or blocked easily by objects within the operating environment (such as metallic gas cylinders); this would effectively lead to assets not being identified and being invisible in the supply chain. However, if the cylinders can communicate directly with each other and relay data to some reader device then this problem needn't arise. This principle is illustrated in Figure 6.

Inventory Management in the Packaged Gas Industry Using WSN

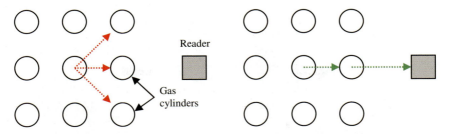

Fig. 6. Gas cylinders with RFID tags (left) and mote devices (right) attached

Fig. 7. Mica2 (left) and MicaZ (right) motes

In order to achieve this idea, a prototype system was developed using miniature wireless hardware based on the popular Berkeley Mica2 and MicaZ mote platforms [17] – see Figure 8 for an illustration. These devices incorporate the Atmel ATmega128L [18] programmable microcontroller, which gives access to eight 10-bit analogue to digital channels (ADC), as well as 4KB RAM, and 128KB program memory. Important additional components include a unique identification IC, a battery compartment for 2xAA type batteries, an MMCX connector for external antennas and also a 51-pin expansion connector which allows the addition of peripheral devices to the motes. Such devices could include sensors, some of which are shown later in this chapter for specific use in the packaged gas industry. Finally, incorporated into the motes are wireless transceivers for communication; the Mica2 operates at 915MHz, whilst the MicaZ operates near the ISM 2.4GHZ band. Please note that both types of sensor node are used as part of this work, and so will be referred to collectively as the Micax motes future references.

The ability of sensor nodes to communicate and form networks is an essential part of their function within a Wireless Sensor Network (WSN). They require this ability in order to communicate with their neighbours on the network, as well as with the base station which serves as a gateway between a WSN and a workstation or the Internet. The nodes within a WSN typically form a mesh topology [19]. This allows the WSN to route data between nodes on the network by relaying, or

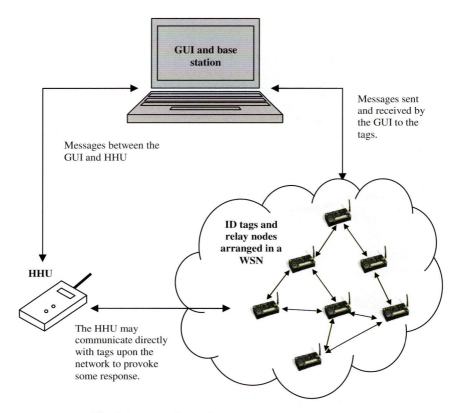

Fig. 8. Prototype inventory management system overview

hopping, data from node to node until an appropriate destination is reached. This is often referred to as multihop communication [20] or multihop routing [21].

Connections within a mesh network may be configured and maintained throughout the life of the network, or there may be no connections at all – that is – data is sent in the hope that it will arrive at the target destination. These are described as connection orientated and connectionless networks [22] respectively. Importantly, the connections in mesh networks can self heal in the event of a failure by simply routing data around the point of failure. This increases the reliability of the network in the event of hardware failure or movement, since connectivity is preserved.

3 Prototype System Implementation

3.1 System Overview

In order to demonstrate the feasibility of wireless sensor devices in the packaged gas industry it was necessary to create a prototype system which could be shown

Inventory Management in the Packaged Gas Industry Using WSN 83

to interested parties. This system is made up of a number of different key components; ID tags and relay nodes, a Graphical User Interface (GUI) with an associated base station, and a Hand Held Unit (HHU). An overview of each of these components is given in this section, and Figure 8 gives an idea of how they are linked together to form this prototype inventory management system.

Each gas cylinder requires unique identification and so should be individually tagged. The tags in this prototype system are Micax motes, which are programmed with a TinyOS [23] application allowing the unique identification of any asset to which they are attached – this is facilitated using a serial silicon number IC (the Dallas Semiconductor DS2401 [24]) which is included as part of the mote design. Each tag can communicate both directly and indirectly (i.e. – via multihop communication) with its neighbouring tags, the GUI and the HHU. Relay nodes are also included with the purpose of providing additional network coverage between the GUI and the tags. In practise the relay nodes are identical to the tags in this system.

A HHU has been created to complement the tagging system. This portable device allows a user to interact with the tags remotely – it has controls, a Liquid Crystal Display (LCD) and its own power supply. In the current prototype the HHU can be used to locate user defined assets via range determination using the received signal strength indicator (RSSI) included as part of the mote radio transceivers.

Finally, a GUI is included to give the user a clear representation of the assets; it allows the tracking of assets as they move, as well as keeping a record of the current inventory at each appropriate location. The GUI instigates interrogation of the tags, as well as allowing communication with the HHU. The GUI interfaces with the tags and the HHU via a base station. The two are inseparable, as the base station serves as the interface (or gateway) between the GUI and the other components. As a result, the remainder of this document refers to data as being sent or received by the GUI for simplicity.

3.2 System Operation

The tags are required to be attached to gas cylinders in order to provide unique identification. Many modern cylinders have screw fitted plastic collars, as shown in Figure 9. The space between the inside of the collar and the gas cylinder is large enough to fit a Micax mote PCB; the battery pack must be detached however. The fit is rather tight and the final part of this chapter considers the future likelihood of much smaller motes becoming available which might be better suited. Motes such as the Mica2Dot [17] are already available, and would easily fit – they are just not as practical to work with as an everyday research platform. They are very similar to the Micax motes in terms of hardware however, hence requiring little or no modification to apply the TinyOS tag application to these devices.

The tags are designed such that they respond to requests made by the GUI with their identification number. As described previously, the tags form part of a mesh network which means that data can be passed directly and indirectly to the GUI. The GUI itself (shown in Figure 10) can scan for tags in order to create an inventory, as well as communicate directly with an individually discovered tag.

Fig. 9. Gas cylinder collar

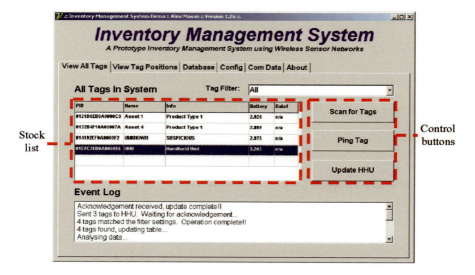

Fig. 10. Inventory list shown on GUI

This later feature is used in the prototype to sound a piezoelectric speaker on the tag as well as flash an LED, which could possibly be used to assist locating a particular asset.

When creating an inventory list, the GUI records the last relay node that a response message from a tag passes through before being interpreted by the GUI. This means that assets can be positioned in real-time relative to their nearest relay nodes, assuming that the positions of the relay nodes are known. In terms of a gas cylinder storage facility, there may be relay nodes in particular areas such as entrance and exit gates, as well as in different areas designated for different types of cylinder. As a result, each relay node could be representative of a particular position in Figure 11, and the resolution of the system is limited only by the number of relay nodes used during deployment of the system.

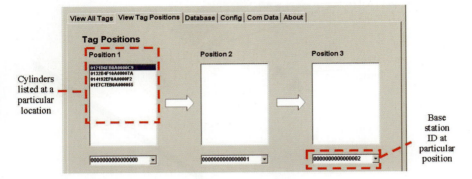

Fig. 11. Location discovery of assets

Figure 10 also shows that the GUI can be used to transfer data to the HHU, the unit which can be used in this scenario to locate assets (see Figure 12). The data transferred is the current inventory which has been detected by the GUI; all tag serial numbers are uploaded to the HHU and buttons on it allow the user to select a tag to actively locate, as shown in Figure 13. The device is powered by a PP3 9V battery, and contains a Micax mote interfaced to a PIC (programmable interface controller [26]) driven LCD. An external antenna is also provided to ensure reasonable range can be achieved – during experimentation this was found to be in the region of 20-30m. Figure 14 shows an example readout from the box; this readout gives the current signal strength of the selected tag, which is identified via its silicon serial number.

3.3 Key Features

The prototype system is rather extensive in its implementation [27], and so rather than attempt to describe it in great detail there will be a description here of its key features – particularly those which allow the system to operate effectively.

Since the multihop communication method of WSN is a key reason for applying them to asset tracking, the first discussion is regarding the messaging mechanism implemented. Typical RFID includes a simple interrogation command which demands tags respond with their serial number. This is a quick and effective way of finding the presence of tags local to a reader. Carrying forward this idea,

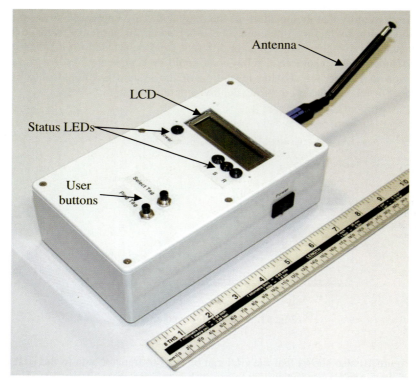

Fig. 12. The hand held unit (HHU) used for asset positioning

Fig. 13. LCD readout from the HHU, showing the currently selected tag (top line) and its serial number (bottom line)

Fig. 14. The HHU LCD showing the serial number of the selected tag (top line), as well as the signal strength currently being experienced (bottom line)

since it is well established, the prototype GUI transmits a request message to the sensor network. All tags which overhear this message are required to respond as well as rebroadcast the message exactly once. This means that nodes multiple hops from the GUI device can receive the message and also respond.

The mechanism used for the response is similar to the reverse routing implemented in AODV (Ad-hoc On-demand Distance Vector) routing [28] – an illustration of the mechanism is shown in Figure 15. AODV, as the name states, is an on-demand mechanism, which means that the tags only need to respond when requested to; for the rest of the time they may sleep and therefore save energy. Energy saving is a key design aspect that will be revisited many times within this section, and its importance cannot be understated since the battery life of the tags is a factor in their cost. It is also apparent that using a single return route for data is preferable to broadcasting, since the number of messages generated is significantly reduced. Figure 16 serves to further highlight this issue by simply looking at the case were a single message is sent out on a network where all the nodes are exactly one hop from the source node, and can all communicate with each other, error free. With a network consisting of 10 nodes, broadcasting generates almost five times more messages than the reverse routing technique used in this system; once 50 nodes is reached this increases to nearly 25 times more sent messages! This indicates that broadcasting is acceptable for transmitting a single on-demand request to many tags, but allowing the tags to broadcast their response causes a significant number of messages to be generated for propagation through the network. This leads to a high level of redundant packets on the network which would increase contention between tags as they try to deal with this redundancy. The ultimate result of this is for data collisions to occur because tags are trying to transmit messages simultaneously. This is commonly referred to as the broadcast storm problem [29].

The use of AODV is not only important in terms of saving energy, however, since it is also quite lightweight in terms of memory consumption. Routes are created when they are required, whereas protocols such as DSDV (Destination Sequenced Distance Vector) routing require that all tags know a path to all other tags on the network. Obviously, with the tags having only a small quantity of memory and potentially millions of routes in a large scale system, DSDV would not be a scalable approach.

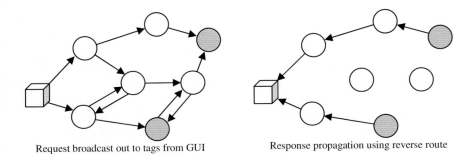

Request broadcast out to tags from GUI   Response propagation using reverse route

Fig. 15. Message propagation using reverse routing

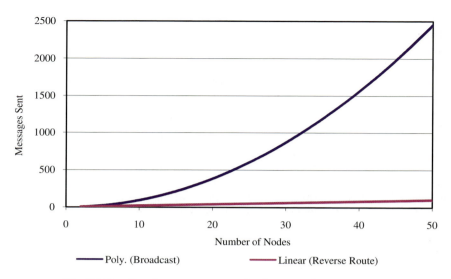

Fig. 16. Broadcasting vs. reverse routing in terms of messages sent

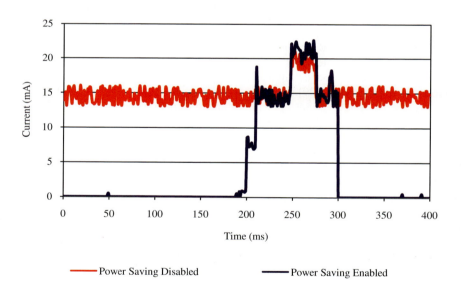

Fig. 17. The effect of enabling power saving features

Another key feature of the system, in particular the design of the tag software, is that the tags are programmed to save energy as much as possible through the use of sleep states. The microcontroller and radio transceiver of the Micax motes support numerous sleep states, with some allowing the tag to consume < 10μA

Inventory Management in the Packaged Gas Industry Using WSN

[17] of power; this is compared to 10mA of power when they tags are awake and processing, or approximately 30mA of power when transmitting data. In Figure 17 one can see the difference between a tag utilising power saving mode, and one which is not; unfortunately the Instek GDS-810S [30] digital oscilloscope used does not have the resolution available to show the power consumption whilst sleeping.

When a tag is switched on, it performs numerous self tests, checking battery voltage and memory status; it then enters a sleep state where the microcontroller, transceiver, and other components are either turned off completely or in a low power state. Every 100ms, the transceiver powers up to test the wireless channel for activity, and goes back to sleep if there is none. In order to prevent message transmission from being missed by a sleeping tag, initial interrogation messages from the GUI are very long (> 200ms) in order to ensure that tags do not miss the request for information. Figure 18 shows a flow diagram of how the tag software operates; experimental results later in this chapter show the effect of such power saving features.

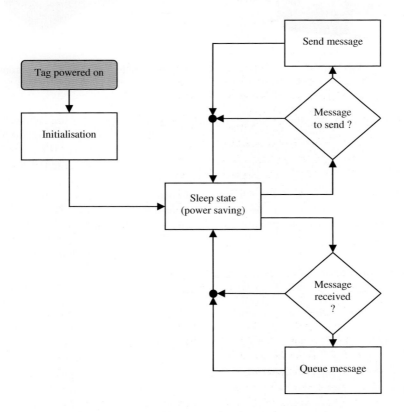

Fig. 18. TinyOS tag application flow diagram

3.4 Sensors for Asset Tracking

The Micax mote platform is highly expandable, as already demonstrated with the HHU, which has a Micax mote at its heart. In addition this work also demonstrates how WSNs could add value to inventory management systems through the provision of additional and highly useful information. Two such examples are presented here in the form of a pressure sensor and also a 2-axis accelerometer; these are shown in Figure 19.

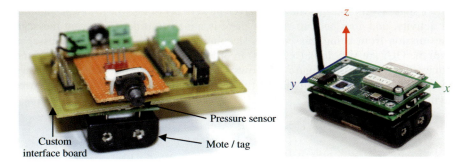

Fig. 19. Pressure sensor (left) and sensor board including 2-axis accelerometer incorporated into the MTS420 [31] sensor board (right)

Dealing with pressure sensing initially, a gauge type pressure sensor was chosen. However, rather than use an expensive sensor [31] capable of reading the high pressures that a gas cylinder can maintain, a cheaper option [33] (< £10) was found to achieve proof of principle. This sensor is capable of reading 0-100psi (0-6.9bar) and operates using the piezoelectric effect [34]; pressure at the sensor input port causes stress on the material contained within and results in an electric signal. It was attached to a custom made mote interface board. The sensor was calibrated using an air compressor, and then the calibration data integrated into the GUI. A cylinder pressure sensor could be useful for a number of purposes such as allowing automatic checking of leaks as well as automated alerts when a gas cylinder is running low; the later alert could even be triggered at the customer location in order to prompt a new delivery of gas. Figure 20 shows how the pressure sensor data can be integrated into the prototype system.

Whilst a pressure sensor could add value to a packaged gas inventory management system, it is not the only sensor which could be useful. Acetylene, a highly flammable gas which is used in oxyacetylene cutting and welding has strict storage guidelines [35]. The gas is unstable when pressurised, and so for storage in a pressurised cylinder it must be dissolved in acetone. The cylinder is also filled with a porous material which acts like a sponge to soak up the acetone. As a result of this arrangement, the cylinder must be kept upright in order to prevent acetone leaking from the cylinder – if such a leak does occur then a build up of acetylene gas may occur which could result in an explosion.

Inventory Management in the Packaged Gas Industry Using WSN

PID	Note	Gas Type	Battery	Pressure
0132B4F10A00007A	Ready for delivery	Argon	2.899	214.4 bar
014192EF0A0000F2	NNODE		2.968	n/a
0121D6EB0A0000C9	Ready for delivery	Argon	2.968	214.4 bar
01E7C7EB0A000055	Empty cylinder	Argon	2.968	0 bar

Fig. 20. Integration of pressure sensor into the GUI; note values are scaled by a factor of 32 in order to make them appear to be realistic gas cylinder pressures.

In order to provide an additional safety feature the tag could have an integrated accelerometer [36] (see Figure 19) so that the orientation of a gas cylinder could be determined. This sensor is a micro electro-mechanical system (MEMS) which measures the acceleration and gravity that it experiences. Any change in gravity or acceleration repositions capacitive plates within the device, which are suspended on springs – this alters the voltage output of the sensor for the axis affected.

Since the accelerometer covers only 2 axes, it can only detect rotations in the *xz* and *yz* planes. In order to demonstrate this sensor in a packaged gas scenario, it was assumed that the mote would be orientated such that its antenna socket would be perpendicular to the earth and be pointing skyward. In this position the *x*-axis output is 1.31V and the y-axis output 1.65V. It was assumed that a deviation greater than ±0.05V on either axis would indicate that the gas cylinder was not upright. As such, this logic was incorporated into the prototype system software. Once the GUI has readings from both axes, it determines whether the gas cylinder is upright or not, as shown in Figure 21.

PID	Note	Gas Type	Battery	Upright?
0132B4F10A00007A	Ready for delivery	Acetylene	2.899	Yes
014192EF0A0000F2	NNODE		2.968	n/a
0121D6EB0A0000C9	Ready for delivery	Acetylene	2.968	NO!!
01E7C7EB0A000055	Empty cylinder	Acetylene	2.968	Yes

Fig. 21. Integration of orientation data into the GUI

4 Experimental Results

4.1 Battery Life

Previously, power saving was discussed as part of the prototype application's key features because it could significantly reduce the consumption of the tag's finite energy resources. To investigate this, two tags were employed, but only one had the power saving enabled. This experiment ran for a little over 6 days (148 hours) – the time it took for the battery voltage of the tag without power saving to drop to 2.5V. It is reported that the Micax motes will operate to 2.1V; the datasheets for the microcontroller and transceivers support this report, however the ADC is not capable of accurately reading voltages this low.

The results of the experiment are shown in Figure 22. The tags each used two type AA alkaline batteries (1500mAh capacity), and the duty cycle was 1%. Studying Figure 22, there is an obvious difference in the gradient of the two lines; in the time it took the mote with the power saving disabled to drop 0.74V, the mote with it enabled drops just 0.09V. This initial result would indicate that the later mote would operate 8 times longer due to power saving being enabled, but this does not take into account the faster decay rate shown between approximately 3.2V and 3.0V when the power management was disabled. Therefore, if the graph is analysed in a linear way and split into two parts (i.e. 3.2-3.0V and 3.0-2.5V), it can be shown that the voltage decay rate is approximately five times greater in the first part than the second. Using this result it is possible to estimate that power saving enables a mote to operate for 254 days from a voltage of 3.2-2.5V. Using the same linear relationship and the assumption that the motes will operate to a voltage of 2.1V it is possible to estimate a maximum lifespan for the motes of 1.24 years which is comparable with the manufacturer's claim of 1.5 years considering the batteries used in these experiments were relatively low capacity. Batteries rated in excess of 2000mAh are easily available, and would increase the mote lifespan significantly; as an example, 2500mAh batteries should be capable of providing over 2 years of operation based on the experimental results obtained in this work. Reducing duty cycle further would further increase this life expectancy, as could the use of energy scavenging techniques.

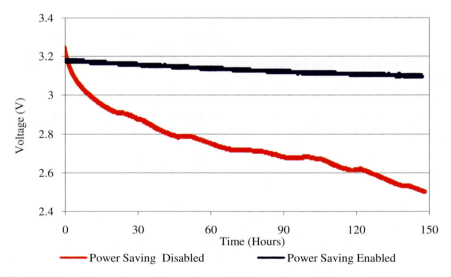

Fig. 22. Comparison of tag voltage decay with time when enabling or disabling power saving features

Inventory Management in the Packaged Gas Industry Using WSN 93

4.2 Using the Hand Held Unit for Asset Discovery

An experiment to test the HHU was set up inside a metal shipping container (5m long, 2.5m wide and 3m tall). This environment was chosen in order to give an ideal situation such that external interference would have no bearing on the measurements taken, as the container acts as a Faraday cage. Since sufficient spare gas cylinders were not available for use, 24 plastic crates (stacked in four columns, each six crates high) were used – one of the crates contained a tag, which was moved at random from crate to crate. This setup is shown in Figure 23.

Fig. 23. Experimental setup of 24 crates in a metal shipping container for testing the HHU

To perform the tests, firstly the HHU was moved along the length of the container in 0.5m increments from the back wall to the door. At the point where the lowest RSSI value was found the HHU was then moved toward the stacked crates in 0.2m increments, maintaining a fixed height of 1m above the ground. Finally the HHU was moved vertically from the floor to the top of the crates (0.3m increments from 0-1.5m) – the aim of these sweeps was to see if the crate with the mote in could be discovered reliably through a procedural method. Due to time limitations imposed upon utilising the container, ten iterations of this test were recorded and it was found that 80% of these tests resulted in the correct crate being discovered.

It was noted that the vertical movement of the HHU was useful, particularly because the HHU was so close to the crates. It was found that if the correct column of crates was discovered through moving along the length of the container, then the vertical movement of the HHU next to the crates always resulted in a high RSSI reading at the crate where a tag was located. Therefore in order to capitalise on this feature further tests were completed which involved holding the HHU close (~10cm) to the front of each crate, which are numbered in

Figure 23, and taking a RSSI reading. It was found that this was 100% reliable in identifying the crate containing the tag. Figure 24 shows an example of two tests where the mote was in different crates. The highest RSSI value in each test represents the position of the mote, and it is noteworthy that this RSSI value is approximately 3-4dBm up on the next highest value in each case.

Given the large error experienced in the first set of tests, it is likely that the HHU would be most suitable for use in close proximity to gas cylinders, rather than as a long range device. This can still be useful however, as the GUI can narrow down the location of a cylinder to the nearest relay node, and therefore a human operator will have a good indication of where to begin with the HHU.

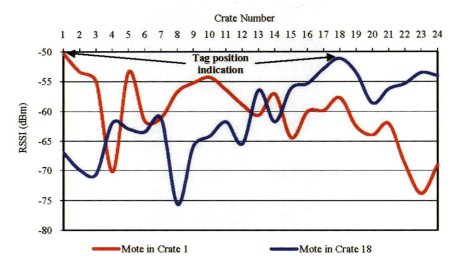

Fig. 24. Results of HHU experimentation

4.3 Industrial Demonstration

A small scale demonstration was set up for the packaged gas industry using the prototype system. The packaged gases used in this demonstration were metallic gas cylinders stored outdoors in a metallic cage (see Figure 25). Three cylinders were used, with each having a tag securely tied around its collar – these cylinders did not have the plastic collar fitting as shown in Figure 9. A relay node was placed approximately 10m away from the gas cylinders. A PC with the prototype software and the base station were placed in an adjacent building. This setup is demonstrated graphically in Figure 26.

In addition to the metallic storage cage, a mote on each cylinder was arranged so that it provided the worst possible operating conditions; non-line of sight conditions were present, antennas were touching the cylinders, and they were not matched in terms of polarisation. One hundred scans were performed with the software and the number of tags recognised upon each scan was recorded. Table 1

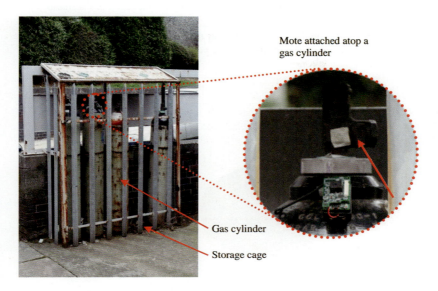

Fig. 25. Gas cylinders used for prototype demonstration

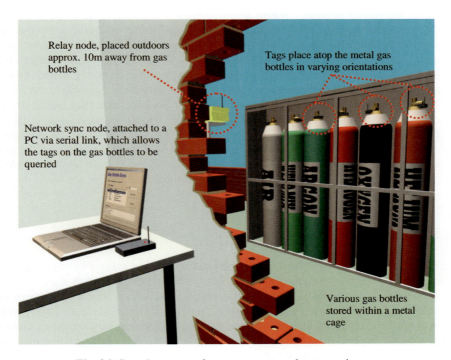

Fig. 26. Complete setup of prototype system demonstration

Table 1. Results of testing the prototype system in a packaged gas scenario

Iterations	Successfully discovered tags (%)
1	94.5%
2	99.75%
3	100%

Fig. 27. System on chip integrated circuit for future motes

shows the results obtained. Whilst it is still possible to achieve 100% reads using the prototype system in this scenario, it requires further scanning iterations in order to achieve this. As a result a reliable delivery system may be required as a modification to the current routing mechanism – this is a topic which is discussed in the final part of this chapter. It is also important to note that the Mica2 motes are used during these experiments since a large quantity were available at the time. These motes operate at 915MHz, so it is also possible that they suffer from cellular network interference, since UK mobile phones operate around these frequencies and is the reason that, legally, 915MHz devices could not be deployed in the UK. In addition, medium access control (MAC) is implemented via the tag software to accommodate the Mica2 motes; the MicaZ motes have hardware MAC implementation as part of their Zigbee [37] compliance and so are likely to be much more effective in collision avoidance.

5 Conclusions and Future Work

The prototype systems utilises a mesh network in order to attempt to overcome the issue of attenuation suffered by RF devices. The motes have been shown to be scalable through the AODV like reverse routing mechanism which was shown to exhibit a linear increase in traffic (see Figure 16) when an increase in network nodes is experienced, which is far more suitable than the traffic increase exhibited by a purely broadcast based network. A positive side effect of this routing method is that, because it generates less traffic, it also saves energy.

Further energy saving has been shown with the use of power saving features implemented by the mote hardware and the operating system, such that high capacity batteries could possibly achieve a two year lifespan for a Micax mote. Additional lifespan gains could be achieved by reducing the duty cycle of the motes, a feasible option since gas cylinders may lay dormant for many months at a time. It follows that the longer a tag battery lasts, the lower the maintenance costs of the system and the more attractive it is to industry. As part of the prototype system, all tags report their battery voltage so that the system can alert operators to tags which may require maintenance.

The intelligent nature of WSN has been shown through the implementation of sensors – not only can the tags be used to identify assets uniquely they can also determine some characteristic of the asset. This has been shown with the gas cylinder application, where the tags have had both pressure and accelerometer sensors attached. This means that the tags can tell how much gas is left in a cylinder – this could be used for leak detection as well as allowing for automatic deliveries when a customer's gas supply is running out. The accelerometer sensor can be used to determine how an acetylene cylinder is orientated, and so could be a useful safety feature.

Expansion is demonstrated in multiple ways – the HHU was created as an application specific tool which could be developed further for other tasks. It was found that using the HHU as a location device was 80% reliable in limited testing, but was 100% reliable at identifying an item if it was held approximately 10-20cm away from it, leading to the conclusion that it may be most useful in its current state for close range asset discovery and authentication. Future work could consider other possible uses of the HHU; for example it could be used to discover assets within its vicinity rather than having a preloaded database of nodes available.

In terms of testing the prototype system it was found that it was reliable provided some redundancy was added to the system through repeated scan requests. It is thought that the small amount of tags which are not read is due to data collisions which appear to occur despite MAC implementation and random messaging offsets.

It is proposed that future work should consider more extensive testing and development of the prototype system in terms of both simulation and industrial trials. For network devices there is simulation software such as ns-2 [38] which can assist evaluation and aid the improvement of communication logic. An interesting comparison could be to see whether the service orientated structure that is in place now would be served better by a harvesting structure, which would remove the on-demand nature of the system presented in this work. Another aspect that could be evaluated is the effect asset movement has on the reliability of communication. It might also be useful to calculate the optimum arrangement of relay nodes, which provide additional network coverage.

Industrial trials will also be important however, since simulations may not take into account all situations which can occur in an industrial situation. Furthermore, industrial experience will allow for increased accuracy in the simulations. In order for such theoretical and practical testing to be feasible there needs to be industrial support and funding, as well as collaboration with standards bodies in order to ensure that any technological advancement complies with recognised standards to avoid the stigma currently attached to RFID use.

Some might argue that this technology is not ready for deployment, however is untrue; many organisations now make money from the sale of WSN hardware which could be used for industrial purposes [39-40]. This proves that there is a market for such devices, so the aim now should be to reduce the cost so that they are feasible for large scale applications. One way in which this could be achieved is through removal of some unnecessary components from the motes – the EEPROM memory on the Micax motes, for example, is not used in this work. In addition, there are system on chip (SoC) devices which combine both the microcontroller and the transceiver into a single package (see Figure 27) [41] – this may not only cut the cost of the device but also assist in size reduction and make them suitable for easy integration into the plastic collars found on many gas cylinders.

References

[1] The Oxford Dictionary and English Usage Guide, Oxford University Press, Oxford (1996) ISBN 0198613253
[2] Waters, C.D.J.: Definition of Inventory Terms. In: Inventory Control and Management, ch. 1, p. 4. John Wiley & Sons, Chichester (1992)
[3] Waters, C.D.J.: Just-In-Time Systems. In: Inventory Control and Management, ch. 8, pp. 295–322. John Wiley & Sons, Chichester (1992)
[4] McAllister, E.W.: Hydrostatic testing for pipelines. In: Pipeline Rules of Thumb Handbook, 6th edn., p. 151 (2005) ISBN: 0750678526
[5] Air Products PLC: ACTS – Cylinder Tracking System. Proprietary cylinder tracking system, http://www.airproducts.co.uk/services/acts.htm
[6] Pollution Control Industries: Gas Cylinder Management, Cylinder tracking and management system, http://www.pollutioncontrol.com/services/cylinders/GasCylinders.html
[7] Symbol technologies reference
[8] Shepard, S.: Bar codes...Up Close and Personal. In: RFID: Radio Frequency Identification, ch. 1, pp. 26–41. McGraw Hill, New York (2005)
[9] Shepard, S.: A History of Bar codes. In: RFID: Radio Frequency Identification, ch. 1, pp. 12–26. McGraw Hill, New York (2005)
[10] Visich Jr, M.: Bar code Symbologies. In: Bar codes and their Applications, Monograph Series of the New Liberal Arts Program, Department of Technology and Society, State University of New York as Stony Brook, (10/03/2008) http://www.math.dartmouth.edu/~mqed/NLA/Barcodes/Barcodes.phtml
[11] Finkenzeller, K.: Bar code Systems. In: RFID Handbook, ch. 1, p. 3. Wiley & Sons, Chichester (2003)
[12] Piasecki, D.: RFID Update: The Basics, The Wal-Mart Mandate, EPC, Privacy Concerns and More. RFID News Website, http://www.inventoryops.com/RFIDupdate.htm

[13] Lin, C.-Y., Ho, Y.-H.: RFID technology adoption and supply chain performance: an empirical study in Chinas logistics study. Supply Chain Manangement: An International Journal, 369–378 (2009) ISSN: 1359-8546
[14] Ulaby, F.T.: Plane Wave Propagation in Lossy Media. In: Fundamentals of Applied Electromagnetics, ch. 7, pp. 277–279 (1999) ISBN: 0130115541
[15] GS1, Blue Tower, 326 Avenue Louise, BE 1050, Brussels, Belguim, http://www.gs1.org/
[16] save
[17] Crossbow: MPR/MIB Users Manual, Crossbow Technology Inc., 4145N First Street, San Jose, California, http://www.xbow.com/
[18] ATMEL: 8-bit AVR Microcontroller with 128K Bytes In-System Programmable Flash, ATmega128L datasheet, http://www.atmel.com/
[19] Zhao, F., Guibas, L.: Roles of Sensor Nodes and Utilities. In: Wireless Sensor Networks: An Information Processing Approach, ch. 5, p. 139. Elsevier Inc., Amsterdam (2004)
[20] Zhao, F., Guibas, L.: Energy-Minimizing Broadcast. In: Wireless Sensor Networks: An Information Processing Approach, vol. 3, p. 83. Elsevier Inc., Amsterdam (2004)
[21] Woo, A., Tong, T., Culler, D.: Taming the Underlying Challenges of Reliable Multihop Routing in Sensor Network. In: Sensys 2003, November 5-7 (2003)
[22] Tanenbaum, A.S.: Connection Oriented and Connectionless Services. In: Computer Networks, 4th edn.,ch. 1, pp. 32–33. Prentice Hall, Englewood Cliffs (2003)
[23] TinyOS Homepage, http://www.tinyos.net/
[24] Dallas Semiconductor: DS2401 Silicon Serial Number, Data sheet, http://www.maxim-ic.com/getds.cfm?qv_pk=2903
[25] http://www.everbouquet.com.tw/
[26] Microchip: PICF87xx 28/40 pin 8-bit CMOS Flash, Data Sheet, http://www.microchip.com/
[27] Mason, A.: Wireless Sensor Networks and their Industrial Applications, PhD Thesis, Liverpool John Moores University, Liverpool (August 2008)
[28] Basagni, S., Conti, M., Giordano, S., Stojmenovic, I.: Ad Hoc On-Demand Distance Vector Routing. In: Mobile Ad Hoc Networking, ch. 10, pp. 281–283 (2004) ISBN: 0471373133
[29] Basagni, S., Conti, M., Giordano, S., Stojmenovic, I.: Destination Sequenced Distance Vector Routing. In: Mobile Ad Hoc Networking, ch. 10, pp. 277–278 (2004) ISBN: 0471373133
[30] Instek: GDS-810 Oscilloscope, Test and Measurement Product Suppliers, 3661 Walnut Avenue, Chino, United States, http://www.instek.com/
[31] Crossbow: MTS/MDA Sensor Board Users Manual, Crossbow Technology Inc, 4145N First Street, San Jose, California, http://www.xbow.com/
[32] Farnell: Pressure Transmitter 0-1000 bar, http://uk.farnell.com/7056473/industrial-controls-automation/product.us0?sku=vdo-imt-3396088001
[33] Farnell: Pressure Sensor 0-100 PSI, http://uk.farnell.com/731729/industrial-controls-automation/product.us0?sku=honeywell-s-c-24pcffm6g
[34] Breithaupt, J.: Piezoelectricity. In: Understanding Physics for Advanced Level, 3rd edn., vol. 13, p. 178 (1995) ISBN: 0748715797

[35] Air Products: Acetylene, Safetygram-13 (1994),
http://www.airproducts.com/Responsibility/EHS/ProductSafety/ProductSafetyInformation/Safetygrams/safetygram13.htm
[36] Analog Devices: ADXL202E Low-cost 2G Dual-Axis Accelerometer, Product Datasheet (2000),
http://www.analog.com/en/prod/0,,764_800_ADXL202,00.html
[37] Zigbee Alliance, Official Website, http://www.zigbee.org/en/about/ Cited 01/03/2008
[38] The Network Simulator: ns-2 (2008),
http://nsnam.isi.edu/nsnam/index.php/User_Information
[39] Crossbow: Wireless Sensor Networks, 4145 N. First Street, San Jose, CA 95134, USA (2008), http://www.xbow.com/Home/wHomePage.aspx
[40] Microstrain: Wireless Sensors, 310 Hurricane Lane, Suite 4, Williston, VT 05495, USA (2008), http://www.microstrain.com/
[41] Texas Instruments: CC2431 System on Chip for 2.4GHz Zigbee / IEEE 802.15.4 with Location Engine, CC2341 PRELIMINARY Datasheet, Rev. 1.01 (2006)

An EM-IMM Method for Simultaneous Registration and Fusion of Multiple Radars and ESM Sensors

Dongliang Huang and Henry Leung

Department of Electrical and Computer Engineering
University of Calgary, Calgary, AB T2N 1N4, Canada
dl_huang@ieee.org, leungh@ucalgary.ca

Abstract. In a multiple sensor tracking scenario, measurements originated from the target of interest are necessarily aligned and fused to provide accurate information about the target. The process known as registration and fusion is generally casted as a joint parameter and state estimation problem for a single target case. The standard solution to this problem is the augmented Kalman filter (AKF) which takes the parameters as variables in the state vector. Despite of its easy implementation, the AKF is not favorable for large number of unknown parameters, as is the case for multiple sensors. Moreover, it is prone to numerical inaccuracy or divergence in application. In this paper, we evaluate the divergence problem of the AKF in simultaneous registration and fusion for the Radar/ESM sensors. Furthermore, we propose an expectation-maximization (EM) method in the maximum likelihood estimation (MLE) framework. In particular, to account for the maneuverability of the target, the interacting multiple model (IMM) filter implemented either by an extended Kalman filter (EKF) or by an unscented Kalman filter (UKF) is embedded into the conditional expectation evaluation in the E-step. The proposed joint registration and fusion method is thus called EM-IMM. Analysis shows that the EM method is convergent and furthermore leads to asymptotically unbiased estimate in an approximation sense. To evaluate the estimation performance, a direct inverse computation algorithm of Fisher information matrix (FIM) in posterior Cramer-Rao bound (PCRB) is also developed. Simulation results are given to demonstrate the effectiveness of the proposed method.

1 Introduction

Multiple sensors are prevalently deployed in various systems such as surveillance system [1], robotics system [2], air traffic control [3], etc. Other than improved reliability and increased coverage, one significant advantage is that higher accuracy can be achieved by fusing the complementary information from various similar/dissimilar sensors [4]. The fusion process requires coordinate system transformation of sensor data from individual platform to a common reference system. The problem is that the systematic biases inherent in the sensors will be nonlinearly entered into the fusion process during the transformation. As a result, these biases will lead to false tracks and ghost targets in the yielded global surveillance picture [4, 5]. Therefore, actions must be taken to eliminate the effect caused by the sensor biases. Sensor registration is such a step in the fusion process to correct the system biases [6].

In the past decade, a large number of registration algorithms have been proposed. The earlier efforts have been made to isolate the estimation of biases from the fusion. For example, the least squares (LS) method is proposed by applying stereographic projection to transform the sensor measurements to the regional plane from which the sensor biases are estimated [6, 7, 8]. Alternatively, to account for the measurement noise or sensor bias characteristics, maximum likelihood methods are proposed [9, 10]. The essence behind these methods is that the estimation problem is treated as a pure parameter identification problem after decoupling the registration with the fusion process. One main disadvantage of the decoupling methods is that it is not desirable for on-line fusion process, in particular, when dealing with the temporal biases between sensors.

Simultaneous registration and fusion has the merit of on-line implementation. Besides, it can deal with the sensor asynchronization due to time biases between sensors. In the previous study, the EKF method is proposed for different specific scenarios to estimate the sensor biases and state variables jointly [11, 12, 13]. The technique is carried out by augmenting the state variables with the unknown sensor biases which is recursively estimated by the filter. The EKF method is easy to implement and is suitable for the small number of biases since the Jacobian matrix is readily computed. However, it is prone to divergence under some situations such as highly nonlinear system, weak observability, and improper initialization procedure, etc., [14, 15, 16]. The UKF method, recently proposed in [17], is a good alternative to the EKF method to achieve improved performance with similar computational cost. Application of UKF to simulation registration and fusion for radar and electronic support measures (ESM) sensors is recently proposed in [18]. Another type of method for simultaneous registration and fusion is the two-stage Kalman filter [19], which decouple the sensor bias estimation from the state estimation. Along this line, registration and fusion approach is developed for multiple asynchronous sensors [20]. To reduce the complexity of the Kalman filter associated with large number of object states, Doorn and Blom proposed to decoupling the large Kalman filter into separate small-sized filters [21].

In this paper, we formulate the problem in the MLE framework for which the EM method [22] can be applied to obtain the ML estimation of the target tracking trajectory as well as the constant sensor biases. Unlike the AKF approach, the sensor biases are estimated from the system states separately. Thus, it avoids the potential numerical instability and divergence problems. Although the EM method has been proposed for target association and tracking problem [23, 24, 25, 26], the sensor biases are not considered in those applications. More recently, the EM method implemented via particle filter is proposed for state estimation of a single target under unknown model parameters and non-Gaussian measurement noise in [27]. But this method did not take the target maneuverability into account. To deal with the maneuvering target, we proposed an EM method combined with IMM filter for car platooning in cooperative driving system [28]. A similar idea is applied here for the target tracking using radar and ESM sensors. It should be noted that besides differences in algorithm development, our primary goal in this paper is to gain certain analytical understanding of the proposed method for the specific application.

In this paper, we assume that the systematic errors are time-constants. For simplicity, data association is not considered at this stage, though the developed EM method can be extended to this case. Following the formulation in [18], we first develop the sensor registration model and multiple model representation for fusion. The EM algorithm is then developed for simultaneous sensor biases and state estimation. In the proposed method, the conditional expectation of the states for the multiple models is approximately evaluated by the IMM filter which provides a good balance between computation complexity and accuracy [5]. In our IMM implementation, both EKF and UKF are considered. The UKF does not require Jacobian matrix computation and directly approximates the distribution of the state vector via the unscented transformation instead of the cumbersome functional approximation. We further apply perturbation analysis to the proposed EM-IMM method. This provides justification on why the augmented Kalman filtering method is prone to biased estimation and divergence while the proposed method is asymptotically convergent and approximately unbiased. To evaluate the estimation performance, PCRB based on a novel computation scheme of the inverse FIM is developed here.

This paper is organized as follows. The multiple active and passive sensor models, in particular, radar and ESM, are described in Section 2. In Section 3, the EM algorithm is developed for simultaneous registration and fusion based on the multiple model representation in which the IMM filter is employed for conditional expectation evaluation of the state. In Section 4, theoretical analyses are given on the augmented Kalman filter approach and the proposed method. Performance evaluation by the developed PCRB is given in Section 5. Section 6 presents the simulation results. Finally, concluding remarks are given in Section 7.

2 Sensor Network Model for Radars and ESM Sensors

Without loss of generality, a single target is used in our formulation. Assume that there are a total number of M radars and K ESM sensors located in a common Cartesian coordinate system. For illustration, a sensor configuration with $M = 2$ and $K = 2$ is shown in Fig. 1 in which $\{r_*\}$ and $\{e_*\}$ represent radars and ESMs, respectively. t_g denotes the target, and r_*, e_* are the corresponding radar and ESM measurements of t_g. It should be noted that this setting can be extended to three-dimension coordinate system by including an additional vertical variable.

For simplicity, we first consider a simple scenario that both the radars and the ESMs are assumed to be in a constant velocity (CV) motion. The discrete-time model description of the j-th radar or ESM is

$$\xi_{k+1}^j = A_\xi \xi_k^j + \Gamma w_k^j, \quad j = \begin{cases} 1, \cdots, M, & \text{for radar} \\ 1, \cdots, K, & \text{for ESM} \end{cases} \tag{1}$$

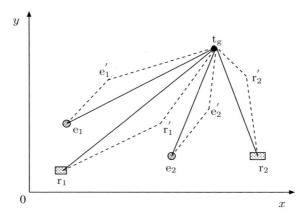

Fig. 1. Radar/ESM sensor geometry

where the system and noise matrices are defined as

$$A_\xi = \begin{bmatrix} 1 & T_s & 0 & 0 \\ 0 & 1 & 0 & 0 \\ 0 & 0 & 1 & T_s \\ 0 & 0 & 0 & 1 \end{bmatrix}, \quad \Gamma = \begin{bmatrix} T_s^2/2 \\ T_s \\ T_s^2/2 \\ T_s \end{bmatrix}, \tag{2}$$

where T_s is the sampling period. Note that all models adopt the same sampling period. The state vector $\boldsymbol{\xi}_k^j = [x_k^{e,j}, \dot{x}_k^{e,j}, y_k^{e,j}, \dot{y}_k^{e,j}]^{\mathrm{T}}$ (ESMs) or $\boldsymbol{\xi}_k^j = [x_k^{r,j}, \dot{x}_k^{r,j}, y_k^{r,j}, \dot{y}_k^{r,j}]^{\mathrm{T}}$ (radars) denotes the position and velocity components in x- and y-axes, respectively, where T is the matrix transpose. The disturbance w_k^j is assumed to be a zero-mean white Gaussian process with a variance of $(\sigma_w^j)^2$.

From a practical viewpoint, it is inefficient to describe the target dynamics by a single model as the target motion may change. To overcome this problem, using multiple models is found to provide an effective solution [29]. In this paper, we adopt three commonly used models to represent the different motions of a target. They are the CV model, the constant acceleration (CA), and the coordinated turn (CT) models. Assume that the dynamical model of the target is uniformly represented by

$$\boldsymbol{\zeta}_{k+1} = \Lambda \boldsymbol{\zeta}_k + \boldsymbol{\epsilon}_k, \tag{3}$$

where $\boldsymbol{\zeta}_k$ denotes the state vector of a target. In the case of CV motion, the system matrix $\Lambda = A_\xi$, and the covariance matrix of ϵ_k is $\Gamma \Gamma^{\mathrm{T}} \sigma_\epsilon^2$ where σ_ϵ^2 is the noise parameter. In the case of CA motion, the system transition matrix becomes $\Lambda = \mathrm{blkdiag}(\Lambda_a, \Lambda_a)$ where blkdiag is the diagonal concatenation of block matrices and the noise covariance matrices $Q_a = \mathrm{blkdiag}(\Gamma_a, \Gamma_a) q_a$ with [4]

$$\Lambda_a = \begin{bmatrix} 1 & T_s & T_s^2/2 \\ 0 & 1 & T_s \\ 0 & 0 & 1 \end{bmatrix}, \quad \Gamma_a = \begin{bmatrix} T_s^5/20 & T_s^4/8 & T_s^3/6 \\ T_s^4/8 & T_s^3/3 & T_s^2/2 \\ T_s^3/6 & T_s^2/2 & T_s \end{bmatrix}, \tag{4}$$

where q_a is a noise parameter. The system state in this case includes the acceleration as its component and is represented by $\zeta_k = [x_k, \dot{x}_k, \ddot{x}_k, y_k, \dot{y}_k, \ddot{y}_k, z_k, \dot{z}_k, \ddot{z}_k]^T$ where $\{\ddot{x}_k, \ddot{y}_k, \ddot{z}_k\}$ denotes the acceleration.

When the target is performing a turn, correlation between orthogonal coordinate axes will occur. In this case, the CT model can be used to describe the corresponding dynamics [4]. The system transition matrix is related to the angular rate and it is represented by $\Lambda = \text{blkdiag}(\Lambda_t, \Lambda_t)$

$$\Lambda_t = \begin{bmatrix} 1 & \sin(\omega T_s)/\omega & (1-\cos(\omega T_s))/\omega^2 \\ 0 & \cos(\omega T_s) & \sin(\omega T_s)/\omega \\ 0 & -\omega \sin(\omega T_s) & \cos(\omega T_s) \end{bmatrix}, \tag{5}$$

where ω is the angular rate which is determined by the cross-product between velocity and acceleration [4]. The noise covariance Q_t in the CT model has the same form as that in the CA model except that the noise parameter becomes q_t.

Assume that the origin is given by a reference. Let Δt_i denote the i-th radar time bias with respect to t_k. The time index of the i-th radar then becomes $t_k^i = t_k + \Delta t_i$. Let $(\tilde{r}_k^i, \tilde{\theta}_k^i)$ denote the range and angular measurements of the i-th radar at time instant t_k^i in the polar coordinate system while $\{\Delta r^i, \Delta \theta^i\}$ denotes the system bias of the i-th radar in the range and angle. The observed measurements $(\tilde{r}_k^i, \tilde{\theta}_k^i)$ of the i-th radar to the target are then equal to the actual range and angle $\{r_k^i, \theta_k^i\}$ perturbed by the system and time biases plus measurement noise. Converting it to the Cartesian coordinate system and denoting the corresponding measurements of the i-th radar by $(\tilde{x}_k^i, \tilde{y}_k^i)$, we have

$$\begin{aligned} \tilde{x}_k^i &= (r_k^i + \Delta r^i)\sin(\theta_k^i + \Delta \theta^i) + \Delta t_i(\dot{x}_k^{r,i} - \dot{x}_k) + n_{k,x}^i \\ &= (\sqrt{(x_k^{r,i} - x_k)^2 + (y_k^{r,i} - y_k)^2} + \Delta r^i)\sin(\theta_k^i + \Delta \theta^i) \\ &\quad + \Delta t_i(\dot{x}_k^{r,i} - \dot{x}_k) + n_{k,x}^i, \\ \tilde{y}_k^i &= (r_k^i + \Delta r^i)\cos(\theta_k^i + \Delta \theta^i) + \Delta t_i(\dot{y}_k^{r,i} - \dot{y}_k) + n_{k,y}^i \\ &= (\sqrt{(x_k^{r,i} - x_k)^2 + (y_k^{r,i} - y_k)^2} + \Delta r^i)\cos(\theta_k^i + \Delta \theta^i) \\ &\quad + \Delta t_i(\dot{y}_k^{r,i} - \dot{y}_k) + n_{k,y}^i, \quad i = 1, \cdots, M, \end{aligned} \tag{6}$$

$$\tag{7}$$

where $n_{k,x}^i$ and $n_{k,y}^i$ are assumed to be white Gaussian noise processes with zero mean and variances $(\sigma_{k,x}^i)^2$ and $(\sigma_{k,y}^i)^2$ respectively.

The positions of the radars and ESMs are assumed to be measured by the global positioning system. They are represented by

$$\tilde{x}_k^{r,i} = x_k^{r,i} + n_{k,x}^{r,i}, \quad \tilde{y}_k^{r,i} = y_k^{r,i} + n_{k,y}^{r,i}, \quad i = 1, \cdots, M, \tag{8}$$

$$\tilde{x}_k^{e,j} = x_k^{e,j} + n_{k,x}^{e,j}, \quad \tilde{y}_k^{e,j} = y_k^{e,j} + n_{k,y}^{e,j}, \quad j = 1, \cdots, K, \tag{9}$$

where $n_{k,x}^{r,i}$ and $n_{k,y}^{r,i}$ are independent zero-mean white Gaussian processes with variances $(\sigma_{k,x}^{r,i})^2$ and $(\sigma_{k,y}^{r,i})^2$, respectively. Similarly, $n_{k,x}^{e,j}$ and $n_{k,y}^{e,j}$ are zero-mean white Gaussian processes with variances $(\sigma_{k,x}^{e,j})^2$ and $(\sigma_{k,y}^{e,j})^2$, respectively. Since the j-th ESM sensor gives the angular measurement $\tilde{\phi}_k^j$ of the target, we have

$$\tilde{\phi}_k^j = \phi_k^j + \Delta\phi^j + n_k^{e,j}, \tag{10}$$

where ϕ_k^j is the actual azimuth of the target with respect to the j-th ESM at t_k and $\Delta\phi^j$ is the sensor bias of the j-th ESM. The measurement disturbance $n_k^{e,j}$ is assumed to be white Gaussian noise with zero-mean and variance $(\sigma_n^j)^2$. We can further expand (10) as

$$\tilde{\phi}_k^j = \tan^{-1}\left(\frac{x_k - x_k^{e,j}}{y_k - y_k^{e,j}}\right) + \Delta\phi^j + n_k^{e,j}. \tag{11}$$

Let the radar measurement vector be $z_k^{r,i}$ with $z_k^{r,i} = [\tilde{x}_k^i, \tilde{y}_k^i]$, the ESM measurement vector be $z_k^e = [\tilde{\phi}_k^1, \cdots \tilde{\phi}_k^K]$, and the position measurement vector for the radars and ESMs be $z_k^{re} = [\tilde{x}_k^{r,1}, \tilde{y}_k^{r,1}, \cdots, \tilde{x}_k^{r,M}, \tilde{y}_k^{r,M}, \tilde{x}_k^{e,1}, \tilde{y}_k^{e,1}, \cdots, \tilde{x}_k^{e,K}, \tilde{y}_k^{e,K}]$. The augmented measurement vector is defined by $z_k = [z_k^{r,1}, \cdots, z_k^{r,M}, z_k^e, z_k^{re}]^T$, which can be established by combining (6)-(7), (8)-(9) and (11). That is,

$$z_k = h(\bar{\xi}_k, \beta) + n_k, \tag{12}$$

where $h(\cdot, \cdot)$ is a nonlinear vector function defined by the radar and ESM measurement equations and β is the sensor biases with $\beta = [\Delta t^1, \cdots, \Delta t^M, \Delta \theta^1, \cdots, \Delta \theta^M, \Delta r^1, \cdots, \Delta r^M, \Delta \phi^1, \cdots, \Delta \phi^K]$. The noise vector n_k is a white Gaussian process with zero-mean and covariance matrix Q_n. The augmented state space equation for simultaneous radar/ESM registration and fusion can be expressed by $\bar{\xi}_{k+1} = \bar{A}\bar{\xi}_k + w_k$ where $\bar{\xi}_k$ denote the states of the target and ESM sensors. The system matrix $\bar{A}$ is thus in the block diagonal form with $\bar{A} = \text{blkdiag}(\Lambda, A_\xi, \cdots, A_\xi)$ while w_k is the cascaded vector of the noise processes with $w_k = [\epsilon_k^T, (\Gamma w_k^1)^T, \cdots, (\Gamma w_k^K)^T]^T$ and a covariance matrix of Q_w. The initial state $\bar{\xi}_0$ is assumed to be a Gaussian process with mean $\mu_{\bar{\xi}_0}$ and covariance $\Sigma_{\bar{\xi}_0}$ which is independent of the noise processes.

Equation (12) indicates that z_k has a dimension of $2M + 5K$ and β has a dimension of $3M + K$. The augmented UKF provides an improved estimation accuracy compared to augmented EKF. However both approaches will suffer from the problem that as the number of sensors increases, the dimensionality of an augmented filter becomes computational prohibitive.

3 Simultaneous Registration and Fusion Using EM-IMM

The objective of our approach is to obtain the unknown parameters by maximizing the likelihood function expressed by $\eta = \arg\max_\eta \mathcal{L}(\eta|\mathcal{Z})$ where η consists of both sensor biases and unknown noise parameters and $\mathcal{Z}$ is the observed data with $\mathcal{Z} = \{z_1, \cdots, z_N\}$. The likelihood function has the same form of parameter-conditioned probability density function $p(\mathcal{Z}|\eta)$. The closed-form solution is difficult to obtain in general. If the likelihood function is expressed in an alternative way, then the optimization can be executed by the EM method efficiently [22]. The alternative likelihood function is based on the so-called complete data. The complete data include both observed data and missing data. The missing data is taken as the unknown system state, namely, $\mathcal{M} = \{\bar{\xi}_0, \cdots, \bar{\xi}_N\}$. The complete data is denoted by $\mathcal{C}$ with $\mathcal{C} = \{\mathcal{M}, \mathcal{Z}\}$. Subsequently the complete data based likelihood function is $\mathcal{L}(\eta|\mathcal{C})$ instead of $\mathcal{L}(\eta|Z)$. The EM algorithm provides an efficient solution to the optimization of the complete-data based likelihood function. It works in an iterative way which guarantees to converge to a stationary point and increases the likelihood function at each iteration. Based on the complete data, the log likelihood function can be expressed as

$$\log \mathcal{L}(\mathcal{C}|\eta, \mathcal{Z})$$
$$= -\frac{N(m+l)+m}{2} \log 2\pi - \frac{1}{2} \log |\Sigma_{\bar{\xi}_0}| - \frac{1}{2}(\bar{\xi}_0 - \mu_{\bar{\xi}_0})^T \Sigma_{\bar{\xi}_0}^{-1}(\bar{\xi}_0 - \mu_{\bar{\xi}_0})$$
$$- \frac{N}{2} \log |Q_w| - \sum_{k=1}^{N} \frac{1}{2}(\bar{\xi}_k - \bar{A}\bar{\xi}_{k-1})^T Q_w^{-1}(\bar{\xi}_k - \bar{A}\bar{\xi}_{k-1})$$
$$- \frac{N}{2} \log |Q_n| - \sum_{k=1}^{N} \frac{1}{2}[z_k - h(\bar{\xi}_k, \beta)]^T Q_n^{-1}[z_k - h(\bar{\xi}_k, \beta)]. \tag{13}$$

where m and l are the dimensions of the state and measurement vector, respectively. It should be noted that the fourth and fifth terms in the right side of (13) simplify the description of the Markov switching property of the multiple models because only conditional expectation of state is involved and it is estimated by IMM as shown in the following.

The EM algorithm iterates between two steps, i.e., the E-step and the M-step as represented by (1) E-step: $\mathcal{Q}(\eta, \hat{\eta}^{(p)}) = \mathbb{E}[\log \mathcal{L}(\eta)|\hat{\eta}^{(p)}, \mathcal{Z}]$, and (2) M-step: $\hat{\eta}^{(p+1)} = \arg\max_\eta \mathcal{Q}(\eta, \hat{\eta}^{(p)})$, where $\mathbb{E}(\cdot)$ denotes mathematical expectation operator. The E-step computes the conditional expectation, i.e., the $\mathcal{Q}$-function, based on the current parameter estimates and the available measurements. In this stage, the state vector of the target and ESM can be estimated by a Kalman filter. The M-step solves for an optimization problem to obtain the unknown sensor biases and other parameters to increase the likelihood function.

3.1 E-STEP

In the E-step, we define the conditional mean and the covariance functions as $\hat{\bar{\xi}}_{k|j} = \mathbb{E}(\bar{\xi}_k|\mathcal{Z}_j)$, $P_{k|j} = \text{cov}(\bar{\xi}_k|\mathcal{Z}_j)$ and $P_{k,k-1|j} = \text{cov}(\bar{\xi}_k, \bar{\xi}_{k-1}|\mathcal{Z}_j)$, respectively, and where $\mathcal{Z}_j = [z_1, \cdots, z_j]$. Furthermore, we can calculate the following terms as

$$\sum_{k=1}^{N} \mathbb{E}\left((\bar{\xi}_k - \bar{A}\bar{\xi}_{k-1})(\bar{\xi}_k - \bar{A}\bar{\xi}_{k-1})^{\mathrm{T}}\bigg|\mathcal{Z}\right)$$

$$=\sum_{k=1}^{N}\left[\mathbb{E}(\bar{\xi}_k\bar{\xi}_k^{\mathrm{T}}|\mathcal{Z}) - \mathbb{E}(\bar{\xi}_k\bar{\xi}_{k-1}^{\mathrm{T}}\bar{A}^{\mathrm{T}}|\mathcal{Z}) - \mathbb{E}(\bar{A}\bar{\xi}_{k-1}\bar{\xi}_k^{\mathrm{T}}|\mathcal{Z})\right.$$
$$\left.+\mathbb{E}(\bar{A}\bar{\xi}_{k-1}\bar{\xi}_{k-1}^{\mathrm{T}}\bar{A}^{\mathrm{T}}|\mathcal{Z})\right]$$

$$=\Upsilon - \sum_{k=1}^{N}\left[(P_{k,k-1|N} + \hat{\bar{\xi}}_{k|N}\hat{\bar{\xi}}_{k-1|N}^{\mathrm{T}})\bar{A}^{\mathrm{T}} - \bar{A}(P_{k-1,k|N} + \hat{\bar{\xi}}_{k-1|N}\hat{\bar{\xi}}_{k|N}^{\mathrm{T}})\right.$$
$$\left.+\bar{A}(P_{k-1|N} + \hat{\bar{\xi}}_{k-1|N}\hat{\bar{\xi}}_{k-1|N}^{\mathrm{T}})\bar{A}^{\mathrm{T}}\right]$$

$$=\Upsilon + \Psi - \Pi - \Pi^{\mathrm{T}}, \tag{14}$$

$$\mathbb{E}\left\{(z_k - \mathbf{h}(\bar{\xi}_k, \boldsymbol{\beta}))(z_k - \mathbf{h}(\bar{\xi}_k, \boldsymbol{\beta}))^{\mathrm{T}}\bigg|\mathcal{Z}\right\}$$
$$\approx \mathbb{E}\left\{\left[z_k - \mathbf{h}(\boldsymbol{\beta}, \hat{\bar{\xi}}_{k|N}) - H_k^{\mathrm{T}}(\bar{\xi}_k - \hat{\bar{\xi}}_{k|N})\right]\left[z_k - \mathbf{h}(\boldsymbol{\beta}, \hat{\bar{\xi}}_{k|N})\right.\right.$$
$$\left.\left.- H_k^{\mathrm{T}}(\bar{\xi}_k - \hat{\bar{\xi}}_{k|N})\right]^{\mathrm{T}}\bigg|\mathcal{Z}\right\}$$
$$=H_k^{\mathrm{T}}P_{k|N}H_k + \left(z_k - \mathbf{h}(\hat{\bar{\xi}}_{k|N}, \boldsymbol{\beta})\right)\left(z_k - \mathbf{h}(\hat{\bar{\xi}}_{k|N}, \boldsymbol{\beta})\right)^{\mathrm{T}}. \tag{15}$$

Using (14), (15) and neglecting the irrelevant constant term, the conditional expectation of (13) can be computed as

$$\mathcal{Q}(\boldsymbol{\eta}, \hat{\boldsymbol{\eta}}^{(p)}) = -\frac{1}{2}\log|\Sigma_{\bar{\xi}_0}| - \frac{1}{2}\mathrm{Tr}\left\{\Sigma_{\bar{\xi}_0}^{-1}[P_{0|N} + (\hat{\bar{\xi}}_{0|N} - \mu_{\bar{\xi}_0})(\hat{\bar{\xi}}_{0|N} - \mu_{\bar{\xi}_0})^{\mathrm{T}}]\right\}$$
$$-\frac{N}{2}\log|Q_w| - \frac{1}{2}\mathrm{Tr}\left\{Q_w^{-1}(\Upsilon + \Psi - \Pi - \Pi^{\mathrm{T}})\right\} - \frac{N}{2}\log|Q_n|$$
$$-\frac{1}{2}\mathrm{Tr}\left\{Q_n^{-1}\sum_{k=1}^{N}\left(H_k^{\mathrm{T}}P_{k|N}H_k + [z_k - \mathbf{h}(\hat{\bar{\xi}}_{k|N}, \boldsymbol{\beta})]\right.\right.$$
$$\left.\left.[z_k - \mathbf{h}(\hat{\bar{\xi}}_{k|N}, \boldsymbol{\beta})]^{\mathrm{T}}\right)\right\}, \tag{16}$$

where Tr denotes the trace of a matrix, and Ψ, Π, Υ are defined by $\Psi = \bar{A}[\sum_{k=1}^{N}(P_{k-1|N} + \hat{\bar{\xi}}_{k-1|N}\hat{\bar{\xi}}_{k-1|N}^{\mathrm{T}})]\bar{A}^{\mathrm{T}}$, $\Pi = \sum_{k=1}^{N}(P_{k,k-1|N} + \hat{\bar{\xi}}_{k|N}\hat{\bar{\xi}}_{k-1|N}^{\mathrm{T}})\bar{A}^{\mathrm{T}}$, $\Upsilon = \sum_{k=1}^{N}P_{k|N} + \hat{\bar{\xi}}_{k|N}\hat{\bar{\xi}}_{k|N}^{\mathrm{T}}$.

3.2 Expectation Evaluation by IMM Filter

It can be seen that the $\mathcal{Q}$-function is involved with computing the conditional expectation of the system state and its covariances. In fact, they can be obtained by using a Kalman smoother if the whole batch of data are available for a non-maneuvering

An EM-IMM Method for Simultaneous Registration

target. For the maneuvering target, multiple models with a Markovian switching property are frequently used to describe the target dynamics. Correspondingly, the conditional expectations and the covariances of the system states can be effectively computed by the IMM filter, a recursive hybrid filtering technique which provides a good balance between performance and complexity [5].

Here, we employ both the EKF and UKF in the implementation of IMM filter. The UKF can provide an approximation accuracy up to the second order [17]. It has a similar computational cost to that of an EKF but it does not require computing the Jacobian matrix. In the following, we briefly introduce the IMM filter/smoother in computing the Q-function.

Let L denote the number of models. The model jumping process is assumed to be a homogeneous Markov process. The model transition probability is defined by $p_{ij} = \Pr\{M_k^j | M_{k-1}^i\}$, where M_k^j and M_{k-1}^i denote the j-th model at time k and the i-th model at time $k-1$, respectively. IMM computes the mixing probabilities as $p(\bar{\xi}_k | \mathcal{Z}_k) = \sum_{i=1}^{L} p(\bar{\xi}_k | M_k^i, z_k, \mathcal{Z}_{k-1}) \mu_k^i$, where $\mu_k^i = \Pr\{M_k^i | \mathcal{Z}_k\}$ is the model probability subject to $\sum_i \mu_k^i = 1$. It can be seen that if the noise processes are Gaussianly distributed, then the conditional probability $p(\bar{\xi}_k | M_k^i, z_k, \mathcal{Z}_{k-1})$ can be approximately calculated by a bank of Kalman filters with the mixing initial probability states as the initial states. To be specific, the recursive algorithm is described as follows; see [29] for details.

First, mixed initial conditions for the bank of Kalman filters are computed according to mixing probabilities. Let $\hat{\bar{\xi}}_{k-1|k-1}^i$ and $P_{k-1|k-1}^i$ denote the original initial state and covariance matrix for each filter respectively. The mixed initial states and covariance matrices for the j-th Kalman filter are then calculated by

$$\hat{\bar{\xi}}_{k-1|k-1}^{0,j} = \sum_{i=1}^{L} \hat{\bar{\xi}}_{k-1|k-1}^i \cdot \mu_{k-1|k-1}^{i|j}, \qquad (17)$$

$$P_{k-1|k-1}^{0,j} = \sum_{i=1}^{L} \left(P_{k-1|k-1}^i + [\hat{\bar{\xi}}_{k-1|k-1}^i - \check{\bar{\xi}}_{k-1|k-1}^i][\hat{\bar{\xi}}_{k-1|k-1}^i - \check{\bar{\xi}}_{k-1|k-1}^i]^T \right)$$
$$\times \mu_{k-1|k-1}^{i|j}, \qquad (18)$$

where the mixing probabilities $\mu_{k-1|k-1}^{i|j}$ are computed as $\mu_{k-1|k-1}^{i|j} = p_{ij} \mu_{k-1}^i / \sum_{i=1}^{L} p_{ij} \mu_{k-1}^i$, and where μ_{k-1}^i is *a priori* known probability evaluated in the previous recursion. Subsequently, the state $\hat{\bar{\xi}}_{k|k}^i$ and its covariance $P_{k|k}^i$ for each of the L filters are estimated by an EKF or UKF with the initial state and covariance setting as $\bar{\xi}_{k-1|k-1}^{0,j}$ and $P_{k-1|k-1}^{0,j}$, respectively. The model probabilities μ_k^i are updated by $\mu_k^i = \alpha_i \sum_{j=1}^{L} p_{ji} \mu_{k-1}^j / \sum_{i=1}^{L} \left(\alpha_i \sum_{j=1}^{L} p_{ji} \mu_{k-1}^j \right)$, where α_i is the conditional probability denoted by $p(z_k | M_k^i, \mathcal{Z}_{k-1})$, which can be computed simply by using the mixed initial state and covariance associated with model M_k^i [29]. The state estimates of $\hat{\bar{\xi}}_{k|k}$ and its covariance are then obtained by mixing various states estimates according to the mixing probabilities. That is, $\hat{\bar{\xi}}_{k|k} = \sum_{i=1}^{L} \hat{\bar{\xi}}_{k|k}^{i} \mu_{k}^{i}$ and $P_{k|k} = \sum_{i=1}^{L} \left(P_{k|k}^{i} + [\hat{\bar{\xi}}_{k|k}^{i} - \hat{\bar{\xi}}_{k|k}][\hat{\bar{\xi}}_{k|k}^{i} - \hat{\bar{\xi}}_{k|k}]^{T} \right) \mu_{k}^{i}$.

It should be noted that the above IMM filter is based on the current measurements. If the batch measurements are available, then the Kalman filter in the IMM can be replaced by a fixed-interval Kalman smoother. This Kalman smoother based IMM method is called SIMM which is proposed in [30]. Since the smoother computes with the whole set of measurements, the evaluated conditional expectation is more accurate. But the tradeoff is a higher computational cost. In SIMM, the forward and backward Kalman filter banks are employed to obtain the smoothed state estimates. Assume that the state and the corresponding estimation error covariances are obtained from the forward and backward filters, i.e., $\hat{\bar{\xi}}_{k|k}^{i}$, $\hat{\bar{\xi}}_{k|k+1}^{b,j}$ and $P_{k|k}^{i}$, $P_{k|k+1}^{b,j}$, respectively. The SIMM method is briefly described as follows. The first step is to obtain $\hat{\bar{\xi}}_{k|k}^{i}$, $\hat{\bar{\xi}}_{k|k+1}^{b,j}$ and $P_{k|k}^{i}$, $P_{k|k+1}^{b,j}$ based on the forward and backward EKF or UKF. Then the individual smoothed estimates from the forward and backward filtering are obtained by

$$\hat{\bar{\xi}}_{k|N}^{ij} = P_{k|N}^{ji} \left[(P_{k|k}^{j})^{-1} \hat{\bar{\xi}}_{k|k}^{j} + (P_{k|k+1}^{b,i})^{-1} \hat{\bar{\xi}}_{k|k+1}^{b,i} \right], \tag{19}$$

$$P_{k|N}^{ji} = \left[(P_{k|k}^{j})^{-1} + (P_{k|k+1}^{b,i})^{-1} \right]^{-1}. \tag{20}$$

The smoothed estimates are then computed based on the following equations.

$$\hat{\bar{\xi}}_{k|N} = \sum_{j=1}^{L} \mu_{k}^{s,j} \sum_{i=1}^{L} \mu_{k+1|k+1}^{i|j} \hat{\bar{\xi}}_{k|N}^{ji}, \tag{21}$$

$$P_{k|N} = \sum_{j=1}^{L} \mu_{k}^{s,j} \left[P_{k|N}^{j} + (\hat{\bar{\xi}}_{k|N}^{j} - \hat{\bar{\xi}}_{k|N})(\hat{\bar{\xi}}_{k|N}^{j} - \hat{\bar{\xi}}_{k|N})^{T} \right], \tag{22}$$

$$P_{k,k-1|N} = \sum_{j=1}^{L} \mu_{k}^{s,j} \left[P_{k,k-1|N}^{j} + (\hat{\bar{\xi}}_{k|N}^{j} - \hat{\bar{\xi}}_{k|N})(\hat{\bar{\xi}}_{k-1|N}^{j} - \hat{\bar{\xi}}_{k-1|N})^{T} \right], \tag{23}$$

$$P_{k,k-1|k}^{i} = (I - K_{k} H_{k}) A_{k-1}^{(i)} P_{k-1|k-1}^{i}. \tag{24}$$

where the derivation of $P_{k,k-1|k}^{i}$ can be found in [31], $\mu_{k}^{s,j}$ and $\mu_{k+1|k+1}^{i|j}$ are updated as given in [30].

3.3 M-STEP

Since the output measurement function is linear with respect to the registration biases. We have

$$\begin{aligned} I_{z} &= \mathbb{E}\left\{ (z_{k} - \mathbf{h}(\bar{\xi}_{k}, \beta))(z_{k} - \mathbf{h}(\bar{\xi}_{k}, \beta))^{T} \Big| \mathcal{Z} \right\} \\ &\approx \mathbb{E}\left\{ [z_{k} - H_{k}\bar{\xi}_{k} - B(\bar{\xi}_{k})\beta][z_{k} - H_{k}\bar{\xi}_{k} - B(\bar{\xi}_{k})\beta]^{T} \Big| \mathcal{Z} \right\} \end{aligned} \tag{25}$$

Hence, we can obtain

$$\operatorname{Tr}\sum_{k=1}^{N} I_z = \sum_{k=1}^{N} \mathbb{E}\left\{ [z_k - H_k\bar{\xi}_k - B(\bar{\xi}_k)\beta]^{\mathrm{T}} [z_k - H_k\bar{\xi}_k - B(\bar{\xi}_k)\beta] \big| \mathcal{Z} \right\}$$

$$\approx \beta^{\mathrm{T}} \mathbb{E}\left\{ \sum_{k=1}^{N} B^{\mathrm{T}}(\bar{\xi}_k) B(\bar{\xi}_k) \big| \mathcal{Z} \right\} \beta - \mathbb{E}\left\{ \sum_{k=1}^{N} (z_k - H_k\bar{\xi}_k)^{\mathrm{T}} B(\bar{\xi}_k) \big| \mathcal{Z} \right\} \beta$$

$$- \beta^{\mathrm{T}} \mathbb{E}\left\{ \sum_{k=1}^{N} B^{\mathrm{T}}(\bar{\xi}_k)(z_k - H_k\bar{\xi}_k) \big| \mathcal{Z} \right\} + \mathbb{E}\left\{ \sum_{k=1}^{N} (z_k - H_k\bar{\xi}_k)^{\mathrm{T}} \right.$$

$$\left. \times (z_k - H_k\bar{\xi}_k) \big| \mathcal{Z} \right\}$$

$$= (\beta - B_b^{-1} Z_b^{\mathrm{T}})^{\mathrm{T}} B_b (\beta - B_b^{-1} Z_b^{\mathrm{T}}) + Z_1 - Z_b B_b^{-\mathrm{T}} Z_b^{\mathrm{T}}, \quad (26)$$

where

$$Z_1 = -\beta^{\mathrm{T}} \mathbb{E}\left\{ \sum_{k=1}^{N} B^{\mathrm{T}}(\bar{\xi}_k) \right.$$

$$\left. \times (z_k - H_k\bar{\xi}_k) \big| \mathcal{Z} \right\} + \mathbb{E}\left\{ \sum_{k=1}^{N} (z_k - H_k\bar{\xi}_k)^{\mathrm{T}} (z_k - H_k\bar{\xi}_k) \big| \mathcal{Z} \right\},$$

$$Z_b = \mathbb{E}\left\{ \sum_{k=1}^{N} (z_k - H_k\bar{\xi}_k)^{\mathrm{T}} B(\bar{\xi}_k) \big| \mathcal{Z} \right\},$$

$$B_b = \mathbb{E}\left\{ \sum_{k=1}^{N} B^{\mathrm{T}}(\bar{\xi}_k) B(\bar{\xi}_k) \big| \mathcal{Z} \right\}.$$

Therefore, we can get an estimate of β as

$$\beta = \left[\mathbb{E}\left\{ \sum_{k=1}^{N} B^{\mathrm{T}}(\bar{\xi}_k) B(\bar{\xi}_k) \big| \mathcal{Z} \right\} \right]^{-1} \mathbb{E}\left\{ \sum_{k=1}^{N} B^{\mathrm{T}}(\bar{\xi}_k)(z_k - H_k\bar{\xi}_k) \big| \mathcal{Z} \right\}. \quad (27)$$

From (26), we have

$$Q_n = Z_1 - Z_b B_b^{-\mathrm{T}} Z_b^{\mathrm{T}}. \quad (28)$$

According to (16), we can obtain

$$Q_w = \frac{1}{N}(\Upsilon + \Psi - \Pi - \Pi^{\mathrm{T}}). \quad (29)$$

In the M-step, parameter estimates are obtained by an approximated algebraic solution via setting the derivative of the expected log-likelihood function to be zero. That is, $\hat{\boldsymbol{\mu}}_{\xi_0} = \hat{\bar{\boldsymbol{\xi}}}_{0|N}$, $\Sigma_{\bar{\xi}_0} = P_{0|N}$, $\hat{\beta} = B_b^{-1} Z_b^{\mathrm{T}}$, $Q_w = (\Upsilon + \Psi - \Pi - \Pi^{\mathrm{T}})/N$, and $Q_n = Z_1 - Z_b B_b^{-\mathrm{T}} Z_b^{\mathrm{T}}$.

4 Bias Analysis of the EM Method

We investigate why the augmented Kalman filtering method is prone to biased estimation and divergence. It is generally difficult to analyze this problem in a rigorous way in [16]. Instead, we resort to an intuitive way using the perturbation technique. The idea is to derive an approximately closed-form formula to describe the propagation of the estimation error on the parameter and state vectors, i.e., $\Delta\beta_0 \rightarrow \Delta x_1 \rightarrow \Delta\beta_1 \rightarrow \Delta x_2 \rightarrow \cdots$. This is reasonable since the parameter estimation and state estimation are separately conducted in the EKF and the proposed EM method. It should be noted that this analysis is also valid for non-maneuvering target case where the conditional expectation of the state is directly evaluated by the Kalman filter instead of the IMM filter.

Assume that an EKF is used, we have [16]

$$\hat{\bar{\xi}}_{k+1} = \hat{\bar{\xi}}_k + K_k^1[z_k - h(\hat{\bar{\xi}}_k, \hat{\beta}_k)], \qquad (30)$$

$$\hat{\beta}_{k+1} = \hat{\beta}_k + K_k^2[z_k - h(\hat{\bar{\xi}}_k, \hat{\beta}_k)], \qquad (31)$$

where K_k^1 and K_k^2 are Kalman gains with appropriate dimensions, respectively.

Assume that the true parameter vector and state are denoted by β^* and $\bar{\xi}^*(k)$, respectively, we have

$$\delta\beta_{k+1} = \delta\beta_k - K_k^2[z_k - h(\bar{\xi}^*, \beta^*)] + K_k^2\left[\frac{\partial h}{\partial \bar{\xi}}\delta\bar{\xi} + \frac{\partial h}{\partial \beta}\delta\beta\right]. \qquad (32)$$

Taking expectation of the above equation under the measurements, it becomes

$$\mathbb{E}[\delta\beta(k+1)|\mathcal{Z}] = \mathbb{E}\left[(I - K_k^2\frac{\partial h}{\partial \beta})\delta\beta(k)|\mathcal{Z}\right] - \mathbb{E}\left[K_k^2\frac{\partial h}{\partial \bar{\xi}}\delta\bar{\xi}_k|\mathcal{Z}\right]$$

$$= [I - K_k^2\frac{\partial h}{\partial \beta})|\mathcal{Z}]\mathbb{E}[\delta\beta_k] - \mathbb{E}\left[K_k^2\frac{\partial h}{\partial \bar{\xi}_k}\delta\bar{\xi}_k|\mathcal{Z}\right]. \qquad (33)$$

Similarly, we have

$$\mathbb{E}[\delta\bar{\xi}(k+1)|\mathcal{Z}] = \mathbb{E}\left[(I - K_k^1\frac{\partial h}{\partial \bar{\xi}})\delta\bar{\xi}_k|\mathcal{Z}\right] - \mathbb{E}\left[K_k^1\frac{\partial h}{\partial \beta}\delta\beta_k|\mathcal{Z}\right]$$

$$= \left[I - K_k^2\frac{\partial h}{\partial \bar{\xi}}\right]\mathbb{E}[\delta\bar{\xi}_k|\mathcal{Z}] - \mathbb{E}\left[K_k^1\frac{\partial h}{\partial \beta}\delta\beta_k|\mathcal{Z}\right]. \qquad (34)$$

It follows from (33) and (34) that

$$\mathbb{E}\left[\begin{bmatrix}\delta\bar{\xi}_{k+1}\\ \delta\beta_k\end{bmatrix}\bigg|\mathcal{Z}\right] = \left\{I - \underbrace{\begin{bmatrix}K_k^1\frac{\partial h}{\partial \bar{\xi}} & K_k^1\frac{\partial h}{\partial \beta}\\ K_k^2\frac{\partial h}{\partial \bar{\xi}} & K_k^2\frac{\partial h}{\partial \beta}\end{bmatrix}}_{A_h}\right\}\mathbb{E}\left[\begin{bmatrix}\delta\bar{\xi}_k\\ \delta\beta_k\end{bmatrix}\bigg|\mathcal{Z}\right]. \qquad (35)$$

When $k \rightarrow \infty$, K_k^1 and K_k^2 will converge to constant matrices K^1 and K^2, respectively. From (35), in order to retain convergence and to reduce estimation error, the

An EM-IMM Method for Simultaneous Registration

singular value of $I - A_h$ should be smaller than 1. Thus the matrix A_h must be non-singular. However, in cases such as one dimensional status of state and parameter, A_h becomes singular, and hence the parameter and state estimates are biased. In fact, the matrix A_h is singular if the dimension of measurement vector is smaller than the dimension of state plus that of parameter as in the case of the radar/ESM.

For the proposed algorithm, using a first-order approximation of the radar measurement, we have

$$\tilde{x}_k^i = x_k^i + \dot{x}_k^i \Delta t^i + y_k^i \Delta \theta^i + \frac{x_k^i}{\sqrt{(x_k^i)^2 + (y_k^i)^2}} \Delta r^i + n_{k,x}^i, \tag{36}$$

$$\tilde{y}_k^i = y_k^i + \dot{y}_k^i \Delta t^i - x_k^i \Delta \theta^i + \frac{y_k^i}{\sqrt{(x_k^i)^2 + (y_k^i)^2}} \Delta r^i + n_{k,y}^i. \tag{37}$$

Similarly, for the ESM measurement equation, we have

$$\tilde{x}_k^{i,j} = x_k^{e,j} + \dot{x}_k^{e,j} \Delta t^i + y_k^{e,j} \Delta \theta^i + \frac{x_k^{e,j}}{\sqrt{(x_k^{e,j})^2 + (y_k^{e,j})^2}} \Delta r^i + n_{k,x}^{e,j}, \tag{38}$$

$$\tilde{y}_k^{i,j} = y_k^{e,j} + \dot{y}_k^{e,j} \Delta t^i - x_k^{e,j} \Delta \theta^i + \frac{y_k^{e,j}}{\sqrt{(x_k^{e,j})^2 + (y_k^{e,j})^2}} \Delta r^i + n_{k,y}^{e,j}. \tag{39}$$

The function $h(\cdot, \cdot)$ is approximated by the first-order Taylor series expansion, i.e., $h(\bar{\xi}_k, \beta) \approx h(\hat{\bar{\xi}}_{k|j}, \beta) + H_k^T(\bar{\xi}_k - \hat{\bar{\xi}}_{k|j})$, where $H_k^T = \left.\frac{\partial h(\bar{\xi}_k, \beta)}{\partial \bar{\xi}_k}\right|_{\bar{\xi}_k = \hat{\bar{\xi}}_{k|j}}$. Since the output measurement function is linear with respect to β, we have $h(\bar{\xi}_k, \beta) \approx H_k \bar{\xi}_k + B(\bar{\xi}_k)\beta$, where H_k is a constant matrix, and $B(\bar{\xi}_k)$ is a nonlinear matrix function with respect to $\bar{\xi}_k$. Thus we have $h(\bar{\xi}_k, \beta) \approx H_k \hat{\bar{\xi}}_{k|j} + B(\hat{\bar{\xi}}_{k|j})\beta + H_k^T(\bar{\xi}_k - \hat{\bar{\xi}}_{k|j})$.

At iteration t the state estimation error $\Delta \bar{\xi}_k^t$ is generated due to an biased estimation of β with an error of $\Delta \beta_{t-1}$. As a result, it yields an error in the state, $\Delta \bar{\xi}_k^t$, which can be computed by

$$\Delta \bar{\xi}_k^t = -K_k B(\bar{\xi}_k) \Delta \beta_{t-1}. \tag{40}$$

Let β^* denote the correct parameter value, we have

$$\beta^* \approx \left[\sum_{k=1}^N B^T(\bar{\xi}_k) B(\bar{\xi}_k)\right]^{-1} \cdot \left[\sum_{k=1}^N B^T(\bar{\xi}_k)(z_k - H_k \bar{\xi}_k)\right]. \tag{41}$$

Similarly, we can derive β_t as follows.

$$\beta_t \approx \left[\sum_{k=1}^N B^T(\hat{\bar{\xi}}_k) B(\hat{\bar{\xi}}_k)\right]^{-1} \cdot \left[\sum_{k=1}^N B^T(\hat{\bar{\xi}}_k)(z_k - H \hat{\bar{\xi}}_k)\right], \tag{42}$$

where $\hat{\bar{\xi}}_k = \bar{\xi}_k - \Delta\bar{\xi}_k^t$ and $\bar{\xi}_k$ is the true state. Thus $B(\hat{\bar{\xi}}_k) = B(\bar{\xi}_k - \Delta\bar{\xi}_k^t) = B(\bar{\xi}_k) + \Delta B$ where ΔB is the small error matrix with $\Delta B = B'(\bar{\xi}_k) \otimes \Delta\bar{\xi}_k^t$ and B' is the derivative of B with respect to $\hat{\bar{\xi}}_k$. Therefore, we can simplify β_t as

$$\beta_t = \left[(B(\bar{\xi}_k) + \Delta B)^{\mathrm{T}}(B(\bar{\xi}_k) + \Delta B)\right]^{-1}$$

$$\times \left[\sum_{k=1}^{N}(B(\bar{\xi}_k) + \Delta B)^{\mathrm{T}}(z_k - H_k\bar{\xi}_k + H_k\Delta\bar{\xi}_k^t)\right]$$

$$= \left[\underbrace{\sum_{k=1}^{N} B^{\mathrm{T}}(\bar{\xi}_k)B(\bar{\xi}_k)}_{B_b}\right.$$

$$\left.+ \underbrace{\sum_{k=1}^{N} B^{\mathrm{T}}(\bar{\xi}_k)\Delta B + \sum_{k=1}^{N} \Delta B^{\mathrm{T}} B(\bar{\xi}_k) + \sum_{k=1}^{N} \Delta B^{\mathrm{T}} \Delta B}_{B_1}\right]^{-1}$$

$$\times \left[\underbrace{\sum_{k=1}^{N} B^{\mathrm{T}}(\bar{\xi}_k)(z_k - H_k\bar{\xi}_k)}_{B_\nu}\right.$$

$$\left.+ \underbrace{\sum_{k=1}^{N} \Delta B^{\mathrm{T}}(z_k - H_k\bar{\xi}_k + H_k\Delta\bar{\xi}_k^t) + \sum_{k=1}^{N} B^{\mathrm{T}}(\bar{\xi}_k)H_k\Delta\bar{\xi}_k^t}_{B_h}\right]$$

$$= [B_b^{-1} - B_b^{-1}B_1(I + B_b^{-1}B_1)B_b^{-1}][B_\nu + B_h]$$

$$= \beta^* + B_b^{-1}\sum_{k=1}^{N} B^{\mathrm{T}}(\bar{\xi}_k)H_k\Delta\bar{\xi}_k^t. \tag{43}$$

In (43), higher-order term of the perturbation matrix multiplication is neglected in the last equality. Using (40), the estimation error in β at iteration t can be expressed by

$$\Delta\beta_t = B_b^{-1}\sum_{k=1}^{N} B^{\mathrm{T}}(\bar{\xi}_k)H_k\Delta\bar{\xi}_k^t$$

$$= \left[\sum_{k=1}^{N} B^{\mathrm{T}}(\bar{\xi}_k)B(\bar{\xi}_k)\right]^{-1} \cdot \sum_{k=1}^{N} B^{\mathrm{T}}(\bar{\xi}_k)\Xi_k B(\bar{\xi}_k)\Delta\beta_{t-1}, \tag{44}$$

where $\Xi_k = H_k K_k$. Therefore, we obtain the following norm expression by using Holder inequality.

$$||\Delta\beta_t|| \leq ||\max_k \Xi_k|| \cdot ||\Delta\beta_{t-1}||$$
$$\leq ||\Delta\beta_{t-1}|| \tag{45}$$

An EM-IMM Method for Simultaneous Registration

The following lemma gives the property of Ξ_k.

Lemma 1. *The matrix Ξ_k is positive semi-definite and $||\Xi_k|| \leq 1$.*

Proof: From the Kalman filter, we further obtain Ξ_k as

$$\Xi_k = HPH^T(HPH^T + R)^{-1} \tag{46}$$

If H is full rank, then HPH^T is non-singular. Thus (46) can be rewritten as $\Xi_k = (I + R(HPH^T)^{-1})^{-1}$. Since $P > 0$, we have $HPH^T > 0$. From $R > 0$, it follows that $\Xi_k^{-1} > I$ and therefore $||\Xi_k|| \leq 1$.

If H is not full rank, we apply Schur decomposition to HPH^T, i.e.,

$$HPH^T = U_h D_h U_h^T \tag{47}$$

where U_h is a unitary matrix and D_h is a block diagonal matrix with some zero diagonal elements given by

$$D_h = \begin{bmatrix} D_h^{11} & \\ & \mathbf{0} \end{bmatrix}. \tag{48}$$

Substituting (47) and (48) into (46), we have

$$\begin{aligned}\Xi_k &= U_h D_h (D_h + R_1)^{-1} U_h^T \\ &= U_h^{11} D_h^{11} (D_h^{11} + R_1^{11})^{-1} (U_h^{11})^T \\ &= U_h^{11} (I + R_1^{11} (D_h^{11})^{-1})^{-1} (U_h^{11})^T,\end{aligned} \tag{49}$$

where $R_1 = U_h^T R U_h$. Since $D_h^{11} > 0$ and $R_1^{11} > 0$, it follows that $||\Xi_k|| < 1$. ∎

From Lemma 1 and (45), we conclude that the proposed method gradually converges and the yielded estimation error does not increase during the iteration. In fact, in most cases, $||\Xi_k||$ will be smaller than 1 due the measurement matrix H is usually not full rank. Therefore, the estimates are asymptotically unbiased.

5 Performance Evaluation by PCRB

To evaluate the proposed method, it is beneficial to obtain a lower bound of the estimation performance. This bound is useful to evaluate the best achievable performance of an algorithm. For a discrete-time nonlinear dynamic state-space system, a recursive algorithm is recently proposed to compute the PCRB of the states [32]. The bound computation algorithm is further extended to target tracking in [33,34,35,36]. In this section, we first compute the PCRB for a radar/ESM system. Application to multiple model case is then introduced. In addition, we also present a novel inverse computation algorithm to obtain the inverse of FIM directly.

Let $p_{z,s}(\mathcal{Z}, S)$ be the joint probability density of the estimated vector s and the observations z with $s_k = [\bar{\xi}_k \ \beta_k]^T$. The PCRB is represented by

$$\mathbb{E}\{[\varpi(z) - s][\varpi(z) - s]^T\} \leq J^{-1}, \tag{50}$$

where $\varpi(z)$ is an estimator of s based on the observation z, and J is the FIM with the ij-th element defined by

$$J_{ij} = \mathbb{E}\left\{-\frac{\partial^2 \log p_{Z,S}(\mathcal{Z},\mathcal{S})}{\partial S_i \partial S_j}\right\}. \tag{51}$$

If S_k and J_k are the information sub-matrices for $[\bar{\xi}_{k-1}, \beta_k]$ and for s_k respectively, then we obtain the following recursions [32].

$$S_{k+1} = \begin{bmatrix} J_k^{11} + H_k^{11} & H_k^{13} & J_k^{12} + H_k^{12} \\ (H_k^{13})^{\mathrm{T}} & H_k^{33} & (H_k^{23})^{\mathrm{T}} \\ (J_k^{12} + H_k^{12})^{\mathrm{T}} & H_k^{23} & J_k^{22} + H_k^{22} \end{bmatrix}, \tag{52}$$

$$J_{k+1} = \begin{bmatrix} S_{k+1}^{22} & S_{k+1}^{23} \\ S_{k+1}^{32} & S_{k+1}^{33} \end{bmatrix} - \begin{bmatrix} S_{k+1}^{21} \\ S_{k+1}^{31} \end{bmatrix} (S_{k+1}^{11})^{-1} \begin{bmatrix} S_{k+1}^{12} & S_{k+1}^{13} \end{bmatrix}. \tag{53}$$

From (52) and (53), a recursive algorithm for computing the FIM is given as follows [32].

$$J_{k+1}^{11} = H_k^{33} - (H_k^{13})^{\mathrm{T}}(J_k^{11} + H_k^{11})^{-1} H_k^{13}, \tag{54}$$

$$J_{k+1}^{12} = (H_k^{23})^{\mathrm{T}} - (H_k^{13})^{\mathrm{T}}(J_k^{11} + H_k^{11})^{-1}(J_k^{12} + H_k^{12}), \tag{55}$$

$$J_{k+1}^{21} = (J_{k+1}^{12})^{\mathrm{T}}, \tag{56}$$

$$J_{k+1}^{22} = J_k^{22} + H_k^{22} - (J_k^{12} + H_k^{12})^{\mathrm{T}}(J_k^{11} + H_k^{11})^{\mathrm{T}}(J_k^{12} + H_k^{12}). \tag{57}$$

From [32], we know that $H_k^{11}, H_k^{12}, H_k^{13}, H_k^{22}, H_k^{23}, H_k^{33}$ are obtained from the second-order derivatives of the probability density function $\bar{p}_k$ which is defined by $\bar{p}_k = p(\bar{\xi}_{k+1}|s_k)p(z_{k+1}|s_k,\bar{\xi}_{k+1})$. We can then obtain these matrices as follows:

$$H_k^{11} = \mathbb{E}\{-\nabla_{\bar{\xi}_k,\bar{\xi}_k}(\log \bar{p}_k)\} = \bar{A} Q_w^{-1} \bar{A}, \tag{58}$$

$$H_k^{12} = \mathbb{E}\{-\nabla_{\bar{\xi}_k,\beta_k}(\log \bar{p}_k)\} = \mathbf{0}, \tag{59}$$

$$H_k^{13} = \mathbb{E}\{-\nabla_{\bar{\xi}_k,\bar{\xi}_{k+1}}(\log \bar{p}_k)\} = \bar{A}(Q_w^{-\mathrm{T}}), \tag{60}$$

$$H_k^{22} = \mathbb{E}\{-\nabla_{\beta_k,\beta_k}(\log \bar{p}_k)\} = \mathbb{E}\{H_2^{\mathrm{T}} Q_n^{-1} H_2\}, \tag{61}$$

$$H_k^{23} = \mathbb{E}\{-\nabla_{\beta_k,\bar{\xi}_{k+1}}(\log \bar{p}_k)\} = \mathbb{E}\{H_1^{\mathrm{T}} Q_n^{-1} H_2\}, \tag{62}$$

$$H_k^{33} = \mathbb{E}\{-\nabla_{\bar{\xi}_{k+1},\bar{\xi}_{k+1}}(\log \bar{p}_k)\} = Q_w^{-\mathrm{T}} + \mathbb{E}\{H_2^{\mathrm{T}} Q_n^{-1} H_1\}, \tag{63}$$

where $H_1 = \frac{\partial \mathbf{h}(s_k)}{\partial \bar{\xi}_k}$, and $H_2 = \frac{\partial \mathbf{h}(s_k)}{\partial \beta_k}$.

Now we derive a direct inverse FIM recursive algorithm. This algorithm is equivalent to the previous one with the same computation cost. However, it is more convenient for PCRB evaluation. We first compute the inverse of S_{k+1}^{-1}. The J_{k+1}^{-1} will then be at the right-lower block of S_{k+1}^{-1}. In our setting, we can rewrite (52) as

$$S_{k+1} = \Phi_k^{-\mathrm{T}} J_h \Phi_k^{-1}, \tag{64}$$

An EM-IMM Method for Simultaneous Registration

where

$$\Phi_k = \begin{bmatrix} I & 0 & 0 \\ 0 & 0 & I \\ 0 & I & 0 \end{bmatrix}, \tag{65}$$

$$J_h = \underbrace{\begin{bmatrix} J_k & 0 \\ 0 & H_k^{33} \end{bmatrix}}_{C_k} + \underbrace{\begin{bmatrix} H_k & \begin{bmatrix} H_k^{13} \\ H_k^{23} \end{bmatrix} \\ \left[(H_k^{13})^{\mathrm{T}} \ (H_k^{23})^{\mathrm{T}} \right] & 0 \end{bmatrix}}_{L_k}. \tag{66}$$

It should be noted that Φ_k is a symmetric unitary matrix, i.e., $\Phi_k = \Phi_k^{-1} = \Phi_k^{\mathrm{T}}$.

It follows from the matrix inverse lemma that we can compute the inverse of J_h as $J_h^{-1} = (I + C_k^{-1} L_k)^{-1} C_k^{-1}$ and after some algebraic manipulations, we obtain

$$J_h^{-1} = \begin{bmatrix} I + J_h^{-1} H_k & J_k^{-1} \begin{bmatrix} H_k^{13} \\ H_k^{23} \end{bmatrix} \\ (H_k^{33})^{-1} \left[(H_k^{13})^{\mathrm{T}} \ (H_k^{23})^{\mathrm{T}} \right] & I \end{bmatrix}^{-1}$$

$$\times \begin{bmatrix} J_k^{-1} & 0 \\ 0 & (H_k^{33})^{-1} \end{bmatrix}$$

$$= \begin{bmatrix} S_D^{-1} & -S_D^{-1} J_k^{-1} \bar{H}_k^{\mathrm{T}} \\ -(H_k^{33})^{-1} \bar{H}_k S_D^{-1} & I + (H_k^{33})^{-1} \bar{H}_k^{\mathrm{T}} S_D^{-1} J_k^{-1} \bar{H}_k \end{bmatrix}$$

$$\times \begin{bmatrix} J_k^{-1} & 0 \\ 0 & (H_k^{33})^{-1} \end{bmatrix}$$

$$= \begin{bmatrix} S_D^{-1} J_k^{-1} & -S_D^{-1} J_k^{-1} \bar{H}_k^{\mathrm{T}} (H_k^{33})^{-1} \\ -(H_k^{33})^{-1} \bar{H}_k^{\mathrm{T}} S_D^{-1} J_k^{-1} & S_s \end{bmatrix}$$

$$= \begin{bmatrix} J_d & -J_d \bar{H}_k^{\mathrm{T}} (H_k^{33})^{-1} \\ -(H_k^{33})^{-1} \bar{H}_k J_d & S_s \end{bmatrix}, \tag{67}$$

where

$$\bar{H}_k = \left[(H_k^{13})^{\mathrm{T}} \ (H_k^{23})^{\mathrm{T}} \right] \tag{68}$$

$$S_s = \left[I + (H_k^{33})^{-1} \bar{H}_k S_D^{-1} J_k^{-1} \bar{H}_k^{\mathrm{T}} \right] (H_k^{33})^{-1}, \tag{69}$$

$$S_D = I + J_k^{-1} D_s, \tag{70}$$

$$D_s = H_k - \bar{H}_k^{\mathrm{T}} (H_k^{33})^{-1} \bar{H}_k, \tag{71}$$

$$J_d = S_D^{-1} J_k^{-1} = (I + J_k^{-1} D_s)^{-1} J_k^{-1}. \tag{72}$$

Let

$$J_d \triangleq \begin{bmatrix} J_d^{11} & J_d^{12} \\ J_d^{21} & J_d^{22} \end{bmatrix}.$$

Inserting (67) into (64), we obtain

$$S_{k+1}^{-1} = \Phi_k J_h^{-1} \Phi_k$$
$$= \begin{bmatrix} J_d^{11} - (J_d^{11} H_k^{13} + J_d^{12} H_k^{23})(H_k^{33})^{-1} J_d^{12} \\ J_s & S_s & \bar{J}_s \\ J_d^{21} - (J_d^{21} H_k^{13} + J_d^{22} H_k^{23})(H_k^{33})^{-1} J_d^{22} \end{bmatrix}, \quad (73)$$

where $J_s = -(H_k^{33})^{-1}[(H_k^{13})^T J_d^{11} + (H_k^{23})^T J_d^{21}]$ and $\bar{J}_s = -(H_k^{33})^{-1}[(H_k^{13})^T J_d^{21} + (H_k^{23})^T J_d^{22}]$. Thus, the inverse FIM at $k + 1$ can be calculated as

$$J_{k+1}^{-1} \triangleq \begin{bmatrix} \bar{J}_{k+1}^{11} & \bar{J}_{k+1}^{12} \\ \bar{J}_{k+1}^{21} & \bar{J}_{k+1}^{22} \end{bmatrix}$$
$$= \begin{bmatrix} S_s & \bar{J}_s \\ -(J_d^{21} H_k^{13} + J_d^{22} H_k^{23})(H_k^{33})^{-1} & J_d^{22} \end{bmatrix}. \quad (74)$$

Since J_d can be obtained from block matrices of J_k^{-1} via (72), all block matrices in J_{k+1}^{-1} can be computed from J_k^{-1} as given by (74).

Proposition 1. *The sequences $\{J_{k+1}^{11}, J_{k+1}^{13}, J_{k+1}^{31}, J_{k+1}^{33}\}$ of the inverse FIM can be recursively calculated from $\{J_k^{11}, J_k^{13}, J_k^{31}, J_k^{33}\}$ by (72) and (74).*

The inverse FIM is obtained directly for performance evaluation. However, the resulted PCRB is calculated only for a single model in effect associated with the maneuvering target. To account for the maneuverability, the enumeration PCRB is developed [33] where the unconditional PCRB is obtained by taking the weighted sum of the model-conditioned PCRB^i with respect to the model sequence $M_i(k)$. That is,

$$\text{PCRB} = \mathbb{E}\{\text{PCRB}^i\} = \sum_{i=1}^{L^k} \Pr\{M_i(k)\} \cdot (J_k^i)^{-1}, \quad (75)$$

where J_k^i is the FIM conditioned on the sequence $M_i(k)$. The main difficulty with this method is the computational cost since the number of the model sequence is exponentially grown with the time. Even though a limited number of terms are used in (75), it is intensive in the computation. Recently, Hernandez, et. al., elegantly designed a best-fitting Gaussian approximation technique to replace the the Markov switch model with a single Gaussian model [37] in the PCRB computation. This can improve the accuracy of the bound and remove the difficult procedure of taking average over the model history sequence. However, the method requires generation

An EM-IMM Method for Simultaneous Registration

of a large number of particles in Gaussian approximation in each iteration. In this paper, similar to [35], we adopt the known model sequence and parameters in the PCRB computation. This corresponds to adopt only one optimal sequence in (75). Similar to (75), the resulted PCRB is also conservative [35, 37]. However the computational burden is significantly reduced.

6 Simulation Results

In this section, we study the performance of the proposed method via computer simulations. The scenario of two radars and two ESMs observing one target is considered here. For simplicity, all motions are assumed to be in two-dimensional space.

The target dynamics is described by three models: CV, CA, CT. More precisely, the CA and CT models are described by (76) and (77) respectively as follows.

$$x_{k+1} = \Lambda_a x_k + \Gamma_a w_k, \tag{76}$$

$$x_{k+1} = \begin{bmatrix} 1 & S_w T_s & 0 & 0 & -C_w T_s & 0 \\ 0 & \cos(\omega T_s) & 0 & 0 & -\sin(\omega T_s) & 0 \\ 0 & -T_s \omega^2 & 1 & 0 & 0 & 0 \\ 0 & C_w T_s & 0 & 1 & S_w T_s & 0 \\ 0 & \sin(\omega T_s) & 0 & 0 & \cos(\omega T_s) & 0 \\ 0 & 0 & 0 & 0 & -T_s \omega^2 & 1 \end{bmatrix} x_k + \begin{bmatrix} \Gamma_a \\ \Gamma_a \end{bmatrix} w_k. \tag{77}$$

In (76), Λ and Γ_a are given in (4) where the sampling period T_s is assumed to be 1s. In (77), the turn rate ω is set to -0.05 arc per second, S_w and C_w are then given by $S_w = \sin(\omega T_s)/(\omega T_s)$ and $C_w = (1 - \cos(\omega T_s))/(\omega T_s)$, respectively. It should be noted that the state dimension of the CV is 4 and that of the CA and CT is 6. Thus, necessary dimensional changes both in the state vector and in its covariance matrices from one model to another model are required in the IMM filter [29]. The standard deviations of measurement noises of the CV, CA, CT models are set to 0.001, 0.001, 0.002, respectively.

The target is initially located at $(10\sin(\pi/6), 10\cos(\pi/6))$km, and the initial velocity is set to $(2.78\sin(\pi/6), 2.78\cos(\pi/6))$km/s. The target first moves with a constant velocity during the first 220s. After that it accelerates according to the CA model for the subsequent 50s. At this stage, the accelerations in x-axis and y-axis are set to 0.88km/s^2 and 0.8km/s^2, respectively. Finally the target performs a turn (in CT model) for the final 30s.

The initial positions of the first radar, the second radar, the first ESM and the second ESM are set to $(-25, -1000)$km, $(-1000, 1000)$km, $(1200, 2500)$km, $(1500, 10)$km, respectively. Correspondingly their respective initial velocities are set to $(7.66, 1.35)$km/s, $(3.92, 1.43)$km/s, $(4.81, 1.39)$km/s, and $(7.22, 2.08)$km/s. The radars and ESMs run according to the CV model for the whole observation period. Fig. 2 plots the true trajectories of the target, the radars and the ESMs.

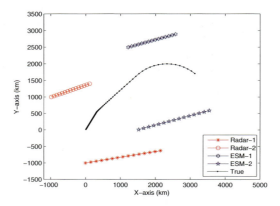

Fig. 2. True trajectory of the target, the radars and and the ESMs

The time biases, the radar angle biases, the radar range biases for the first and second radars are $\{\Delta t_1 = 1\text{s}, \Delta \theta_1 = -0.05\text{rad}, \Delta r_1 = 30\text{m}\}$ and $\{\Delta t_2 = 0.5\text{s}, \Delta \theta_2 = 0.1\text{rad}, \Delta r_2 = 60\text{m}\}$, respectively. The angular biases for the first and the second ESMs are set to $\Delta \phi_1 = 0.2\text{rad}$ and $\Delta \phi_2 = 0.1\text{rad}$, respectively.

In the implementation of the IMM smoother, either EKF or UKF is employed which is hence named as EM-IMM-EKF and EM-IMM-UKF methods, respectively. A total of 10 Monte-Carlo runs are performed in the simulation. In the IMM, the initial forward mode probability is set to [0.9 0.1 0] and the backward mode probability [1/3 1/3 1/3]. The initial state is randomly chosen according to the actual measurements. The initial sensor biases vector is randomly chosen such as around [1.2s, 0.02rad, 15km, 0.03rad]. In the implementation of the IMM smoother, the mode transition probability is given as

$$p_{ij} = \begin{bmatrix} 0.95 & 0.09 & 0.01 \\ 0.05 & 0.9 & 0.05 \\ 0.01 & 0.09 & 0.9 \end{bmatrix}.$$

For the proposed method, iteration is stopped when the norm of consecutive parameter estimation error reaches the threshold of $1e\text{-}3$. Our simulation shows that a few iterations such as 5-10 iterations, are sufficient to obtain the estimates.

The estimation result of the target state is shown in Figs. 3 and 4 for the x-axis and y-axis, respectively. The RMSE in position estimation of the target is given by Fig. 3a and 4a. It is observed that the EM-IMM-EKF method yields a smaller position estimation error than the EM-IMM-UKF method during the CV, CA modes in x-axis. But it is inferior to the EM-IMM-UKF in position estimation along the y-axis. In the x-axis, when the target is in the CT mode, the EM-IMM-EKF shows a larger position estimation error than the EM-IMM-UKF. This is because the EKF could not quickly responds to the changing dynamics after the CA model is switched to the CT model, while the UKF is capable of providing a more approximation in the

An EM-IMM Method for Simultaneous Registration

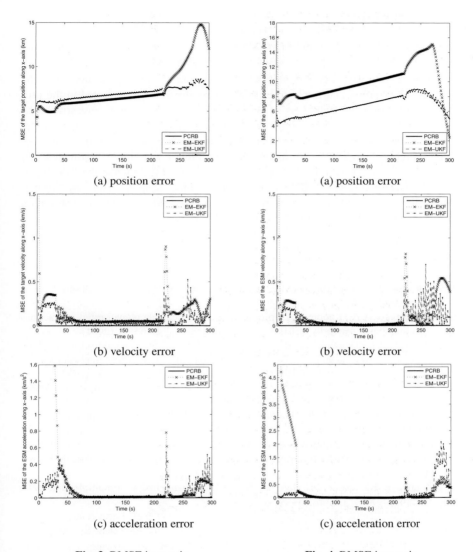

Fig. 3. RMSE in x-axis   **Fig. 4.** RMSE in y-axis

propagation of the mean and covariance matrices. Fig. 3b and 4b show the velocity estimates by both methods in the x- and y-axis, respectively. It is seen that the EM-IMM-EKF has very large fluctuations at the beginning and the model changes from CA to CT. This hence leads to larger position estimation error. Similar observation is also found in acceleration estimation along the x- and y-axis. As UKF has the same complexity level as EKF, the proposed EM-IMM-UKF is considered to be more practical.

7 Conclusions

In this paper, we propose an EM-IMM method for simultaneous registration and fusion of multiple active and passive sensors. The IMM smoother/filter is embedded in the E-step to evaluate conditional expectation of the target state. Since the parameter estimation and the state estimation are separately conducted in the proposed algorithm, this enables us to apply perturbation analysis to investigate the performance. The analysis explains why the augmented Kalman filter is prone to divergence but the proposed method is gradually convergent. Furthermore, the proposed method is shown to give asymptotically unbiased parameter estimates. For performance evaluation, a novel computation algorithm for inverse the FIM is derived. And a conservative PCRB is employed to reduce computational cost. Simulation results demonstrated that the proposed method provides satisfactory estimation performance as well as good convergence capability.

References

[1] Bar-Shalom, Y.: Multiarget-Multisensor Tracking: Advanced Applications. Artech House, Norwood (1990)
[2] Makarenko, A.A., Kaupp, T., Durrant-Whyte, H.F.: Scalable human-robot interactions in active sensor networks. IEEE Pervasive Computing 2(4), 63–71 (2001)
[3] Blom, H.A.P., Hogendoorn, R.A., Doorn, B.A.V.: Design of a multisensor tracking system for advanced air traffic control. In: Bar-Shalom, Y. (ed.) Multitarget-Multisensor Tracking: Applications and Advances, vol. II, pp. 31–63. Artech House (1992)
[4] Blackman, S., Pop, R.: Design and Analysis of Modern Tracking Systems. Artech House, Norwood (1999)
[5] Bar-Shalom, Y., Li, X.R.: Multitarget-Multisensor Tracking: Principles and Techniques. YBS, Storrs (1995)
[6] Dana, M.P.: Registration: a prerequisite for multiple sensor tracking. In: Bar-Shalom, Y. (ed.) Multiarget-Multisensor Tracking: Advanced Applications, Artech House, Norwood (1990)
[7] Leung, H., Blanchette, M., Harrison, C.: A least squares fusion of multiple radar data. In: Proc. of RADAR, pp. 364–369 (1994)
[8] Zhou, Y., Leung, H., Blanchette, M.: Sensor alignment with Earth-centered Earth-fixed (ECEF) coordinate system. IEEE Trans. Aerospace and Electronic Systems 35(2), 410–418 (1999)
[9] Zhou, Y., Leung, H., Yip, P.C.: An exact maximum likelihood registration algorithm for data fusion. IEEE Trans. Signal Processing 45(6), 1560–1572 (1997)
[10] Okello, N., Ristic, B.: Maximum likelihood registration for multiple dissimilar sensors. IEEE Trans. Aerospace and Electronic Systems 39(3), 1074–1083 (2003)
[11] Cruz, E.D., Alouani, A., Rice, T., Blair, W.: Sensor registration in multisensor systems. In: Proceedings of SPIE, vol. 1698, pp. 382–393 (1992)
[12] Zhou, Y., Leung, H., Bosse, E.: Registration of mobile sensors using the parallelized extended Kalman filter. Optical Engineering 36(3), 780–788 (1997)
[13] Nabaa, N., Bishop, R.H.: Solution to a multisensor tracking problem with sensor registration errors. IEEE Trans. Aerospace and Electronic Systems 35(1), 354–363 (1999)

14. Tenney, R., Hebbert, R., Sandell, N.J.: A tracking filter for maneuvering sources. IEEE Trans. Automatic Control 22(2), 246–251 (1977)
15. Aidala, V.J.: Kalman filter behavior in bearings-only tracking applications. IEEE Trans. Aerospace and Electronic Systems 15(1), 29–39 (1979)
16. Ljung, L.: Asymptotic behavior of the extended Kalman filter as a parameter estimator for linear systems. IEEE Trans. Automatic Control 24(1), 36–50 (1979)
17. Wan, E.A., van der Merwe, R.: The unscented Kalman filter. In: Haykin, S. (ed.) Kalman filter and Neural networks, ch. 7. Wiley Publishing, Chichester (2001)
18. Li, W., Leung, H., Zhou, Y.: Space-time registration of radar and ESM using unscented Kalman filter. IEEE Trans. Aerospace and Electronic Systems 40(3), 824–836 (2004)
19. Friedland, B.: Treatment of bias in recursive filtering. IEEE Trans. Automatic Control 14(4), 359–367 (1969)
20. Zhou, Y., Mickeal, J., Ford, B.: A neural network based approach for ESM/Radar track association. In: Proc. of the 7th Int. Conf. on Information Fusion, vol. 3, pp. 1238–1244 (2004)
21. Doorn, B.A.V., Blom, H.A.P.: Systematic error estimation in multisensor fusion systems. In: Proc. SPIE Symp. Signal and Data Processing of Small Targets, vol. 1954, pp. 450–461 (1993)
22. Dempster, A.P., Laird, N.M., Rubin, D.B.: Maximum likelihood from incomplete data via the EM algorithm. J. Royal Statiscal Soc., Ser. B 39(1), 1–38 (1977)
23. Logothetis, A., Krishnamurthy, V., Holst, J.: A Bayesian EM algorithm for optimal tracking of a maneuvering target in clutter. Signal Processing 82, 473–490 (2002)
24. Molnar, K.J., Modestino, J.W.: Application of the EM algorithm for the multitarget/multisensor tracking problem. IEEE Trans. Signal Processing 46(1), 115–129 (1998)
25. Pulford, G.W., Scala, B.F.L.: MAP estimation of target manoeuvre sequence with the expectation-maximization algorithm. IEEE Trans. Aerospace and Electronic Systems 38(2), 367–377 (2002)
26. Krishnamurthy, V., Dey, S.: Reduced spatio-temporal complexity MMPP and image-based tracking filters for maneuvering targets. IEEE Trans. Aerospace and Electronic Systems 39(4), 1277–1291 (2002)
27. Zia, A., Kirubarajan, T., Reilly, J.P., Yee, D., Punithakumar, K., Shirani, S.: An EM algorithm for nonlinear state estimation with model uncertainties. IEEE Trans. Signal Processing 56(3), 921–936 (2008)
28. Huang, D., Leung, H.: An expectation-maximization based interacting multiple model approach for cooperative driving systems. IEEE Trans. Intelligent Transportation Systems 6, 206–228 (2005)
29. Bar-Shalom, Y., Li, X.R., Kirubarajan, T.: Estimation with Applications to Tracking and Navigation. Wiley, New York (2001)
30. Helmick, R.E., Blair, W.D., Hoffman, S.A.: Fixed-interval smoothing for Markovian switching systems. IEEE Trans. Information Theory 41(6), 1845–1855 (1995)
31. Huang, D., Fujiyama, N., Sugimoto, S.: Blind image identification and restoration for noisy blurred images based on discrete sine transform. IEICE Trans. Information and Systems E86-D, 727–735 (2003)
32. Tichavsky, P., Muravchik, C.H., Nehorai, A.: Posterior Cramér-Rao bounds for discrete-time nonlinear filtering. IEEE Trans. Signal Processing 46(5), 1386–1396 (1998)
33. Bessell, A., Ristic, B., Farina, A., Wang, X., Arulampalam, M.S.: Error performance bounds for tracking a manoeuvring target. In: Proc. of the 6th Conf. Information Fusion, vol. 2, pp. 903–910 (2003)

[34] Farina, A., Ristic, B., Timmoneri, L.: Cramér-Rao bound for nonlinear filtering with $P_d < 1$ and its application to target tracking. IEEE Trans. Signal Processing 50(8), 1916–1924 (2002)
[35] Ristic, B., Arulampalam, M.S.: Tracking a manoeuvring target using angle-only measurements; algorithms and performance. Signal Processing 83, 1223–1238 (2003)
[36] Ristic, B., Farina, A., Hernandez, M.: Cramér-Rao lower bound for tracking multiple targets. IEE Proc.-Radar Sonar Naving 151(3), 129–134 (2004)
[37] Hernandez, M., Ristic, B., Farina, A., Sathyan, T., Kirubarajan, T.: Performance measure for Markovian switching systems using best-fitting Gaussian distributions. IEEE Trans. Aerospace and Electronic Systems 44(2), 724–747 (2008)

Locatable, Sensor-Enabled Multistandard RFID Tags

D. Brenk, J. Essel, J. Heidrich, and R. Weigel

University of Erlangen-Nuremberg, Institute for Electronics Engineering, Germany
{brenk,essel,heidrich,r.weigel}@ieee.org

Abstract. This chapter covers all aspects that have to be considered for building highly efficient passive RFID tags with special abilities. Within a concrete project we are currently implementing such a tag. Besides the mandatory RF frontend and digital baseband processing it incorporates functionalities for multistandard communication, localization and sensing. Ultra low-power techniques enable these circuits to be implemented in a single chip without any other power supply than the RF field. The possibility to produce such RFID tags in high volumes and at low costs is a main objective of our project. Manifold monitoring solutions would profit from maintenance-free and ubiquitous sensor-data acquisition with position information using cheap, disposable RFID tags.

1 Introduction

Radio Frequency Identification (RFID) is a key technology for several industry sectors. Logistic chains can be optimized using central management of worldwide acquired RFID data via Enterprise Resource Planning Tools (ERP). Medical implants are readable and configurable via RFID and thus enhancing the effectiveness and quality of medical attendance. The automotive industry uses RFID-labled supply parts for quick checking which configuration is built into the vehicle without the need of connecting to its CAN bus. RFID-based immobilizer-, passive entry- and tire pressure monitoring-systems give a security and comfort benefit for the actual driver of such a car. These and a lot of other applications are already implemented using RFID tags at different frequencies and different power levels. LF-tags that make use of inductive coupling and load modulation have a range of centimeters (bidirectional, passive) to several meters (unidirectional, active). HF-tags have a range of centimeters to one meter. LF- and HF-tags are commonly used for applications, where a short communication range is required, e.g. in security- or payment-systems. Active UHF-tags work with conventional transceiver concepts. At 868 MHz they reach communication ranges of up to 100 meters [1]. Passive UHF RFID devices have an operating range of up to 9 meters at 868 MHz [2] and up to 3.5 meters at 2.45 GHz [3].

Today, complex energy demanding tasks can only be accomplished by active tags or tags that make use of near field inductive coupling. The energy, that can be gained out of the far field is not sufficient for operating microcontrollers, crypt algorithms or sensors. Two approaches can be combined to solve the dilemma of using more features within passive tags at higher reading ranges. First, the efficiency of the energy harvesting analog frontend has to be improved and second, the power consumption of the tag has to be minimized. Only a few µW are available as supply for all functional blocks [4]. For example a sensor integration requires an ultra low-power data acquisition circuit. For an FMCW[1] radar-based positioning feature a short-time stable tag oscillator is needed. A reference source like a bandgap circuitry must be integrated to enable both functionalities. The need of these three blocks points out the importance of an extremely energy saving design. The baseband processing has to work at the same time as well. In the following sections we present methods and techniques for an implementation of locatable, sensor-enabled multistandard RFID tags with high reading ranges. The benefit of adding more features to passive UHF tags while preserving or even expanding their reading range is in a higher cost efficiency. These tags are absolutely maintenance-free and even disposable if produced in high volumes. Applications like cold chain surveillance, tire pressure monitoring or humidity monitoring of living spaces become feasible at a much lower price.

2 System Architecture

Fig. 1 shows the system architecture of the integrated RFID tag presented in this article. The only external component of this multistandard tag is the antenna which can be used at UHF and HF. The two combined highly efficient power rectifiers convert the incoming electromagnetic RF wave at UHF or the magnetic wave at HF to DC for powering all active circuits on the transponder. If the DC supply voltage reaches a defined level, the power-on-reset generates a reset signal which places the digital logic into a known state. The demodulator recovers the information, which is ASK modulated on the carrier by the interrogator and delivers the data to the digital logic. Because of different modulation standards at UHF and HF it was necessary to implement two independent ASK demodulators. The backscatter modulator at UHF and the load modulator at HF enables the tag-to-reader communication. This is achieved by varying the impedance of the RF node in order to modulate the reflected power (UHF) or by changing the load of the inductively coupled transformer network. The integrated voltage limiter protects the analog frontend at high RF levels. The current source has been designed to drive the different filters and references. When the digital logic decodes an EPC custom command for sensor reading (see section 5), the sensor interface is turned on. It selects the ADC input source via the multiplexer and

---

[1] Frequency Modulated Continuous Wave.

Locatable, Sensor-Enabled Multistandard RFID Tags

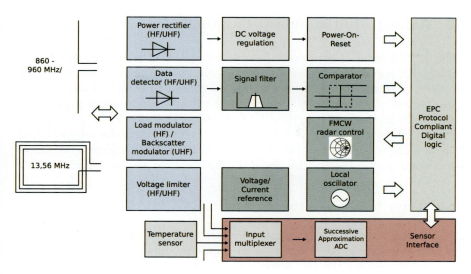

Fig. 1. System architecture of a locatable, passive multistandard RFID transponder with sensor interface

performs a data conversion. The acquired data is sent back to the reader on either the HF- or UHF-communication channel. In case of decoding an EPC custom command for positioning, the FMCW radar control module performs a constant backscattering for a predefined duration.

3 Technology Aims

Because passive RFID transponder devices are high-volume products for mass applications, the use of a low-cost technology is essential. To obtain cheaper tags in the cent domain with more operational reliability, a lot of scientific work is in progress in the area of low cost technology development, e.g. polymer electronics [5].

However, typical state of the art RFID transponder devices, especially for UHF applications, are fabricated in bulk CMOS technologies. The advantages in comparison to other possible technologies are the low static power consumption, the relatively low production costs, the controllable yield and a possible high complexity. Furthermore, it is state of the technology to combine analog and digital circuit blocks with a non-volatile-memory (NVM) like EEPROM and other novel techniques [6]. Poly EEPROM is the established technology for ultra low-cost RFID applications with only small memory arrays [7]. The used technologies for passive RFID transponders typically feature gate lengths of 0.13 µm. A further shrinkage is not reasonable, due to the unproportional increase of the static power consumption. In recent process generations the leakage losses are no longer negligible and therefore ultra low-power design techniques are absolutely required.

4 Analog Multistandard Frontend

The RF frontend of integrated passive transponders represents one of the most complex analog devices used in the wireless communication area. This is because in analog RF frontends non-linear and time-variant components are coupled in a single node. The implemented frontend topology depends on the requirements given by a specific application. Thus, for an ultra low-power frontend design, it is essential to know the behavior and the interaction of the non-linear devices and the field of application. In this section, the most important analog building blocks of the RF frontend are presented.

4.1 Ultra Low-Power Rectification

Fig. 2 shows the topology of a single ended voltage multiplier for the use at ultra low-power RF levels in UHF. This modified Dickson topology is frequently used for the DC power supply generation from the incident RF signal [8, 9]. The functionality of this rectifying structure is based on simple non-linear devices like Schottky-, pn- or MOS-diodes. RF power rectifiers are usually compared by the overall efficiency (η_T). This figure of merit is defined by

$$\eta_T = \frac{P_{DC}}{P_q - P_{refl}^{RF}}, \qquad (1)$$

where P_{DC} is the available output power, P_q is the incident RF power and P_{refl}^{RF} is the reflected RF power. The threshold voltage at the low power domain of the used rectifying diodes leads to the problem of a high input impedance, even at the fundamental frequency [10]. For a highly efficient rectifier design, devices with a low forward voltage drop are essential. To achieve this requirement, Schottky diodes are used in many frontend designs. These diodes have, beside the low forward voltage drop, a low series resistance and a low Schottky junction capacitance [2, 8]. As mentioned in section 3 the bulk CMOS technology is often used for low-cost RFID transponder designs. In a standard CMOS process Schottky diodes are often not available and therefore standard MOS devices have to be used. As simple MOS diodes have a higher forward voltage drop the efficiency of such rectifiers is lower. Fig. 3(a) shows the measured efficiency of a 2-stage rectifier implemented with diode-connected NMOS transistors (as shown in Fig. 2). For comparison, Fig. 3(b) shows the measured efficiency of a 2-stage rectifier implemented with Schottky diodes. In the area of highly efficient rectifier design with MOS devices for ultra low-cost RFID transponders the research activities are still ongoing [11]. The optimal number of rectifying stages depends amongst others on the connected load, the available technology and the used non-linear devices. The conversion efficiency of a multi-stage rectifier results from the matching between the source and the load impedance. Finding the optimum numbers of rectification stages is a typical task during the design [12, 13].

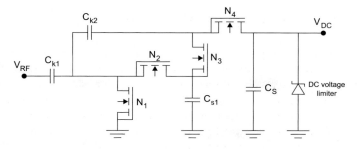

Fig. 2. Single ended two-stage voltage multiplier (modified Dickson topology) for the UHF mode

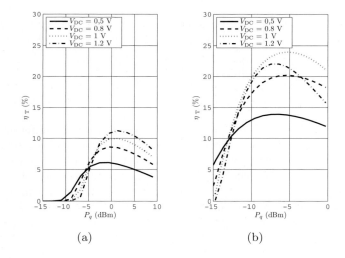

Fig. 3. Efficiency of a 2-stage NMOS (a) and a 2-stage Schottky (b) rectifier

Fig. 4 shows the temperature dependency for a UHF frontend with an integrated 2-stage Schottky rectifier at an input power level (P_q) of approx. -10 dBm. This diagram is particularly important for sensor-enabled RFID tags with an integrated temperature sensor. The efficiency drop at a DC voltage of 1.2 V is mainly caused by the DC voltage limiter shown in Fig. 2. This DC voltage limiter protects the auxiliary devices as well as the digital logic against damage caused by high voltage. The maximum value of the supply voltage can be easily adjusted (e.g. higher than 1.2 V even at higher temperatures) by redesigning the DC voltage limiter.

Fig. 5 shows the simplified schematic of the implemented differential CMOS HF rectifier. The transistors M_1 and M_2 are defining the ground

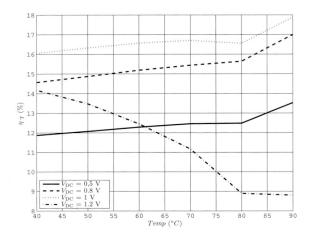

Fig. 4. Temperature dependent efficiency of an UHF frontend with a 2-stage Schottky rectifier at approx. -10 dBm input power P_q.

potential and the transistor pair M_3 and M_4 rectifies the differential input signal at L_{A_HF} and L_{B_HF}. Because of the higher input voltage at HF high-threshold MOS transistors were used to protect the rectifying devices against overvoltage. The lack of efficiency due to high-threshold devices is no problem because near field inductive coupling is in general much more suitable for energy harvesting. To protect the auxiliary devices and the digital logic against overvoltage in HF a DC voltage limiting structure has been implemented as well.

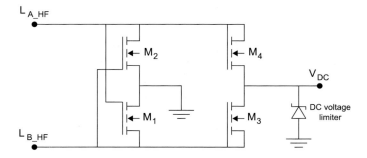

Fig. 5. Power rectifier for the HF mode

4.2 Tag-to-Reader Communication

Backscattering of the incident RF power is the communication principle used in passive UHF RFID systems. The tag-to-reader communication is implemented by varying the impedance of the RF node's analog frontend in order to modulate the reflected power. There are two types of modulation allowed by the EPC UHF Class1 Gen2 standard for RFID backscattering communication: amplitude shift keying (ASK) and phase shift keying (PSK) [14]. Both modulate the complex impedance $\underline{Z}_m = R_m + jX_m$ (m = 1,2) between two different states. Fig. 6 shows four examples of changing the magnitude and/or the phase of the backscattered signal. As Fig. 6(a) and 6(c) show an ideal ASK modulation can only be achieved by a switching resistance if the imaginary part of $\underline{Z}_m$ is zero. Otherwise a more complex modulation circuit is needed. It is demonstrated in [2] and [8] that the PSK modulation would be more efficient than ASK because the available power to the tag is kept almost constant during both modulation states. Analogous to the ideal ASK modulation Fig. 6(b) depicts that the PSK modulation can also be achieved by a complex modulation circuit only. Due to the fact, that most RFID

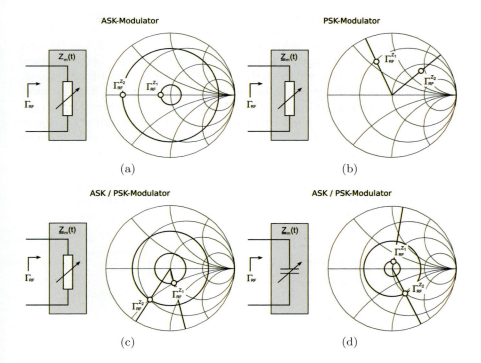

Fig. 6. Change of the magnitude by varying the real part (a), of the phase by varying the real and the imaginary part (b), of the magnitude and phase by varying the real part (c), of the magnitude and phase by varying the imaginary part (d).

transponders are using a switched resistance or a switched capacitance and that several parasitic effects get more important at high frequencies, common integrated backscatter modulators use a combination of PSK and ASK (Fig. 6(c), 6(d)). Therefore, the most reasonable modulation type depends on the available solid-state device technology.

In passive HF RFID Systems an integrated load modulator is used for the tag-to-reader communication. For this type of communication the transponder antenna has to be in the reactive near field of the reader antenna. Because of the transformational coupling between the coil of the transponder and the coil of the reader, a change of the transponder impedance can be detected by the reader. This change can be realized by a carefully designed MOS transistor between the input pins L_{A_HF} and L_{B_HF} (compare Fig. 5) which is driven by the digital baseband signal.

4.3 Reader-to-Tag Communication

The ASK demodulator is an important circuit block for the bi-directional data transmission between the RFID transponder and the reader. It recovers the information, which is ASK modulated on the carrier by the interrogator and delivers the data to the digital logic. To achieve a higher reading range, the reception of data has to be ensured at low power levels. Because of different modulation depths and different input power levels for UHF and HF two demodulation blocks are used [14]. Fig. 7 shows the simplified block diagram of the ASK demodulator for the UHF mode. The input signal $Env_{In,UHF}$ of the demodulator is the envelope of the carrier. This signal is initially level shifted by the non-linear resistors and then filtered by either the current source I_1 (upper path) or the current source I_2 and the capacitor C_1 (lower path). To receive the digital baseband signal, these two values with different time constants and slightly different voltage levels are compared by a low-power comparator with hysteresis. The different voltage levels result from the different dimensions of the current sources I_1 and I_2 and assure a robust demodulation. Because these signals are generated from the same input signal the filters of both paths can be designed with a very low power consumption.

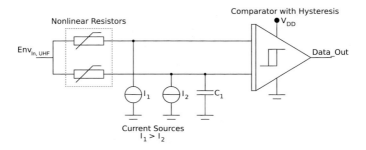

Fig. 7. ASK data detector (UHF)

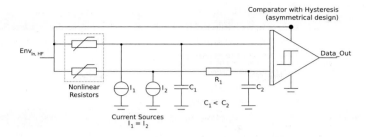

Fig. 8. ASK data detector (HF)

Fig. 8 shows the simplified schematic of the implemented HF demodulator. It looks very similar to the demodulator which is used for UHF. Due to the lower modulation depth in HF (HF: 10 - 30 % ASK, UHF: 80 - 100 % ASK) the used filter structures have to be designed more complex. To demodulate the very small change in the carrier's amplitude, the current sources have the same dimension to prevent different voltage levels. Only the time constant of the signals are adjusted by the capacitors C_1, C_2 and the resistor R_1. To assure a robust demodulation in HF the hysteresis behavior of the comparator is amplified by an unsymmetric design and the use of the modulation state dependent signal $Env_{In,HF}$ as supply voltage.

5 Digital Baseband Processing

When implementing locatable, sensor-enabled multistandard RFID tags, all analog abilities of the tag have to be controlled by a digital logic. First a communication protocol stack is needed. In our case the EPC protocols were chosen: EPC HF draft2 [15] for the HF communication protocol and EPC UHF class1 gen2 [14] for the UHF communication protocol. The advantage of choosing EPC standards for both frequency bands is founded on a very similar baseband implementation, thus saving a lot of logic gates during synthesis. The next point is the integration of a sensor control unit. First a method of sensor data reading has to be defined. This can be achieved by two kind of ways: using a read command on a special memory location or defining a custom command out of the reserved range of the EPC standard. The decision mainly depends on the abilities of the interrogator used. Although being standard-conform, the definition of new *user custom commands* are not supported by all reader products on the market. However, for a higher flexibility the implementation using custom commands was chosen instead of using the failsafe all-interrogator compliant memory-read method. The sensor- and the locate-command are implemented using the same custom command structure. In Fig. 9 an exemplary protocol sequence for both custom commands is displayed.

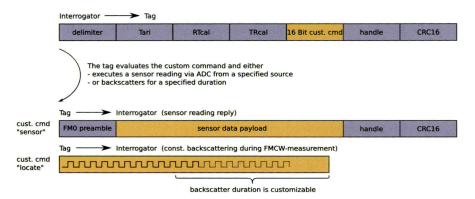

Fig. 9. Sensor command enhancement on the EPC RFID protocol using custom commands

The Interrogator-to-Tag communication is standard-compliant using the delimiter-, Tari-, RTcal- and TRcal-values specified in the communication before invoking the custom command. The handle also has to be the negotiated one from an established EPC session. This way several RFID sensor nodes might be queried in the same session. The 16 Bit custom command contains the command code – either *sensor* or *locate* – and corresponding parameters. For the sensor command this is the input source that should be switched to the ADC and for the locate command this is the duration of the constant backscatter signal.

5.1 Energy Efficiency of Digital Circuits

When implementing this multistandard sensor-enhanced control logic, one has to consider the amount of energy, that is consumed by the logic. For example a standard implementation of the EPC UHF Class1 Gen2 protocol stack might consume up to 9 µW [16]. If also considering the other power consuming design blocks of the tag as the ADC or the oscillator, some measures have to be taken to ensure an acceptable reading range while operating in passive UHF mode [17]. The switching power of logic gates can be expressed as

$$P_{sw} = \alpha C_L V_{DD}^2 f_{clk} \qquad (2)$$

where C_L is the capacitive load at the gate's output, f_{clk} is the switching frequency and α is a factor that accounts for clock gating. In low-power digital designs the clock gating method is always used. Parts of the design that are not needed at a certain time are cut off the clock tree. With the same method slower clocks can be applied to a single unit, i.e. the duty cycle of the clock signal can be minimized. The clock gating method does not address the standby current issue wich is caused by leakage. This power dissipation can be eliminated by using the power gating technique, giving the possibility to cut off whole units of the design from the supply rails. When switching these units back to the supply, a known state has to be ensured to avoid errors. Another method for saving energy is to keep the supply voltage V_{DD} as low as the used CMOS technology allows, since P_{sw} is direct proportional to the square of V_{DD}. A well-considered implementation of units using all techniques in combination results in an energy saving design.

6 Localization of UHF Labels

In radar engineering it is distinguished between primary and secondary radar systems. Primary radars have passive targets which only reflect the radar signal dependent on their shape without changing it. Secondary radars use active targets which amplify the signal [18]. Therefore the operating range is increased because the amplified signal must only overcome the distance tag-to-reader, assumed that the transponder receives enough power from the RF signal of the reader. Furthermore an explicit target identification is possible. This can be achieved by a characteristic modulation or with a codeword appended to the signal. Assigned to passive RFID systems the radar signal is just modulated and backscattered with a lower frequency f_{mod} without being amplified. This technique is called *Modulated Backscattering* [19].

The standard signaling between reader and UHF transponder is based on varying the reflection coefficient of the transponder antenna. To implement positioning for a UHF RFID system the available backscatter modulator of the tranponder can be used for both, basic data transmission and distance measurement. All frequency ranges must be covered by a shared antenna structure on the transponder, but contrary requirements exist for them. Power matching is desired at the HF and the UHF range. At the FMCW frequency range most power should be reflected at the transponder antenna back to the FMCW reader and should not be absorbed by the chip to achieve a distance measurement up to the highest operating range [20]. To fulfill this requirement, specific reflection networks for positioning and data transmission could be implemented. Fig. 10 illustrates this in the Smith

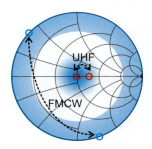

Fig. 10. Ideal modulation states in the Smith chart

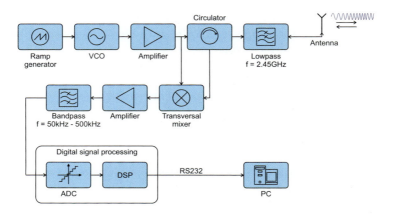

Fig. 11. Block diagram of the FMCW reader

chart. While for power matching the region around the matching point is desired, the reflection coefficient should be close to the border of the Smith chart for the FMCW region. Therefore it should mainly vary in phase (PSK modulation) between the two states.

The use of microwave frequencies (e.g. 2.45 GHz) for the positioning system offers several advantages in comparison to optical and ultrasonic applications: the conditions of the propagation medium have only a small influence on propagation characteristics of the electromagnetic waves. The wave length of microwaves is usually greater than waste particles on sensors. Therefore they are much better suited for positioning in harsh environments than optical systems. Microwaves can also be radiated at a large angle so the coverage area in which targets could be detected can also be large. This depends on the selected antenna architecture and its directional characteristic. However the larger the coverage area the larger is the probability of cancellation by multipath effects. Another advantage is that several objects can be detected simultaneously by the use of different modulation techniques [19, 21].

6.1 Distance Measurement with FMCW Radar

Fig. 11 shows the block diagram of the FMCW reader which is used for this system. It transmits a continous signal which is additionally frequency modulated. The voltage controlled oscillator (VCO) can be periodically tuned by a voltage function. Here a ramp generator is used so the frequency value increases from a base frequency f_0 up to a final value f_0+B within a modulation interval T (Fig. 12). The resulting signal is then radiated by a transmitting and receiving antenna. The lowpass filter behind the antenna assures that only signals in the desired band around 2.45 GHz are processed.

The signal is reflected at the antenna of the UHF transponder at the distance d but at any other target close to the reader, too. Therefore the

Locatable, Sensor-Enabled Multistandard RFID Tags

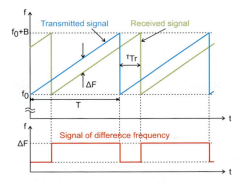

Fig. 12. Frequency curve of transmitted and received FMCW signal

reflection coefficient of the transponder antenna is characteristically modulated after activating the distance measurement with the related custom command. Thereby it can be distinguished from the false targets in the operating range of the reader. A square wave signal at a frequency of about 300 kHz is used as the modulation signal. This modulation should be performed for the duration of a modulation interval T of the FMCW reader to use its whole bandwidth B. This is important for a good resolution of the distance measurement [18]. The backscattered signal of the transponder arrives at the antenna of the reader delayed by the run-time τ_{Tr}. This signal is then mixed with the transmitted signal [21].

The time signal at the output of the mixer for the duration of one ramp interval T is then digitized and the further data processing is done in the digital domain. The related baseband spectrum is not dependent on time. It consists of two mainlobes around a center frequency f_{mod} (Fig. 13). Its value is given by the modulation frequency f_{mod} of the transponder. The distance ΔF of the two mainlobes depends on the distance d between reader and transponder and the known ramp frequency. In [18] a detailed insight into radar theory is given. Fig. 12 shows the frequency curves of the transmitted and received FMCW

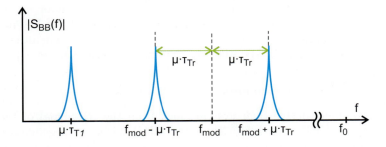

Fig. 13. Baseband spectrum with modulation

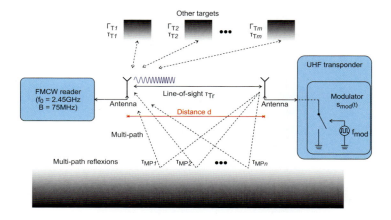

Fig. 14. Channel of the distance measurement

signal. They have the same waveform and are just displaced by the run-time τ_{Tr} from reader to transponder and back. Therefore a characteristic frequency shift ΔF is given which is used for the distance measurement (Eq. 4). Fig. 14 shows the common channel for the positioning function. The desired line-of-sight signal is directly backscattered from the transponder with the run-time τ_{Tr}. The signal can also be reflected again at any target before arriving at the reader antenna. These multi-path signals have always greater run-times τ_{MPn} than the direct one and they are meaningless for the distance measurement but they can corrupt it. Backscattered signals at other targets than the transponder with the common run-times τ_{Tm} are not modulated and they can clearly be separated in the spectrum. Therefore the signal components of the receiving signal can be commonly expressed by Eq. 3.

$$s_r(t) = s_{Tr}(t) + \sum_{i=1}^{n} s_{MPi}(t) + \sum_{j=1}^{m} s_{Tj}(t) \qquad (3)$$

Besides the undesired backscattered FMCW signals the distance measurement could also be influenced by other effects. At the ISM band of 2.45 GHz operate several commercial applications like WLAN systems or microwave ovens. Fig. 13 shows the resulting baseband spectrum for the line-of-sight signal and one (unmodulated) other target, where $\mu = B/T$. The other components have been left for the sake of a general view [20, 21].

6.2 Common Requirements for the Oscillator

HF transponders derive their system clock from the RF signal of the HF reader. At the UHF range (860 MHz to 960 MHz) this is not possible because of the high nominal frequency. The clock dividers for generating a lower system clock would consume large amounts of power. So UHF transponders

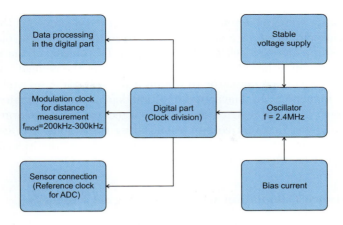

Fig. 15. Clock distribution of the transponder

feature an own oscillator for data processing in the digital part [14]. Regarding an economic power consumption this oscillator should be used for all advanced features of the chip, as there are the modulation signal f_{mod} for the distance measurement, the clock for the sensor interface and the system clock of the control logic. The adjustment of the backscatter frequency value is performed by clock dividers within the digital control logic. The whole setup is shown in Fig. 15.

Besides a proper circuit design which is optimized for low supply voltages and power further requirements must be fulfilled by the whole oscillator system: the oscillator signal must be short-time stable for the duration of a modulation interval T of the FMCW reader (see Fig. 12). Otherwise the modulation signal cannot utilize the whole bandwidth B of the FMCW signal. Fig. 16 shows all necessary components to generate the system clock. The oscillator needs a bias current. With regard to the sensor interface it could be

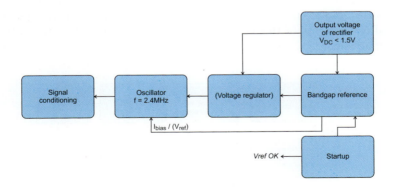

Fig. 16. Components of the oscillator system

derived from a voltage reference which provides the reference for the ADC. The reference circuit needs a startup circuit to settle at the right bias point after power up. The bias cell is supplied by the DC output of the rectifier. The oscillator can either also be directly supplied by the recified voltage or connected to a voltage regulator to obtain a more stable V_{DD} voltage and therefore a higher PSRR. The actual selection depends on the architecture of the oscillator. The digital signal should have a rectangular waveform to minimize the undesired current consumption during transitions. The generated waveform has not sufficiently high slopes which must be conditioned by the succeeding driver stages.

6.3 Distance Calculation with Digital Signal Processing

The digital spectrum analysis offers several advantages compared to its analog counterpart, like lower costs and a short processing time. Fig. 17 demonstrates the different steps which must be performed for the distance calculation. The digitized baseband signal of the FMCW reader contains a constant component. Therefore the mean value of the time signal (one ramp period T) is calculated and then subtracted from it. The limitation of the time signal for one ramp duration T corresponds to a multiplication with a rectangular window. This in turn corresponds to a convolution of the spectrum with the spectrum of a rectangular window which is a sinc(x) function. The first sidelobes of a sinc(x) function are only 13.5 dB below its mainlobe. This has a negative impact on the spectral analysis when the sidelobes of a signal superpose with the main peak of another signal which would falsify the maximum detection (Fig. 13). Using a specific window function (for example the Hamming window) allows a better suppression of sidelobes in the spectrum. Unfortunately this leads to a larger -3 dB bandwidth of the mainlobe which influences the resolution (Fig. 18). The next step is to increase the length of the time signal by adding zeros at its end. This method is called zero-padding which increases synthetically the number of FFT points. Because of the resulting smaller frequency steps the peaks in the spectrum can be better interpolated so the measurement of the distance is more precise. Then the spectral components of this conditioned time signal are calculated with the FFT algorithm.

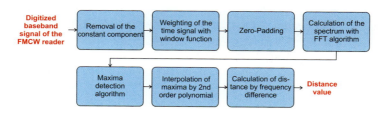

Fig. 17. Steps of distance calculation

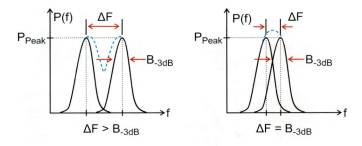

Fig. 18. Achievable resolution of distance measurement

The actual distance can be evaluated with the following algorithm [21]: the calculation begins at the current modulation frequency f_{mod}. Starting from here the algorithm searches the next both peaks in the spectrum above and below f_{mod} which fulfill further boundary conditions: their magnitude must exceed a certain threshold value to avoid the detection of noise peaks. The magnitude of the spectrum must decline by a certain value on both sides of the respective peak. Otherwise the first value which is beyond the threshold would be regarded as the maximum. After that the difference of the related FFT points is calculated and then converted to its equivalent frequency difference ΔF. The desired distance value d is then given by

$$d = \frac{\Delta F \cdot T \cdot c_0}{4 \cdot B} \qquad (4)$$

where c_0 is the speed of light. The numerical calculation of the spectrum with the FFT algorithm is only performed at equidistant frequency points. This causes discretization errors during the calculation of the frequency difference dependent on the distance of the points. The maximum possible error for the calculation of one peak is half the distance of two consecutive points in the FFT spectrum. However, the distance is measured by the spacing between the two peaks so the maximum error is the interval between two FFT points. The accuracy of the distance value can be increased by interpolating around the desired maximum. The respective frequency points of the calculated absolute maximum and two adjacent points are interpolated with a second-order polynomial. Then the related maximum is calculated again. Whether two signals can be distinguished in the spectrum is determined by the width of their main lobes. If the peaks of two mainlobes with the same magnitude P_{Peak} and bandwidth approximate each other more than their -3 dB bandwidth the respective maxima of the two lobes are merging to one lobe (shown in Fig. 18). In this case no distance measurement is possible [21, 22].

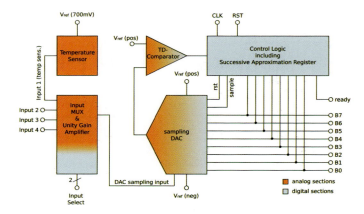

Fig. 19. Block diagram of the sensor interface including an eight Bit successive approximation ADC with DAC-integrated sampling, an input multiplexer and a temperature sensor.

7 Integrated Sensors and Their Interface

Sensors integrated into passive RFID tags have to fulfill low power demands to ensure a sufficient reading range. According to [23] the sensor combined with any circuit to digitize the sensed signal should not consume more than 3 µW. There are specialized RFID sensors as proposed e.g. for temperature in [24]. This can be an advantage from an energy efficiency point of view but it limits the usability of the whole sensor tag to only one purpose. For possible future mass applications an integrated multi-purpose solution is more convenient, where arbitrary sensors can be used with one sensing RFID platform. A generic interface with an analog-to-digital converter (ADC) as digitizing element gives the benefit of connecting arbitrary sensors to it as long as their output voltage resides in the input range of the ADC. The RFID tag presented here, contains a temperature sensor, an ADC and an input multiplexer providing four sensor inputs in sum. It is conceivable, that one or two inputs of the input multiplexer are connected to the HF- and UHF-shunt transistors of the analog frontend. This way RSSI[2] measurements become feasible. The RSSI values might improve or give redundancy to the distance values obtained by the FMCW radar measurements described in section 6.

In Fig. 19 the architecture of our sensor interface is illustrated. On the left the ultra low current temperature sensor is shown, which is connected to the first of four multiplexer inputs. The multiplexer unity gain buffer output has the capacitive DAC array of the ADC as load. The DAC is controlled by the ADC's logic and its analog output is linked to the time-domain comparator which represents the digitizing element of the sensor interface. The next subsections will explain the single design blocks in more detail.

---

[2] Received Signal Strength Information.

7.1 Analog to Digital Converter

The sensor interface mainly consists of a successive approximation (SA) ADC. This SA-ADC-method is predestined for an ultra low-power operation at low clock rates. Fig. 19 shows, that one comparator, a DAC and a digital control logic is needed for this architecture. The comparator is implemented using the time-domain architecture presented in [17]. Main advantage of using a time-domain comparator instead of a classical differential stage is the replacement of the bias current by the charge currents for the capacitors. This nearly eliminates the power dissipation of the comparator while being in idle state. In Fig. 20 the schematic of the time-domain comparator is depicted. When the compare signal is active, both C_1 and C_2 charge over R_1, R_2 and the

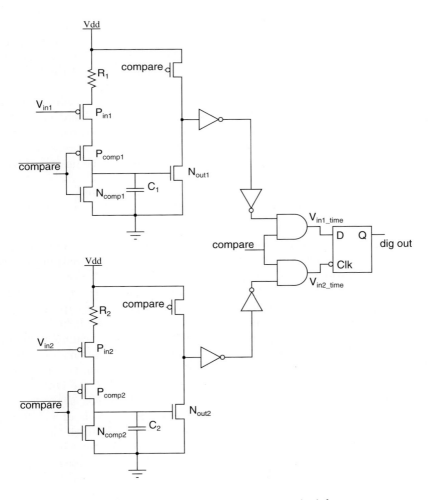

Fig. 20. Time-domain comparator principle

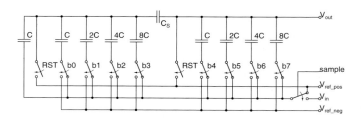

Fig. 21. Schematic of the eight Bit DAC consisting of two four Bit sub-DACs

on-resistance of P_{in1}, P_{in2}. Assumed the input voltage V_{in1} is more positive than V_{in2}, C_1 charges slower than C_2. This results in a later activation of the buffer chain connected to N_{out1} and therefore in a later time of arrival at the D-FlipFlop. Once the compare-signal is switched to "off" the decision which signal (V_{in1_time} or V_{in2_time}) arrived first is stored in the FlipFlop until the compare signal will be activated again. The output state Q of the FlipFlop would be "0" (low) in this exemplary case of V_{in1} exceeding V_{in2}.

The digital to analog converter (DAC) is built using the capacitive charge-redistribution- and charge-scaling-techniques. The input signal is used as voltage offset, thus the DAC is working as a sampling element. Two four Bit sub-DACs form an eight Bit array (Fig. 21). After the reset and the sampling phase, the analog output signal of the DAC can be described as

$$V_{out} = 2V_{ref\_pos} - V_{in} - \left(\frac{1}{16} \frac{\sum_{i=0}^{3} b_i 2^i + 1}{2^4} + \frac{\sum_{i=0}^{3} b_{4+i} 2^i}{2^4} \right) V_{ref\_pos} \quad (5)$$

where $2V_{ref\_pos} - V_{in}$ is the offset caused by the input sampling and the summands in the braces represent the switch states $b_0...b_7$. The output voltage is reciprocally proportional to the bit pattern applied. As the gradient of the temperature sensor output is negative this results in a positive gradient for the temperature codes at the output of the ADC. The scaling capacitor C_S is the demanding element for an implementation of the circuit. It has to match 16/15 of the unity capacitance C. The benefit of using the scaling technique is a lower over-all capacitance in comparison to a straight binary weighted 8 Bit capacitive array. This saves up to 87.9 % of charge when the unity gain buffer of the input multiplexer samples the sensor signal to the DAC.

The digital control logic is implemented with standard cells, that are available in the 0.13 µm CMOS process we used. It mainly consists of a shift register, which performs the successive approximation and stores the current state in latches. Additional elements control the phases for sampling and comparison. Recent measurements in [17] prove that future power improvements should focus on this digital control logic. In this paper we reported a SAR-logic portion of the current consumption beyond 50 % of the whole sensor interface. In section 5.1 we addressed this issue.

7.2 Input Multiplexer

The input multiplexer provides four ports for possible sensor signals. The signals are switched to a unity gain amplifier which delivers the current needed by the sampling input of the capacitive DAC. As only one unity gain amplifier is needed for arbitrary voltage domain sensor signals, the number of sensor input channels can be multiplied easily by adding additional switches. In our implementation we chose four inputs: one originating from the integrated temperature sensor, two to be connected to the HF- and UHF-shunt transistors of the analog frontend (for RSSI) and one as external sensor input.

7.3 Temperature Sensor

The integrated sensor makes use of the temperature chracteristic of a diode-connected NMOS transistor. A constant drain current over temperature is needed to gain a linear relation between voltage drop and temperature for the diode. The circuit in Fig. 22(b) shows how the constant current is generated in an energy efficient manner. The current I_{ref} through N_0 is constant as long as the ZTC[3] bias point for N_0 is met [25]. The temperature characteristic of this bias point is illustrated in Fig. 22(a). For device N_0 a gate source voltage of $V_{ref} = 700\,\text{mV}$ results in a zero deviation of the drain current. V_{ref} is provided by the same bandgap reference source, that is needed for the oscillator and the ADC of the RFID tag. Transistors P_0 and P_1 form a current mirror, that generates I_{const} from I_{ref}. This is necessary, as the supply voltage V_{DD} can vary in a wide range. A direct constant current generation via a ZTC-biased PMOS device connected to N_1 would require a constant gate source voltage, which is difficult to achieve with a fluctuating tag supply voltage.

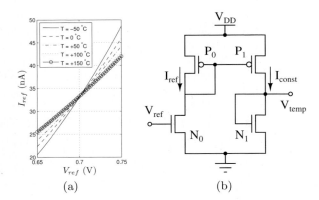

Fig. 22. Temperature sensor: (a) Illustration of the ZTC bias point of transistor N_0. (b) Schematic of the temperature sensor.

---

[3] Zero Temperature Coefficient.

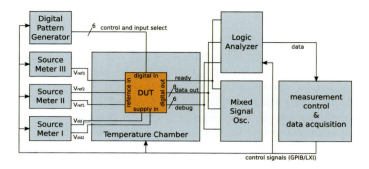

Fig. 23. Sensor interface measurement setup used for verification of the functionality and the ultra low-power characteristics

7.4 Measurements

The sensor interface was implemented in a 0.13 µm CMOS process. The whole interface can be cut off the supply rails by a power switch to minimize leakage current during idle state. The chip was bonded into a CLCC44 package. The package was connected to the measurement instruments using an adaptor PCB interfacing all digital and analog signals. Fig. 23 shows a block diagram of the setup. A digital pattern generator emulates the control signals, that are provided by the sensor-enhanced EPC control logic normally. The source meters provide several reference voltages that are RFID-internally generated by the bandgap circuit. Source meter I provides the supply for the sensor interface (V_{DD}) and the pad buffers (V_{DD2}). At the same time the current consumption is determined by the source meters. The digital output signals are connected to a mixed signal oscilloscope and a logic analyzer. The oscilloscope is used for monitoring purposes while the logic analyzer does the actual sampling of the data for later offline analysis. The whole setup is controlled by a Matlab script which sets the current temperature, measures all relevant parameters, reads the results and then jumps to the next temperature. This way a fully automatic temperature characterization of the sensor interface is obtained.

Fig. 24(a) shows the digital output of the sensor interface when switched to the internal temperature sensor source. A linear relation is given between the environment temperature and the output code. A slight negative offset over supply voltage shows up for supplies below 1.2 V. An important effect is the rising demand for current at higher temperatures shown in Fig. 24(b). When comparing both figures (24(a) and 24(b)), it can be stated that a supply voltage of 1.2 V would be ideal for acquiring temperature codes with low power consumption. Especially a supply of 1.4 V draws a lot of current that is dissipated in leakage. Fig. 4 in section 4.1 shows the power conversion efficiency over temperature for the analog frontend. The efficiency drop at 1.2 V for higher temperatures is caused by the DC voltage limiter. To avoid

Locatable, Sensor-Enabled Multistandard RFID Tags

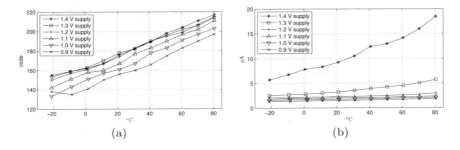

Fig. 24. Measurement results: (a) Temperature codes for different supply voltages. (b) Current consumption of the sensor interface for different supply voltages.

the high current consumptions of the sensor interface at supply voltages beyond 1.2 V, this zener structure has to be adapted. The maximum voltage should be 1.3 V to ensure a good 1.2 V-performance while concurrently avoiding higher supply voltages. Otherwise the reading range would drop at higher temperatures.

8 Summary

All aspects of locatable, sensor-enabled multistandard RFID tags were covered. The different design blocks that were discussed in the last sections have approximate power consumptions as listed in Tab. 1. An available DC power of at least 10 µW is generated by the multistandard analog frontend[4] at an UHF input power of -10 dBm. All values were confirmed by measurements at 1 V supply and room temperature. The value for the EPC compliant digital logic is reported from a predecessor project as first measurements for our digital-only chip are still missing.

The option of a fully passive operation gives the benefit of being absolutely maintenance free. Very cheap, disposable and locatable sensor tags could be developed using the circuits we described. If the energy saving techniques shown here are used to build active RFID tags, the battery lifetime will increase. Based on a total power consumption in active state of 9.8 µW as listed in Tab. 1, a standard AAA battery with 1250 mAh would be sufficient to power such an active tag for more than 10 years. With a much smaller button battery (e.g. a CR2032 with 225 mAh), which are often used for remote keyless entry transmitters, the lifetime would be 2.5 years. If taken into account, that the EPC digital logic is not running during sensor measurements and the whole RFID tag is in idle state for most of the time, *lifetime batteries* for active RFIDs, key fobs and other mobile devices are technically feasible. The only prerequisite is that the tag-to-reader communication has to

---

[4] Schottky diode based rectifier for UHF.

Table 1. Power consumption of all relevant components for locatable, sensor-enhanced multistandard RFID tags

supplier:

component	power consumption	available DC power
UHF&HF Analog Frontend	1.5 μW	> 10 μW

consumers:

component	power consumption	percentage of avail. power
EPC compl. Digital Logic	6 μW	60 %
sensor interface	1.4 μW	14 %
oscillator	1.2 μW	12 %
bandgap reference	1.2 μW	12 %
sum	9.8 μW	98 %

take place via backscattering or load modulation instead of an active transmission by means of a power amplifier. Ultra low-power architectures will enhance reading ranges and add more features to mobile devices in the next few years. Our implementation of a locatable, sensor-enhanced multistandard RFID tag is a first step to demonstrate the development potential for passive, semi-active and active wireless sensor nodes.

References

[1] Active RFID UHF Beacon Tag (Model 137001), GAO Tek Inc., Toronto, Ontario,
http://www.gaotek.com/index.php?main_page=product_info&cPath=63_136&products_id=1457
[2] Karthaus, U., Fischer, M.: Fully integrated passive uhf rfid transponder ic with 16.7-uw minimum rf input power. IEEE J. Solid-State Circuits 38(10), 1602–1608 (2003)
[3] Pillai, V., Heinrich, H., Dieska, D., Nikitin, P., Martinez, R., Rao, K.: An ultra-low-power long range battery/passive rfid tag for uhf and microwave bands with a current consumption of 700 na at 1.5 v. IEEE Trans. Circuits Syst. I 54(7), 1500–1512 (2007)
[4] Seemann, K., Hofer, G., Cilek, F., Weigel, R.: Single-ended ultra-low-power multistage rectifiers for passive rfid tags at uhf and microwave frequencies. In: Proc. IEEE Radio and Wireless Symposium, January 17–19, pp. 479–482 (2006)
[5] Krumm, J., Eckert, E., Glauert, W.H., Ullmann, A., Fix, W., Clemens, W.: A polymer transistor circuit using PDHTT. IEEE Electron Device Letters 25(6), 399–401 (2004)

[6] Glidden, R., Bockorick, C., Cooper, S., Diorio, C., Dressler, D., Gutnik, V., Hagen, C., Hara, D., Hass, T., Humes, T., Hyde, J., Oliver, R., Onen, O., Pesavento, A., Sundstrom, K., Thomas, M.: Design of ultra-low-cost UHF RFID tags for supply chain applications. IEEE Communications Magazine 42(8), 140–151 (2004)

[7] Barnett, R.E., Liu, J.: An EEPROM programming controller for passive UHF RFID transponders with gated clock regulation loop and current surge control. IEEE Journal of Solid-State Circuits 43(8), 1808–1815 (2008)

[8] Pardo, D., Vaz, A., Gil, S., Gomez, J., Ubarretxena, A., Puente, D., Morales-Ramos, R., Garcia-Alonso, A., Berenguer, R.: Design criteria for full passive long range UHF RFID sensor for human body temperature monitoring. In: IEEE International Conference on RFID 2007, March 26–28, pp. 141–148 (2007)

[9] Tran, N., Lee, B., Lee, J.-W.: Development of long-range UHF-band RFID tag chip using schottky diodes in standard CMOS technology. In: 2007 IEEE Radio Frequency Integrated Circuits (RFIC), June 3–5, pp. 281–284 (2007)

[10] McSpadden, J.O., Fan, L., Chang, K.: Design and experiments of a high-conversion-efficiency 5.8-GHzrectenna. IEEE Transactions on Microwave Theory and Techniques 46, 2053–2060 (1998)

[11] Salter, T.S., Metzel, G.M., Goldsman, N.: Parasitic aware optimization of an rf power scavenging. circuit with applications to smartdust sensor networks. In: 2009 IEEE Radio and Wireless Week, January 18–20 (2009)

[12] Seemann, K., Weigel, R.: The system design of integrated passive transponder devices. In: Asia-Pacific Microwave Conference, 2006. APMC 2006, Yokohama, December 12–15, pp. 1114–1117 (2006)

[13] Essel, J., Brenk, D., Heidrich, J., Weigel, R.: A highly efficient UHF RFID frontend approach. In: Proc. IEEE MTT-S International Microwave Workshop on Wireless Sensing, Local Positioning, and RFID IMWS 2009, pp. 1–4 (2009), doi:10.1109/IMWS2.2009.5307866

[14] EPCglobal: Radio-frequency identity protocols - class-1 generation-2 rfid v1.2.0, EPCglobal Standard, EPCglobal Inc. (Mar 2007),
http://www.epcglobalinc.org/standards/uhfc1g2

[15] EPCglobal: Radio-frequency identity protocols - hf version 2 rfid - draft version 0.0.9, EPCglobal Standard (July 2006),
http://www.epcglobalinc.org/standards/hfg2/

[16] Yan, H., Jianyun, H., Qiang, L., Hao, M.: Design of low-power baseband-processor for rfid tag. In: Proc. International Symposium on Applications and the Internet Workshops SAINT Workshops 2006, pp. 4–63 (2006)

[17] Brenk, D., Essel, J., Heidrich, J., Weigel, R.: Ultra low-power techniques for sensor-enhanced RFID tags. In: Proc. IEEE MTT-S International Microwave Workshop on Wireless Sensing, Local Positioning, and RFID IMWS 2009, pp. 1–4 (2009), doi:10.1109/IMWS2.2009.5307885

[18] Barton, D.K.: Radar Systems Analysis and Modeling. Artech House, Boston (2005)

[19] Vossiek, M., Roskosch, R., Heide, P.: Precise 3-d object position tracking using fmcw radar. In: Proc. 29th European Microwave Conference, vol. 1, pp. 234–237 (1999)

[20] Heidrich, J., Brenk, D., Essel, J., Fischer, G., Weigel, R., Schwarze, S.: Local positioning with passive UHF RFID transponders. In: Proc. IEEE MTT-S International Microwave Workshop on Wireless Sensing, Local Positioning, and RFID IMWS 2009, pp. 1–4 (2009), doi:10.1109/IMWS2.2009.5307861
[21] Wiebking, L.: Entwicklung eines zentimetergenauen mehrdimensionalen Nahbereichs-Navigationssystems. VDI Verlag GmbH, Duesseldorf (2003)
[22] Smith, S.W.: The Scientist and Engineers's Guide to Digital Signal Processing. California Technical Publishing, California (2006)
[23] Brenk, D., Essel, J., Heidrich, J., Weigel, R., Hofer, G., Holweg, G.: Wireless sensing by means of passive multistandard RFID tags. In: Proc. IEEE Sensors, October 25-28, pp. 61–64 (2009), doi:10.1109/ICSENS.2009.5398526
[24] Shenghua, Z., Nanjian, W.: A novel ultra low power temperature sensor for uhf rfid tag chip. In: Proc. IEEE Asian Solid-State Circuits Conference ASSCC 2007, pp. 464–467 (2007)
[25] Filanovsky, I., Lim, S.T.: Interaction of threshold voltage and mobility temperature dependencies applied to stabilization of current and voltage. In: Proc. 43rd IEEE Midwest Symposium on Circuits and Systems, August 8–11, vol. 3, pp. 1022–1025 (2000)

Optimal Sensor Network Configuration Based on Control Theory

Takashi Takeda[1] and Toru Namerikawa[2]

[1] Division of Electrical Engineering and Computer Science,
Graduate School of Natural Science and Technology, Kanazawa University,
Kakuma-machi, Kanazawa 920-1192 Japan
ttakashi@scl.ec.t.kanazawa-u.ac.jp

[2] Department of System Design Engineering, Keio University, 3-14-1 Hiyoshi,
Kohoku-ku, Yokohama 223-8522 Japan
namerikawa@sd.keio.ac.jp

1 Optimal Sensor Network Configuration Considering Estimation Error Variance and Communication Energy

1.1 Introduction

Recently wireless sensor networks with the function distribution of memory units, communications, calculations etc. have attracted much attention, and there has been increasing research [1, 2]. In these researches, sensor nodes are connected wirelessly and some local estimates are merged into the common estimate by the wireless communication. It is well known that sensor networks are superior to a observation by a system with single sensor in a fault tolerance, load reduction of operator, collection and application of information etc. and its application to not only a sensing system but also the guidance control system via a sensor network has received attention. For example, traffic control systems, nano-medicines and disaster countermeasures. Meanwhile, each sensor node requires electric power more than a case of only sensing because of communications and calculations, but sensor nodes are generally powered and driven by batteries. Moreover it is difficult to change batteries frequently or charge by power cable because of increasing of costs. Therefore, it is important to utilize the energy efficiently to achieve the energy-saving and prolong sensor nodes life. For this requirement, the sensor scheduling, the optimization of the communication rate and communications traffic and decreasing communication distances by the multi-hop communication are discussed [4,3,5]. Consequently, in this paper, we discuss a sensor scheduling problem selecting available sensor nodes considering the estimation error variance and communication energy in a feedback control system via a sensor network.

Distributed Kalman Filter in sensor networks has been studied in [7,8,9,10]. Each sensor node calculates the local estimate and a sensor network system generates the common estimate by communications between sensor nodes. However, they deal with a sensor network system as a sensing system and do not consider arbitrary control inputs applied to the plant. Thus, it is difficult to apply to the guidance control that the plant receives arbitrary control inputs. Moreover, they do not consider the communication energy. Meanwhile, the network configuration and the sensor scheduling algorithm considering an estimation error variance and communication energy were proposed in [6, 4, 5]. However, each sensor node has only a observation and communication function and does not have a calculation function. The fusion center calculates the estimate and transmits the control input to the plant. In our framework, each sensor node has the calculation, communication and observation functions and the control input is applied to the plant. Thus we can not apply these previous methods.

In this paper, we discuss a sensor scheduling problem considering the estimation error variance and communication energy in a feedback control system via a sensor network. We first propose the estimation algorithm with the unknown input of the plant in the feedback control system via a sensor network. Each sensor node calculates the local estimate without information of the control input and transmits its information to the sensor node applying the control input to the plant. This sensor node calculates the common estimate and control input using received information. Then we show that there is the unique positive definite solution to the discrete algebraic Riccati equation in the error covariance update. Secondly, we propose a sensor scheduling algorithm considering estimation error variance and communication energy. This scheduling algorithm achieves sub-optimal network topology with minimum energy and a desired error variance. Finally, we verify effectiveness of a sensor scheduling algorithm by experiments.

This section is organized as follows. The feedback control system via a sensor network and the network topology are presented and problems are formulated accordingly. Secondly, we describes a novel decentralized estimation algorithm with the unknown input and the unique solution to the discrete algebraic Riccati equation under some assumptions. A sensor scheduling algorithm is proposed. Finally, some experimental results are presented.

1.2 Problem Formulation

Plant and Sensor Nodes

In this paper, we consider the feedback control system via a sensor network illustrated in Fig. 1. This system consists the plant and N sensor nodes S_i, $(i = 1, 2, ..., N)$. We assume all sensor nodes have enough computation capability and can take a measurement of the plant. The process dynamics of the plant and the measurement equation of a sensor node S_i are given by

$$x_{k+1} = Ax_k + Bu_k + w_k, \qquad (1)$$
$$y_k^i = C_i x_k + v_k^i, \qquad (2)$$

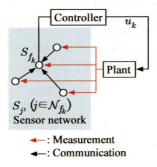

Fig. 1. Sensor network

where $x_k \in \mathbb{R}^n$, $u_k \in \mathbb{R}^m$, $y_k^i \in \mathbb{R}^{q_i}$ are the state, the control input and the measurement output of a sensor node S_i respectively. Additionally, $w_k \in \mathbb{R}^n$, $v_k^i \in \mathbb{R}^{q_i}$ are the process noise and measurement noise respectively. We assume that the control input u_k is given by the following equation and applied from the sensor node $S_{f_k}, (f_k = 1, 2, ..., N)$ to the plant:

$$u_k = L\hat{x}_k^{f_k}, \qquad (3)$$

where $\hat{x}_k^{f_k} \in \mathbb{R}^n$ is the estimate of the sensor node S_{f_k} and L is the feedback gain. Now we assume we can arbitrarily determine which sensor node is the sensor node S_{f_k} at each time step. Moreover, (1) and (2) satisfy following *Assumptions 1-3*.

Assumption 1. w_k, $v_k = \left[(v_k^1)^\mathrm{T} \; (v_k^2)^\mathrm{T} \; \cdots \; (v_k^N)^\mathrm{T}\right]^\mathrm{T} \in \mathbb{R}^q$, $(q = \sum_i^N q_i)$ *are zero mean white Gaussian noise and satisfy equations*

$$\mathrm{E}\left\{\begin{bmatrix} w_k \\ v_k \end{bmatrix} \begin{bmatrix} w_k^\mathrm{T} & v_k^\mathrm{T} \end{bmatrix}\right\} = \begin{bmatrix} Q & 0 \\ 0 & R \end{bmatrix}, \qquad (4)$$

$$\mathrm{E}\left\{w_k x_0^\mathrm{T}\right\} = \mathrm{E}\left\{v_k x_0^\mathrm{T}\right\} = 0, \qquad (5)$$

where Q, $R = \mathrm{diag}(R_1, R_2, ...)$ *are the positive semidefinite and positive definite covariance matrix of noises* w_k, v_k *respectively.*

Assumption 2. *The matrix pair* $(A, Q^{\frac{1}{2}})$ *is reachable.*

Assumption 3. *The matrix pair* (C, A) *is detectable, where*

$$C = \begin{bmatrix} C_1^\mathrm{T} & C_2^\mathrm{T} & \cdots & C_N^\mathrm{T} \end{bmatrix}^\mathrm{T}. \qquad (6)$$

Network Topology

The sensor network consists N sensor nodes and one of them is the sensor node S_{f_k} applying the control input to the plant. We assume the sensor node

S_{f_k} can communicate to other sensor nodes directory and define the set $\mathcal{N}_{f_k}$ containing sensor nodes communicating the sensor node S_{f_k}. Then there is no communication between arbitrary sensor nodes belonging to the set $\mathcal{N}_{f_k}$ at time step k. We assume we can arbitrary determine sensor nodes belonging to the set $\mathcal{N}_{f_k}$ as a case of the sensor node S_{f_k}.

Remark 1. In this paper, a wireless communication between the sensor node $S_j, (j \in \mathcal{N}_{f_k})$ and S_{f_k} means that the sensor node $S_j, (j \in \mathcal{N}_{f_k})$ transmit information to the sensor node S_{f_k}.

This communication pass is unidirectional. In general, if two sensor nodes have a bidirectional communication pass, each sensor node can get and use more information. However, in this paper, we discuss a sensor scheduling problem by determining the sensor node S_{f_k} and set $\mathcal{N}_{f_k}$. Thus, due to different communication ranges of each sensor node or obstacles, two sensor node might not communicate each other [11].

Consequently, all sensor nodes satisfy following *Assumption 4*.

Assumption 4. *Sensor nodes $S_j, (j \in \mathcal{N}_{f_k})$ can communicate to the sensor node S_{f_k} once per time step with a time delay less than a sampling time. Additionally, when the sensor node S_{f_k} applies the control input u_k to the plant and sensor nodes $S_j, (j \in \mathcal{N}_{f_k})$ transmit information to the sensor node S_{f_k}, These sensor nodes use the energy $E_{f_k,p}, E_{j,f_k} \in \mathbb{R}_+$ respectively.*

We define the energy E_k the whole system is using. The energy E_k is described as

$$E_k = E_{f_k, x_k} + \sum_{j \in \mathcal{N}_{f_k}} E_{j, f_k}. \tag{7}$$

Remark 2. The communication energy $E_{i,j}$ generally can be $E_{i,j} = b_{i,j} + a_{i,j}(d_{i,j})^{c_{i,j}}$ and depend on a distance between sensor nodes S_i and S_j, where $b_{i,j}$ is a static part and $a_{i,j}$ is a dynamic part. $c_{i,j}$ is typically from 2 through 6 [6].

Control Problems

In this paper, we discuss an estimation problem with unknown input u_k and a sensor scheduling problem. Problems can be formulated as following *problems 1, 2*.

Problem 1. We assume the plant and all sensor nodes satisfy *Assumption 1-4* and the sensor node S_{f_k} and the set $\mathcal{N}_{f_k}$ is determined. Then compute the optimal state estimate $\hat{x}_k^{f_k}$ that minimizes following estimation error variance.

$$J = \mathrm{E}\left\{(x_k - \hat{x}_k^{f_k})^\mathrm{T}(x_k - \hat{x}_k^{f_k})\right\}. \tag{8}$$

Problem 2. Find the optimal network topology T_k^* satisfying $J \leq \gamma$ and following equation at time step k:

$$T_k^* = \arg\min_{T_k} E_k, \tag{9}$$

where $\gamma > 0$ is a design parameter.

1.3 Estimation Algorithm

In this section, we propose an estimation algorithm in the feedback control system via a sensor network. A novel algorithm is based *Decentralized Kalman Filter* in [9, 10]. Each sensor node S_j, $(j \in \mathcal{N}_{f_k})$ computes the local estimate $\hat{x}_k^j$. Then these sensor nodes can not know the control input because all communication passes are unidirectional. Consequently, a novel estimation algorithm is considering the unknown control input applied from the sensor node S_{f_k} to the plant. In this algorithm, each sensor node S_j, $(j \in \mathcal{N}_{f_k})$ transmits $\hat{x}_k^{j-}$, $\hat{x}_k^j$, P_k^{j-}, P_k^j to the sensor node S_{f_k}. The sensor node S_{f_k} computes estimate $\hat{x}_k^{f_k}$ by information from sensor nodes S_j, $(j \in \mathcal{N}_{f_k})$.

Estimation Algorithm of Sensor Nodes S_j, $(j \in \mathcal{N}_{f_k})$

First, we discuss an estimation algorithm of sensor nodes S_j, $(j \in \mathcal{N}_{f_k})$. Each sensor node S_j, $(j \in \mathcal{N}_{f_k})$ do not have information of the control input because all communication lines are unidirectional. Consequently, we can not apply a existing method to the feedback system via a sensor network. Proposed algorithm satisfies following *Theorem 1*.

Theorem 1. *Consider the system (1) and (2) with Assumption 1-4. Then a estimation algorithm of each sensor node S_j, $(j \in \mathcal{N}_{f_k})$ is given by following equations and the estimate $\hat{x}_k^j$ is minimum variance estimate based measurements of sensor node S_j:*

$$\hat{x}_{k+1}^{j-} = A\hat{x}_k^j + B\hat{u}_k^j, \tag{10}$$

$$\hat{x}_k^j = \hat{x}_k^{j-} + K_k^j(y_k^j - C_j\hat{x}_k^{j-}), \tag{11}$$

$$\hat{u}_k^j = L\hat{x}_k^j, \tag{12}$$

$$P_{k+1}^{j-} = (A+BL)P_k^j(A+BL)^\mathrm{T} + Q$$
$$+ BLP_k^{f_k}L^\mathrm{T}B^\mathrm{T}$$
$$- (A+BL)M_k^j L^\mathrm{T}B^\mathrm{T}$$
$$- BL(M_k^j)^\mathrm{T}(A+BL)^\mathrm{T}, \tag{13}$$

$$P_k^j = \left\{(P_k^{j-})^{-1} + C_j^\mathrm{T}R_j^{-1}C_j\right\}^{-1}, \tag{14}$$

$$M_k^j = (I - K_k^j C_j)M_k^{j-}(I - K_k^{f_k}C_{f_k})^\mathrm{T}, \tag{15}$$

$$M_{k+1}^{j-} = (A+BL)M_k^j A^\mathrm{T} + Q - BLP_k^{f_k}A^\mathrm{T}, \tag{16}$$

where definition of each variable is described as

$$\hat{x}_k^{j-} = \mathrm{E}\left\{x_k|y_{k-1}^j, y_{k-2}^j, \cdots\right\},$$

$$\hat{x}_k^j = \mathrm{E}\left\{x_k|y_k^j, y_{k-1}^j, \cdots\right\},$$

$$P_k^{j-} = \mathrm{E}\left\{(x_k - \hat{x}_k^{j-})(x_k - \hat{x}_k^{j-})^\mathrm{T}\right\},$$

$$P_k^j = \mathrm{E}\left\{(x_k - \hat{x}_k^j)(x_k - \hat{x}_k^j)^\mathrm{T}\right\},$$

$$M_k^j = \mathrm{E}\left\{(x_k - \hat{x}_k^j)(x_k - \hat{x}_k^{f_k})^\mathrm{T}\right\},$$

$$M_k^{j-} = \mathrm{E}\left\{(x_k - \hat{x}_k^{j-})(x_k - \hat{x}_k^{f_k-})^\mathrm{T}\right\}.$$

Proof. Consider the filter equation for (1) and (2) are given by

$$\hat{x}_{k+1}^{j-} = A\hat{x}_k^j + B\hat{u}_k^j, \tag{17}$$

$$\hat{x}_k^j = \hat{x}_k^{j-} + K_k^j(y_k^j - C_j \hat{x}_k^{j-}), \tag{18}$$

$$\hat{u}_k^j = L\hat{x}_k^j. \tag{19}$$

It follows from (1), (2), (17), (18) and (19) that errors $e_k^j = x_k - \hat{x}_k^j$, $e_k^{j-} = x_k - \hat{x}_k^{j-}$ are

$$e_k^j = (I - K_k^j C_j) e_k^{j-} - K_k^j v_k^j, \tag{20}$$

$$e_{k+1}^{j-} = (A + BL) e_k^j + w_k - BL e_k^{f_k}. \tag{21}$$

Thus estimation error covariance matrices P_k^j, P_{k+1}^{j-} are

$$P_k^j = (I - K_k^j C_j) P_k^{j-} (I - K_k^j C_j)^\mathrm{T} + K_k^j R_j (K_k^j)^\mathrm{T}, \tag{22}$$

$$\begin{aligned} P_{k+1}^{j-} =& (A + BL) P_k^j (A + BL)^\mathrm{T} + Q \\ &+ BL P_k^{f_k} L^\mathrm{T} B^\mathrm{T} \\ &- (A + BL) M_k^j L^\mathrm{T} B^\mathrm{T} \\ &- BL (M_k^j)^\mathrm{T} (A + BL)^\mathrm{T}, \end{aligned} \tag{23}$$

where M_k^j is the cross covariance matrix between the estimation errors of the estimate $\hat{x}_k^{f_k}$ and $\hat{x}_k^j$.

Firstly, we consider the covariance matrix (22). It follows from the condition $\frac{\partial}{\partial K_k} \mathrm{tr} P_k^j = \mathbf{0}$ and (22) that the filter gain K_k^j is

$$K_k^j = P_k^j C_j^\mathrm{T} R_j^{-1}. \tag{24}$$

From (24), (22) can be rewritten as follow

$$P_k^j = \left\{(P_k^{j-})^{-1} + C_j^\mathrm{T} R_j^{-1} C_j\right\}^{-1}. \tag{25}$$

Secondly, we consider the cross covariance matrix M_k^j in (23). It follows from the definition of that

$$M_k^j = (I - K_k^j C_j) M_k^{j-} (I - K_k^{f_k} C_{f_k})^{\mathrm{T}}. \tag{26}$$

The sensor node S_{f_k} knows the value of the control input u_k because this sensor node applies the control input to the plant. Thus the estimation error $e_k^{f_k}$ is given as follow

$$e_{k+1}^{f_k-} = A e_k^{f_k} + w_k. \tag{27}$$

It follows from (27) that the cross covariance matrix M_k^{j-} is given as follow

$$M_{k+1}^{j-} = (A + BL) M_k^j A + Q - BL P_k^{f_k} A^{\mathrm{T}}. \tag{28}$$

These equations complete the proof.

Secondly, we consider the estimation algorithm of the sensor node S_{f_k}. The estimation of the sensor node S_{f_k} is based its measurement and received information $\hat{x}_k^{j-}$, $\hat{x}_k^j$, P_k^{j-} and P_k^j from sensor nodes $S_j, (j \in \mathcal{N}_{f_k})$. The sensor node S_{f_k} has information of the control input u_k. In this paper, we use following *Decentralized Kalman Filter* in [9, 10] as the estimation algorithm of the sensor node S_{f_k}.

$$\hat{x}_{k+1}^{f_k-} = (A + BL)\hat{x}_k^{f_k}, \tag{29}$$

$$\bar{x}_k^{f_k} = \hat{x}_k^{f_k-} + K_k^{f_k}(y_k^{f_k} - C_{f_k}\hat{x}_k^{f_k-}), \tag{30}$$

$$K_k^{f_k} = \bar{P}_k^{f_k} C_{f_k}^{\mathrm{T}} R_{f_k}^{-1}, \tag{31}$$

$$P_k^{f_k-} = A P_k^{f_k} A^{\mathrm{T}} + Q, \tag{32}$$

$$\bar{P}_k^{f_k} = \left\{ (P_k^{f_k-})^{-1} + C_{f_k}^{\mathrm{T}} R_{f_k}^{-1} C_{f_k} \right\}^{-1}, \tag{33}$$

$$P_k^{f_k} = \left[(\bar{P}_k^{f_k})^{-1} + \sum_{j \in \mathcal{N}_{f_k}} \left\{ (P_k^j)^{-1} - (P_k^{j-})^{-1} \right\} \right]^{-1}, \tag{34}$$

$$\hat{x}_k^{f_k} = P_k^{f_k} \left[(\bar{P}_k^{f_k})^{-1} \bar{x}_k^{f_k} \right.$$
$$\left. + \sum_{j \in \mathcal{N}_{f_k}} \left\{ (P_k^j)^{-1} \bar{x}_k^j - (P_k^{j-})^{-1} \hat{x}_k^{j-} \right\} \right], \tag{35}$$

where the definition of variables is as follow

$$\bar{x}_k^{f_k} = \mathrm{E}\left\{ x_k | y_k^{f_k}, y_{k-1}^{f_k}, \ldots \right\},$$

$$\hat{x}_k^{f_k} = \mathrm{E}\left\{ x_k | y_k^{f_k}, y_{k-1}^{f_k}, \ldots, y_k^j, y_{k-1}^j \right\}, j \in \mathcal{N}_{f_k},$$

$$\bar{P}_k^{f_k} = \mathrm{E}\left\{(x_k - \bar{x}_k^{f_k})(x_k - \bar{x}_k^{f_k})^\mathrm{T}\right\},$$
$$P_k^{f_k} = \mathrm{E}\left\{(x_k - \hat{x}_k^{f_k})(x_k - \hat{x}_k^{f_k})^\mathrm{T}\right\}.$$

The estimate $\bar{x}_k^{f_k}$ is only based measurements of the sensor node S_{f_k}. However the estimate $\hat{x}_k^{f_k}$ is based on measurements of the sensor node S_{f_k} and some sensor nodes belonging to the set $\mathcal{N}_{f_k}$. Then the covariance matrix $P_k^{f_k}$ satisfies following *Theorem 2*.

Theorem 2. *Consider the (1) and (2) with Assumptions 1-4. If sensor nodes $S_{f_k} = S_f, S_{j_1}, S_{j_2}, ..., (j_1, j_2 \in \mathcal{N}_f)$ are determined and the matrix pair (H_f, A), $H_f = [C_f^\mathrm{T} \; C_{j_1}^\mathrm{T} \; C_{j_2}^\mathrm{T} \cdots]^\mathrm{T}$ is detectable, then the estimate $\hat{x}_k^f$ is the solution of Problem 1 and there is the unique positive definite solution P_∞^f to algebraic Riccati equation satisfying the following equation:*

$$P_\infty^f = AP_\infty^f A^\mathrm{T} + Q \\ - AP_\infty^f H_f^\mathrm{T} \left(H_f P_\infty^f H_f^\mathrm{T} + V_f\right)^{-1} H_f P_\infty^f A^\mathrm{T}, \tag{36}$$

where $V_f = \mathrm{diag}\{R_f, R_{j_1}, R_{j_2}, ...\}$.

Proof. Substituting (14) into (34).

$$P_k^f = \left[\left(P_k^{f-}\right)^{-1} + H_f^\mathrm{T} V_f^{-1} H_f\right]^{-1} \tag{37}$$

From (32) and (37), this is the algebraic Riccati equation. Consequently, it follows from *Assumption 2* and detectability of the matrix pair (H_f, A) that the covariance matrix P_k^f has the unique positive definite solution P_∞^f.

It follows from *Theorem 2* that there is the unique positive definite solution to the algebraic Riccati equation (32)-(34) while sensor nodes S_{f_k} and S_j, $(j \in \mathcal{N}_{f_k})$ is being determined. Additionally, from *Assumption 3*, if we use $N - 1$ sensor nodes as S_j, $(j \in \mathcal{N}_{f_k})$, there is the the unique positive definite solution to the algebraic Riccati equation. In next section, we propose a sensor scheduling algorithm considering the estimation error variance $J = \mathrm{tr} P_\infty^{f_k}$ and communication energy. If we determine the set $\mathcal{N}_{f_k}$ include all sensor nodes, the estimation error variance of estimate is minimized. However the communication energy will increase because all sensor node have to transmit the information to the sensor node S_{f_k}. If on the contrary, we determine the set $\mathcal{N}_{f_k}$ is empty set, the communication energy is zero because there are no communication passes. However the estimation error variance of the estimate will increase. Consequently, there is a trade-off between an estimation accuracy and a communication energy.

1.4 Sensor Scheduling Algorithm

In previous section, we showed that the estimation error variance of the estimate $\hat{x}_k^{f_k}$ is $J = \text{tr}(P_\infty^{f_k})$. In this section, we propose a sensor scheduling algorithm minimizing communication energy in subset of all available network topology under the condition $J \leq \gamma$. The network topology can be determined uniquely by determining sensor nodes S_{f_k} and $S_j, j \in \mathcal{N}_{f_k}$. Then $N2^{N-1}$ network topologies are available. Consequently, we propose a following sub optimal algorithm to reduce computation costs. In a proposed algorithm, $N(N-1)$ network topologies are available. Additionally, E^i, J_i are communication energy of the whole system and the estimation error variance respectively when sensor nodes $S_{f_k} = S_i$ and $S_j, (j \in \mathcal{N}_i)$ are determined.

Sensor Scheduling Algorithm

1: **for** $\alpha = 1$ to N **do**
2: $\quad \mathcal{N}_\alpha = \mathcal{B} = \{1, ..., N\} \backslash \alpha$
3: $\quad$ **repeat** $N - 1$
4: $\quad\quad \beta = \arg\max_{j \in \mathcal{N}_\alpha \cap \mathcal{B}} E_{\alpha, j}$
5: $\quad\quad$ **if** $J_{\alpha'} \leq \gamma, (\mathcal{N}_{\alpha'} := \mathcal{N}_\alpha \backslash S_\beta)$ **then**
$$\mathcal{N}_\alpha := \mathcal{N}_\alpha \backslash S_\beta$$
6: $\quad\quad \mathcal{B} := \mathcal{B} \backslash S_\beta$
7: **return** $S_{i^*}, \mathcal{N}_{i^*}, (i^* = \min_{i=1,...,N} E^i)$

In this algorithm, firstly, we determine the sensor node $S_{f_k} = S_\alpha$. Secondly, we remove the sensor node S_β from the set $\mathcal{N}_\alpha$ in order of decreasing the communication energy $E_{\alpha, \beta}$ under the condition $J_\beta \leq \gamma$. We calculate these subroutine N times ($\alpha = 1, 2, ..., N$). Finally, the sensor node S_{f_k} and set $\mathcal{N}_{f_k}$ minimizing communication energy in subset of all available network topology under the condition $J \leq \gamma$ are determined.

Example 1 is described as follows.

Example 1. Consider 3 sensor nodes ($N = 3$) illustrated in Fig. 2. We assume following conditions:

1) Distances between each sensor nodes are $d_{1,2} = d_{2,3} = 1$, $d_{1,3} = 2$.
2) A communication energy is $E_{i,j} = \epsilon d_{i,j}^2$.
3) The condition $J \leq \gamma$ is satisfied if and only if we use sensor nodes S_1, S_2, S_3 or S_1, S_3.

Now, we examine proposed sensor scheduling algorithm in *Example 1*.

We first define $\alpha = 1$ and $\mathcal{N}_1 = \mathcal{B} = \{2, 3\}$. These definitions mean that we check the communication energy in a case of the sensor node S_{f_k} is S_1. Then

4:, 5: and 6: in a sensor scheduling algorithm are calculated 2 times. we can be $\beta = 3$ at the initial calculation. Then the sensor node S_3 is removed from $\mathcal{N}_\alpha$ because the condition $J_{\alpha'} \leq \gamma$, $(\mathcal{N}_{\alpha'} = \{2\})$ is satisfied. Consequently, $\mathcal{N}_\alpha = \{2\}$, $\mathcal{B} = \{2\}$. After the initial calculation, we can be $\beta = 2$ at the second calculation. Because the condition $J_{\alpha'} \leq \gamma$, $(\mathcal{N}_{\alpha'} = \emptyset)$ is not satisfied, the sensor node S_2 is not removed. Consequently, if we determine the sensor nodes S_{f_k} is S_1, the set $\mathcal{N}_1 = \{2\}$ (see Fig. 5(a)) and communication energy E_k is the following equation:

$$E^1 = E_{1,x_k} + \epsilon d_{1,2}^2 = E_{1,x_k} + \epsilon. \tag{38}$$

Next, we can be define $\alpha = 2$ and $\mathcal{N}_\alpha = \mathcal{B} = \{1,3\}$. We can calculate the communication energy E^2 and the set $\mathcal{N}_2$ by a method similar to above calculation. In this subroutine, because we can not remove sensor nodes from the set $\mathcal{N}_2$ under the condition $J_2 \leq \gamma$, we can be define $\mathcal{N}_2 = \{1,3\}$ (see Fig. 3(b)) and the communication energy is following equation when the sensor node S_{f_k} is S_2

$$E^2 = E_{2,x_k} + \epsilon \left(d_{1,2}^2 + d_{2,3}^2\right) = E_{2,x_k} + 2\epsilon. \tag{39}$$

We lastly define $\alpha = 3$ and $\mathcal{N}_3 = \{1,2\}$. Then we can remove the sensor node S_1 from the set $\mathcal{N}_3$ under the condition. Consequently, $\mathcal{N}_\alpha = \{2\}$ (see Fig. 3(c)) and the communication energy is following equation.

$$E^3 = E_{3,x_k} + \epsilon d_{1,3}^2 = E_{3,x_k} + \epsilon \tag{40}$$

(38)-(40) is the communication energies when the sensor nodes S_{f_k} is S_1, S_2 or S_3 respectively. These energy is depend on the distance between the plant and each sensor node. We consider a case of the distances between sensor nodes S_1, S_2 and S_3 is following equation at step k.

$$d_{1,x_k} = \epsilon, \quad d_{1,x_k} = 2\epsilon, \quad d_{1,x_k} = 3\epsilon. \tag{41}$$

(a) A network topology ($S_{f_k} = S_1$).

(b) A network topology ($S_{f_k} = S_2$).

(c) A network topology ($S_{f_k} = S_3$).

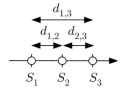

Fig. 2. An example of sensor network

Fig. 3. Network topologies when each case of Example 1

Then the communication energy is given as follow

$$d_{1,x_k} = 2\epsilon, \quad d_{1,x_k} = 4\epsilon, \quad d_{1,x_k} = 4\epsilon. \tag{42}$$

Consequently, we can determine $S_{i^*} = S_1$, $\mathcal{N}_{i^*} = \{2\}$ at time step k.

1.5 Experimental Evaluation

In this section, a effectiveness of a sensor scheduling algorithm is evaluated by experiments. The experiment was carried out on a two-wheeled vehicle, a CCD camera and a computer as shown in Fig. 4. Now Two-wheeled vehicle has the nonholonomic constraint. However two-wheeled vehicle can be defined following framework by virtual structure for feedback linearization [12].

$$A = \begin{bmatrix} 1 & 0 & \delta & 0 \\ 0 & 1 & 0 & \delta \\ 0 & 0 & 1 & 0 \\ 0 & 0 & 0 & 1 \end{bmatrix}, \quad B = \begin{bmatrix} \frac{\delta^2}{2} & 0 \\ 0 & \frac{\delta^2}{2} \\ \delta & 0 \\ 0 & \delta \end{bmatrix}, \tag{43}$$

where $\delta = 0.2$ and $x_0 = [\,1.3\ 0.7\ 0\ 0\,]^\mathrm{T}$ are the sampling time and the initial state respectively. Additionally, we design the feedback gain L by LQG control. There are nine sensor nodes available and each sensor nodes has the following measurement equation and these position is shown in Fig. 5.

$$y_k^i = [\,1\ 0\ 0\ 0\,]\, x_k + v_k^i, \quad (i = 1, 5, 9)$$

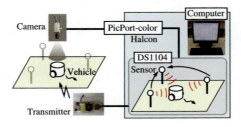

Fig. 4. Experimental setup

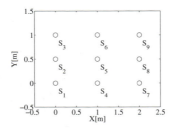

Fig. 5. Position of sensor nodes

$$y_k^i = \begin{bmatrix} 0 & 1 & 0 & 0 \end{bmatrix} x_k + v_k^i, \quad (i = 2, 6)$$
$$y_k^i = \begin{bmatrix} 0 & 0 & 1 & 0 \end{bmatrix} x_k + v_k^i, \quad (i = 3, 7)$$
$$y_k^i = \begin{bmatrix} 0 & 0 & 0 & 1 \end{bmatrix} x_k + v_k^i, \quad (i = 4, 8)$$

Each measurement output is calculated from the image of a CCD camera mounted above the vehicle. The video signals are acquired by a frame grabber board PicPort-color and image processing software HALCON generate nine measurements. Consequently, nine sensor nodes, a network topology and measurement noises exist in the computer. We use DS1104 (dSPACE Inc.) as a real-time calculating for an estimation and sensor scheduling. Additionally, the covariance matrices of noises are $Q = 1 \times 10^{-4} I_4$, $R = 0.1 I_9$ respectively.

Here we define the communication energy between arbitrarily two sensor nodes. We assume that the communication energy between sensor nodes S_i

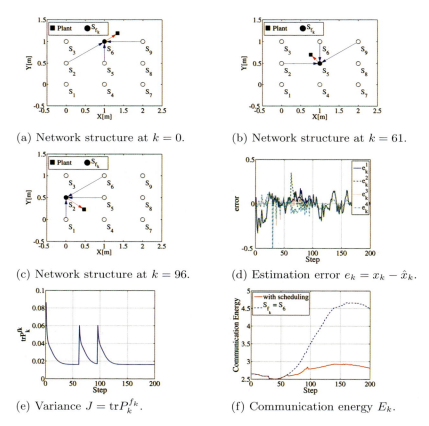

(a) Network structure at $k = 0$.

(b) Network structure at $k = 61$.

(c) Network structure at $k = 96$.

(d) Estimation error $e_k = x_k - \hat{x}_k$.

(e) Variance $J = \mathrm{tr} P_k^{f_k}$.

(f) Communication energy E_k.

Fig. 6. Experimental results

and S_{f_k} is $E_{i,f_k} = \epsilon d_{i,f_k}^2$. d_{i,f_k} is the distance between sensor nodes S_i and S_{f_k} and ϵ is the positive constant.

The experiment was done designing $\gamma = 0.02$. The experimental results are shown in Fig. 6. Fig. 6(a)-(c) show the trajectory of vehicle and network topology. As shown in Fig. 6(a)-(c), sensor nodes are switched while the vehicle is moving. Fig. 6(b) shows the estimation error. As shown in Fig. 6(b), the estimation error is zero mean. Fig. 6(e) shows the estimation error variance $P_k^{f_k}$. As shown in Fig. 6(e), the estimation error variance converge to the solution to algebraic Riccati equation and the solution is less than design parameter γ at all times. Finally, Fig. 6(e) is a comparison between following *Cases 1, 2*.

Case 1. A case that a sensor scheduling algorithm was applied.

Case 2. A case that the sensor node S_{f_k} was S_6 at all times.

In these case, the error variance $\text{tr}P_k^{f_k}$ is same. However from Fig. 6(f) the communication energy is different. This figure shows the energy of the whole system is reduced by a sensor scheduling algorithm. Consequently, by designing γ, a proposed algorithm reduce the communication energy under the condition that the estimation error is smaller than desired value.

1.6 Conclusions

In this paper, we discussed a sensor scheduling problem considering the estimation error variance and communication energy in a feedback control system via a sensor network. We first have proposed the estimation algorithm with the unknown input of the plant in the feedback control system via a sensor network. Each sensor node calculates the local estimate without information of the control input and transmits its information to the sensor node applying the control input to the plant. This sensor node calculates the common estimate and control input using received information. Then we showed that there is the unique positive definite solution to the discrete algebraic Riccati equation in the error covariance update. Secondly, we have proposed a sensor scheduling algorithm considering estimation error variance and communication energy. This scheduling algorithm achieved sub-optimal network topology with minimum energy and a desired error variance. Finally, we have verified effectiveness of a sensor scheduling algorithm by experiments.

2 Optimal Sensor Network Configuration via Multi-hop Communication

2.1 Introduction

Sensor networks contain a large number of sensor nodes connected wirelessly each other and there has been increasing research in the areas of controls, communication technology, computer science etc. [13].

Each sensor node generally has memory units, communications function and calculations function. It is well known that sensor networks are superior in a fault tolerance, sensing in broad area, collection and application of information etc. There are applications to a environmental monitoring, security and intelligent space. Additionally, its application to not only a sensing system but also configuration of the feedback control system via a sensor network and large scale online information processing has received attention in the areas of traffic control, nano-medicines and disaster countermeasures [13].

Meanwhile, each sensor node requires electric power more than a case of only sensing because of communications and calculations, but sensor nodes are generally powered and driven by batteries. Moreover it is difficult to change batteries frequently or charge by a power cable because of increasing of costs. Therefore, it is important to utilize the energy efficiently to achieve the energy-saving and prolong sensor nodes life [5]. For this requirement, the sensor scheduling, the optimization of the communication rate and communications traffic and decreasing communication distances by the multi-hop communication are discussed [14, 5, 6]. Consequently, in this paper, we discuss a multi-hop network configuration problem considering the estimation error variance and communication energy in a feedback control system via a sensor network.

The estimation problem in a sensor network system has been studied in [7,8,9,10]. A distributed Kalman filtering algorithm with a consensus strategy were proposed in [1, 15]. In these methods each sensor node communicates with its neighbors on a network. However, if the plant receives control inputs from fusion center or one of sensor nodes, all sensor node have to obtain its information in real time and it is difficult to develop real system.

In [6], a network configuration problem with a multi-hop communication and a feedback control system considering communication energy and estimation error variance. However amount of information transmitted from each sensor node increase with a number of sensor nodes.

In this paper, we discuss a network configuration problem considering the priori estimation error variance and communication energy in a feedback control system via a sensor network. We first define a sensor network with multi-hop communication. Then we assume that each sensor node transmit same amount of information. In this system, we discuss a estimation problem and a network configuration problem. Then we show that there is the unique positive definite solution to the discrete algebraic Riccati equation in the error covariance update and a trade-off between the estimation error variance and a communication energy. Secondly, we propose a network configuration algorithm considering this trade-off. This network configuration algorithm achieves sub-optimal network topology with minimum energy and a desired error variance. Finally, we verify effectiveness of a sensor scheduling algorithm by experiments.

This section is organized as follows. The feedback control system via a sensor network and the network topology are presented and problems are formulated accordingly. Secondly we describes a novel information fusion algorithm, a estimation algorithm and the unique solution to the discrete algebraic Riccati equation under some assumptions. Finally, a network configuration algorithm is proposed and some experimental results are presented.

2.2 Problem Formulation

Plant and Sensor Nodes

In this paper, we consider the feedback control system via a sensor network illustrated in Fig. 7. This system consists the plant and N sensor nodes S_i, $(i = 1, 2, ..., N)$. We assume all sensor nodes can take a measurement of the plant. The process dynamics of the plant and the measurement equation of a sensor node S_i are given by

$$x_{k+1} = Ax_k + Bu_k + w_k \tag{44}$$

$$y_k^i = C_i x_k + v_k^i \tag{45}$$

where $x_k \in \mathbb{R}^n$, $u_k \in \mathbb{R}^m$, $y_k^i \in \mathbb{R}^{q_i}$ are the state, the control input and the measurement output of a sensor node S_i respectively. Additionally, $w_k \in \mathbb{R}^n$, $v_k^i \in \mathbb{R}^{q_i}$ are the process noise and measurement noise respectively. From (45), each sensor node take a different measurement. Moreover, (44) and (45) satisfy following *Assumptions 5-7*.

Assumption 5. w_k, $v_k = \left[(v_k^1)^\mathrm{T} \ (v_k^2)^\mathrm{T} \ \cdots \ (v_k^N)^\mathrm{T}\right]^\mathrm{T} \in \mathbb{R}^q$, $(q = \sum_i^N q_i)$ *are zero mean white Gaussian noise and satisfy equations*

$$\mathrm{E}\left\{\begin{bmatrix} w_k \\ v_k \end{bmatrix} \begin{bmatrix} w_k^\mathrm{T} & v_k^\mathrm{T} \end{bmatrix}\right\} = \begin{bmatrix} Q & \mathbf{0} \\ \mathbf{0} & R \end{bmatrix}, \tag{46}$$

$$\mathrm{E}\left\{w_k x_0^\mathrm{T}\right\} = \mathrm{E}\left\{v_k x_0^\mathrm{T}\right\} = \mathbf{0}, \tag{47}$$

where Q, $R = \mathrm{diag}(R_1, R_2, ...)$ are the positive semidefinite and positive definite covariance matrix of noises w_k, v_k respectively.

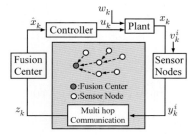

Fig. 7. Sensor network system

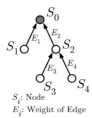

Fig. 8. An example of Network

Assumption 6. *The matrix pair $(A, Q^{\frac{1}{2}})$ is reachable.*

Assumption 7. *The matrix pair (C, A) is detectable, where*

$$C = \begin{bmatrix} C_1^{\mathrm{T}} & C_2^{\mathrm{T}} & \cdots & C_N^{\mathrm{T}} \end{bmatrix}^{\mathrm{T}}. \tag{48}$$

Network Topology

In this paper, we deal multi-hop communication. N sensor nodes and the fusion center S_0 are connected wirelessly and information transmitted from each sensor node are passed on to the fusion center via some relay nodes. The example of a network topology is illustrated in Fig. 8. Let $\mathcal{G} = (\mathcal{V}, \mathcal{E})$ denoted a graph with the set of vertices $\mathcal{V}$ and the set of edges $\mathcal{E}$. Then sensor node S_i and network topology satisfy following Assumption 8, 9.

Assumption 8. *Sensor nodes S_i can transmit $z_k^i \in \mathbb{R}^r$ to the other sensor node once per time step with a time delay less than a sampling time. Additionally, when a sensor node S_i transmit information, this sensor node uses the energy E_i.*

Assumption 9. *A network topology T is a directed spanning tree with root S_0.*

From Assumption 9, a sensor node S_i transmits z_k^i containing information of a measurement of S_i to the other sensor node. the dimension of z_k^i is r in all sensor nodes. Moreover, each sensor node use a energy E_i for transmitting z_k^i to other sensor node. We assume $E = \sum_{i=1}^{N} E_i$ is the energy the whole system is using. The energy E_i is the weight of the edge of the network topology T. In general, the communication energy depend on a length of a communication pass between sensor nodes S_i and S_0. Consequently, if there are some relay node between S_i and S_0, the communication energy to pass to the sensor node S_0 from S_i will be reduced. But all sensor nodes transmit information once per one time step and the time delay between sensor nodes S_i and S_0 will increase. Consequently, there is a trade-off between an estimation accuracy and a communication energy.

Control Problems

In this paper, we discuss an estimation problem and a network configuration problem.

Problems can be formulated as following *problems 3, 4*.

Problem 3. We assume the plant and all sensor nodes satisfy *Assumption 5-9* and the network topology T is determined. Then compute the optimal state estimate $\hat{x}_k^-$ that minimizes following estimation error variance.

$$J = \mathrm{E}\left\{\left(x_k - \hat{x}_k^-\right)^{\mathrm{T}}\left(x_k - \hat{x}_k^-\right)\right\} \tag{49}$$

Problem 4. Find the optimal network topology T^* satisfying $J \leq \gamma$, Assumption 9 and following equation:

$$T^* = \arg\min_T E, \tag{50}$$

where $\gamma > 0$ is a design parameter.

2.3 Proposed Method

Information Merge Method

In this paper, we define the sensor node receiving information from a sensor node S_i as the sensor node $\mathrm{Par}(S_i)$ and the set including sensor nodes transmitting information to a sensor node S_i as the set $\mathcal{N}_i = \{j|\mathrm{Par}(S_j) = S_i\}$. Moreover we define the depth h_i of a sensor node S_i, the height $\bar{h} = \max_i h_i$ of the network topology T. For example, $\mathcal{N}_0 = \{1,2\}$ and $\bar{h} = 2$ in Fig. 8.

A measurement output of each sensor node S_i have to merge via z_k^i with same dimension. Consequently, we propose following information fusion method for each sensor node.

$$z_k^i = C_i^{\mathrm{T}} R_i^{-1} y_{k-\bar{h}+h_i}^i + \sum_{j \in \mathcal{N}_i} z_{k-1}^j, \tag{51}$$

where $y_k^i = y_0^i, (k \leq 0)$. A dimension of $C_i^{\mathrm{T}} R_i^{-1} y_{k+h_i-\bar{h}}^i$ is n and all sensor nodes transmit information with same dimension. Moreover we propose a following information fusion method for fusion center.

$$z_k = \sum_{i \in \mathcal{N}_0} z_k^i, \tag{52}$$

where $z_k^0 = z_k$ is information merged in the fusion center. It follows from $y_{k-\bar{h}+h_i}^i$ and z_{k-1}^j, $(j \in \mathcal{N}_i)$ in (51) that z_k^i delays 1 time step per one relay node. Consequently, in a network topology with Assumption 9, information z_k merged in the fusion center is given by following equation.

$$z_k = \sum_{j \in \mathcal{N}_0} z_k^j = \sum_{j=1}^N C_j^{\mathrm{T}} R_j^{-1} y_{k-\bar{h}+1}^j \tag{53}$$

z_k is calculated in the fusion center at time step k and includes $C_i R_i^{-1} y_{k-\bar{h}+1}^i$ of all sensor nodes. The time step of measurements belonging to z_k depend on $\bar{h}$. The bigger $\bar{h}$ is, the bigger a time delay of measurement belonging to z_k.

State Estimation Algorithm

We showed fusion center calculate z_k including measurements with delay $y^i_{k-\bar{h}+1}$ at time step k. In this section, we propose a estimation algorithm using z_k. Then a estimation algorithm satisfies following Theorem 1 in a sensor network system (44) and (45).

Theorem 1. Consider the system (44), (45) and network topology T with Assumption 5-9. Then a estimation algorithm is given by following equations and the estimate $\hat{x}^j_k$ is minimum variance estimate based measurements of sensor node S_j:

$$\hat{x}^-_k = A^{\bar{h}-1}\hat{x}_{k-\bar{h}+1} + \bar{B}_{\bar{h}}\bar{u}_{k-\bar{h}+1}, \tag{54}$$

$$\hat{x}_{k-\bar{h}+1} = \hat{x}^-_{k-\bar{h}+1} + P_{k-\bar{h}+1}\left(z_k - C^T R^{-1} C \hat{x}^-_{k-\bar{h}+1}\right), \tag{55}$$

$$P^-_k = A^{\bar{h}-1} P_{k-\bar{h}+1} \left(A^{\bar{h}-1}\right)^T + G_{\bar{h}} \bar{Q} G^T_{\bar{h}}, \tag{56}$$

$$P_{k-\bar{h}+1} = \left\{\left(P^-_{k-\bar{h}+1}\right)^{-1} + C^T R^{-1} C\right\}^{-1}, \tag{57}$$

where $\bar{B}_{\bar{h}}, \bar{G}_{\bar{h}}, \bar{Q} \in \mathbb{R}^{n(\bar{h}-1) \times n(\bar{h}-1)}$ is as follows

$$\bar{B}_{\bar{h}} = \begin{bmatrix} B & AB & \cdots & A^{\bar{h}-2}B \end{bmatrix}, \tag{58}$$

$$\bar{G}_{\bar{h}} = \begin{bmatrix} I_n & A & \cdots & A^{\bar{h}-2} \end{bmatrix}, \tag{59}$$

$$\bar{Q} = \text{block diag}\{Q, Q, ..., Q\}. \tag{60}$$

Proof. we first define following fictitious measurement output $y_{k-\bar{h}+1}$.

$$y_{k-\bar{h}+1} = \begin{bmatrix} y^1_{k-\bar{h}+1} \\ y^2_{k-\bar{h}+1} \\ \vdots \\ y^N_{k-\bar{h}+1} \end{bmatrix}$$

$$= C x_{k-\bar{h}+1} + v_{k-\bar{h}+1}. \tag{61}$$

$y_{k-\bar{h}+1}$ include measurements taken at time step $k-\bar{h}+1$ of all sensor nodes. Then we consider the estimation algorithm using $y_{k-\bar{h}+1}$ taken at time step k. (44) can be rewritten as follow

$$x_k = A^{\bar{h}-1} x_{k-\bar{h}+1} + \bar{B}_{\bar{h}} \bar{u}_{k-\bar{h}+1} + \bar{G}_{\bar{h}} \bar{w}_{k-\bar{h}+1}, \tag{62}$$

where $\bar{u}_{k-\bar{h}+1}, \bar{w}_{k-\bar{h}+1}$ is as follows

$$\bar{u}_{k-\bar{h}+1} = \begin{bmatrix} u_{k-1} \\ u_{k-2} \\ \vdots \\ u_{k-\bar{h}+1} \end{bmatrix}, \bar{w}_{k-\bar{h}+1} = \begin{bmatrix} w_{k-1} \\ w_{k-2} \\ \vdots \\ w_{k-\bar{h}+1} \end{bmatrix}. \tag{63}$$

(62) is difference equation of time step k and $k - \bar{h} + 1$. Then we propose following estimation algorithm for (62) and (61).

$$\hat{x}_k^- = A^{\bar{h}-1}\hat{x}_{k-\bar{h}+1} + \bar{B}_{\bar{h}}\bar{u}_{k-\bar{h}+1} \tag{64}$$

$$\hat{x}_{k-\bar{h}+1} = \hat{x}_{k-\bar{h}+1}^- + K_{k-\bar{h}+1}\left(y_{k-\bar{h}+1} - C\hat{x}_{k-\bar{h}+1}^-\right) \tag{65}$$

where $\hat{x}_k^- = E\{x_k|y_0, y_1, ..., y_{k-\bar{h}+1}\}$ and $\hat{x}_{k-\bar{h}+1} = E\{x_{k-\bar{h}+1}|y_0, y_1, ..., y_{k-\bar{h}+1}\}$ are estimations of x_k and $x_{k-\bar{h}+1}$ based all measurements up to time step $k - \bar{h} + 1$. Now, the estimation error variance J is given by following equation.

$$J = E\{(x_k - \hat{x}_k^-)^\mathrm{T}(x_k - \hat{x}_k^-)\} = \mathrm{tr}P_k^- \tag{66}$$

The filter gain minimizing J satisfies following equations.

$$\frac{\partial}{\partial K_k}\mathrm{tr}P_k^- = \mathbf{0} \tag{67}$$

It follows from (62), (64) and (65) that the filter gain K_k and the error covariance matrix P_k satisfying (67) are as follows

$$\begin{aligned}K_{k-\bar{h}+1} &= P_{k-\bar{h}+1}^- C^\mathrm{T}\left(CP_{k-\bar{h}+1}^- C^\mathrm{T} + R\right)^{-1} \\ &= P_{k-\bar{h}+1}C^\mathrm{T}R^{-1}\end{aligned} \tag{68}$$

$$P_{k-\bar{h}+1} = \left\{\left(P_{k-\bar{h}+1}^-\right)^{-1} + C^\mathrm{T}R^{-1}C\right\}^{-1} \tag{69}$$

Meanwhile, error covariance matrix P_k^- is as follow

$$P_k^- = A^{\bar{h}-1}\left\{\left(P_{k-\bar{h}+1}^-\right)^{-1} + C^\mathrm{T}R^{-1}C\right\}^{-1}\left(A^{\bar{h}-1}\right)^\mathrm{T} \\ + G_{\bar{h}}\bar{Q}G_{\bar{h}}^\mathrm{T}, \tag{70}$$

where $\bar{Q}$ is covariance matrix of $\bar{w}_{k-\bar{h}+1}$. Consequently, a estimation algorithm using a measurement output (61).

Secondly, we show this algorithm is a estimation algorithm using z_k in (51).

It follows from (68), (61) and (65) that we can get following.

$$\begin{aligned}\hat{x}_{k-\bar{h}+1} &= \hat{x}_{k-\bar{h}+1}^- \\ &+ P_{k-\bar{h}+1}\left(z_k - C^\mathrm{T}R^{-1}C\hat{x}_{k-\bar{h}+1}^-\right).\end{aligned} \tag{71}$$

(71) is a estimation algorithm using z_k merged in the fusion center. These equations complete the proof.

Relation between an Estimation Error Variance and a Network Topology

In this section, we consider an estimation error variance $\mathrm{tr}P_k^-$ and a network topology. It follows from Assumptions 6, 7 that there is the unique positive definite solution $P_\infty^{\bar{h}}$ to algebraic Riccati equation (56) satisfying following equation:

$$P_\infty^{\bar{h}} = A^{\bar{h}-1}\left\{\left(P_\infty^{\bar{h}}\right)^{-1} + C^\mathrm{T}R^{-1}C\right\}^{-1}\left(A^{\bar{h}-1}\right)^\mathrm{T} + G_{\bar{h}}\bar{Q}G_{\bar{h}}^\mathrm{T}. \tag{72}$$

From (72), the solution $P_\infty^{\bar{h}}$ depend on the depth $\bar{h}$. Now the solution $P_\infty^{\bar{h}}$ satisfies following Theorem 2.

Theorem 2. We assume if $\bar{h} = \alpha, \beta$, $(\alpha > \beta)$, there is the unique positive definite solutions P_∞^α, P_∞^β to algebraic Riccati equation (56) respectively. Then P_∞^α and P_∞^β satisfy following relation:

$$\mathrm{tr}P_\infty^\alpha \geq \mathrm{tr}P_\infty^\beta. \tag{73}$$

Proof. It follows from Assumptions 6 and 7 that the solution to (72) do not depend on initial value. Moreover (56) is different equation between k and $k - \bar{h} - 1$. Consequently it is apparent from these.

From Theorem 2, The smaller $\bar{h}$ is, the smaller priori estimation error is. Consequently, there is trade-off between an estimation error variance and communication energy.

2.4 Network Configuration Algorithm

In this section, we discuss a network configuration algorithm. We have to configurate a network topology satisfying $J = \mathrm{tr}P_\infty^- \leq \gamma$ and Assumption 9. For this purpose, we first need to calculate $\bar{h}$ satisfying $J = \mathrm{tr}P_\infty^{\bar{h}} \leq \gamma$. secondly, we find rooted spanning tree where depths of all sensor node are less than $\bar{h}$ and a communication energy E is minimized. This tree is known as $\bar{h}$-HMST(the minimum-cost $\bar{h}$-hop spanning tree). In several researches, they showed approximation algorithm [18]. In this paper, we propose an algorithm minimizing in a subset of available network topology. We first consider following operation.

- Change destination of sensor nodes receiving information from sensor nodes belonging the set $\mathcal{V}_1$ into sensor nodes belonging the set $\mathcal{V}_2$,

where $\mathcal{V}_1 = \{S_j | h_j > \bar{h}\}$ and $\mathcal{V}_2 = \{S_j | h_j < \bar{h}\}$. It follows from this operation that all sensor nodes have depths with less than $\bar{h}$. We assume the set of all available network topology that we can get from this operation as $\mathcal{T}_s$. We rewrite Problem 2 to following problem.

Problem 5. Find the optimal network topology T^* satisfying $J \leq \gamma$, Assumption 9 and following equation:

$$T = \arg\min_{T \in \mathcal{T}_s} E, \tag{74}$$

where $\gamma > 0$ is a design parameter.

In Problem 2 we find the network topology minimizing a communication energy in all available network topology. However Problem 3 minimize in the subset of all available network topology.

We propose *Network Configuration Algorithm* and it is a solution of Problem 3. In this algorithm, we use Prim's Algorithm finding the minimum spanning tree. In *network configuration algorithm*, $e(S_i, S_j)$ is communication energy between sensor nodes S_i and S_j.

Network Construction Algorithm satisfying following theorem.

Theorem 3. *Network Construction Algorithm minimize a communication energy E in subset $\mathcal{T}_s$ and it is the solution of problem 3.*

Proof. In **3:** of *Network Construction Algorithm*, we select a sensor node with minimum communication energy belonging the set $\mathcal{V}_2$. because the operation are applied these sensor nodes, this algorithm is the solution of Problem 3.

Consequently, by designing γ, we can configurate a network topology what are superior to estimation accuracy or communication energy.

2.5 Experimental Evaluation

In this section, a effectiveness of a sensor scheduling algorithm is evaluated by experiments. The experiment was carried out on a two-wheeled vehicle, a CCD camera and a computer as shown in Fig. 9. Now Two-wheeled vehicle has the nonholonomic constraint. However two-wheeled vehicle can be defined following framework by virtual structure for feedback linearization [12].

$$A = \begin{bmatrix} 1 & 0 & \delta & 0 \\ 0 & 1 & 0 & \delta \\ 0 & 0 & 1 & 0 \\ 0 & 0 & 0 & 1 \end{bmatrix}, \quad B = \begin{bmatrix} \frac{\delta^2}{2} & 0 \\ 0 & \frac{\delta^2}{2} \\ \delta & 0 \\ 0 & \delta \end{bmatrix}, \tag{75}$$

where $\delta = 0.2$ and $x_0 = [\,1.3\ 0.7\ 0\ 0\,]^\mathrm{T}$ are the sampling time and the initial state respectively. Additionally, we design the feedback gain L by LQG control. There are ten sensor nodes available and each sensor nodes has the following measurement equation and these position is shown in Fig. 10.

$$\begin{aligned} y_k^i &= \begin{bmatrix} 1 & 0 & 0 & 0 \end{bmatrix} x_k + v_k^i, & (i = 1, 2) \\ y_k^i &= \begin{bmatrix} 0 & 1 & 0 & 0 \end{bmatrix} x_k + v_k^i, & (i = 3, 4) \\ y_k^i &= \begin{bmatrix} 0 & 0 & 1 & 0 \end{bmatrix} x_k + v_k^i, & (i = 5, 6) \end{aligned}$$

Network Construction Algorithm

1: Compute of $\bar{h}$ satisfying
$$J = \text{tr} P_\infty^{\bar{h}} \leq \gamma.$$

2: Compute rooted minimum spanning tree T by Prim's algorithm and define
$$\begin{aligned}\mathcal{V}_1 &= \{S_j | h_j > \bar{h}\},\\ \mathcal{V}_2 &= \{S_j | h_j < \bar{h}\}.\end{aligned}$$

3: Change $\text{Par}(S_i)$, $(S_i \in \mathcal{V}_1)$
    if $\mathcal{V}_1$ is not an empty set
$$\begin{aligned}\text{Par}(S_i) &:= \arg\min_{S_j \in \mathcal{V}_2} e(S_i, S_j)\\ E_i &:= e(S_i, S_j)\\ h_i &:= h_j + 1\end{aligned}$$
    end

4: **return** T

$$\begin{aligned}y_k^i &= \begin{bmatrix}0 & 0 & 0 & 1\end{bmatrix} x_k + v_k^i, \quad (i = 7, 8)\\ y_k^i &= \begin{bmatrix}1 & 0 & 0 & 0\\ 0 & 1 & 0 & 0\end{bmatrix} x_k + v_k^i, \quad (i = 9, 10)\end{aligned}$$

Each measurement output is calculated from the image of a CCD camera mounted above the vehicle. The video signals are acquired by a frame grabber board PicPort-color and image processing software HALCON generate nine measurements. Consequently, nine sensor nodes, a network topology and measurement noises exist in the computer. We use DS1104 (dSPACE Inc.) as a real-time calculating for an estimation and sensor scheduling. Additionally, the covariance matrices of noises are $Q = 1 \times 10^{-4} I_4$, $R = 0.05 I_{12}$ respectively.

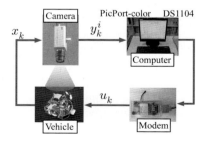

Fig. 9. Experimental setup

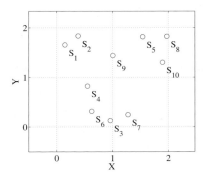

Fig. 10. Position of sensor nodes

Here we define the communication energy between arbitrarily two sensor nodes. We assume that the communication energy between sensor nodes S_i and S_j is $e_{i,j} = \epsilon d_{i,j}^2$. d_{i,f_k} is the distance between sensor nodes S_i and S_j and ϵ is the positive constant.

Additionally, experiments were done following *Case 1* and *Case 2*.

Case 1 : The experiment designing $\gamma = 0.015$
Case 2 : The experiment designing $\gamma = 0.03$

The experimental results of *Case 1* and *Case 2* are shown in Fig. 11, 12. Fig. 11, 12(a), (b), (c) and (d) show a network topology, the state x_k, the estimate $\hat{x}_k$ and a information variable z_k respectively. As shown in Figs. 11(a), 12(a), network topologies satisfying the condition are $\bar{h} = 4, 6$ respectively. Additionally, error variances are $J = 0.0297, 0.0137$ and communication energy are $E = 10.5\epsilon, 3.12\epsilon$ respectively. Consequently, there is a trade-off between an estimation accuracy and a communication energy. As shown in Figs. 11(c), 12(c), a vibration of the estimate in *case 1* is smaller than *Case 2*. As shown in Figs. 11(d), 12(d), z_k has information of weighted measurement. Fig. 13 shows the variance $J = \mathrm{tr} P_k^-$ in *Case 1* and *Case 2* respectively. As shown in Fig. 13, $\mathrm{tr} P_k^-$ converge on $\mathrm{tr} P_k^- = 0.0297, 0.0137$ and it is less than the design parameters respectively.

Consequently, we have showed that we can configurate a network topology what are superior to estimation accuracy or communication energy by designing γ.

2.6 Conclusion

In this paper, we discussed a network configuration problem considering the priori estimation error variance and communication energy in a feedback control system via a sensor network. We first have defined a sensor network with multi-hop communication. Then we have assumed that each sensor

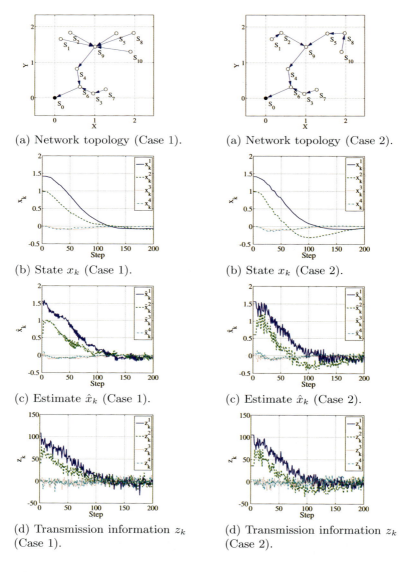

Fig. 11. Experimental results(Case 1)

Fig. 12. Experimental results (Case 2)

node transmit same amount of information for issue resolution of increasing amount of information transmitted. Then we showed that there is the unique positive definite solution to the discrete algebraic Riccati equation in the error covariance update and a trade-off between the estimation error variance and a communication energy. Secondly, we have proposed a network con-

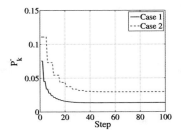

Fig. 13. Variance $\mathrm{tr}P_k^-$ (Case 1, Case 2)

figuration algorithm considering this trade-off. This network configuration algorithm achieves sub-optimal network topology with minimum energy and a desired error variance. Finally, we have verified effectiveness of a sensor scheduling algorithm by experiments.

References

1. Olfati-Saber, R.: Distributed Kalman Filter for Sensor Networks. In: Proc. of IEEE Conference Decision & Control, pp. 5492–5498 (2007)
2. Olfati-Saber, R.: Distributed Kalman Filter with Embedded Consensus Filters. In: Proc. of IEEE Conference Decision & Control, pp. 5492–5498 (2005)
3. Shi, L., Johansson, K.H., Murray, R.M.: Estimation Over Wireless Sensor Networks: Tradeoff between Communication, Computation and Estimation Qualities. In: Proc. of the 17th International Federation of Automatic Control, pp. 605–611 (2007)
4. Sandberg, H., Rabi, M., Skoglund, M., Johansson, K.H.: Estimation over Heterogeneous Sensor Networks. In: Proc. of IEEE Conference Decision & Control, pp. 4898–4903 (2008)
5. Arai, S., Iwatani, Y., Hashimoto, K.: Fast Sensor Scheduling for Spatially Distributed Heterogeneous Sensors. In: Proc. of American Control Conference, pp. 2785–2790 (2009)
6. Shi, L., Johansson, K.H., Murray, R.M.: Change Sensor Topology When Needed: How to Efficient Use system Resource in Control and Estimation over Wireless Network. In: Proc. of IEEE Conference on Decision & Control, pp. 5478–5485 (2007)
7. Casbeer, D.W., Beard, R.: Distributed Information Filtering using Consensus Filters. In: Proc. of American Control Conference, pp. 1882–1887 (2009)
8. Olfati-Saber, R., Sandell, N.F.: Distributed tracking in sensor networks with limited sensing range. In: Proc. of American Control Conference, pp. 3157–3162 (2008)
9. Nebot, E.M., Bozorg, M., Durrant-Whyte, H.F.: Decentralized Architecture for Asynchronous Sensors. Autonomous Robots 6(2), 147–164 (1999)
10. Schlosser, M.S., Kroschel, K.: Communication Issues in Decentralized Kalman Filter. In: Proc. of the Seventh International Conference on Information Fusion, pp. 731–738 (2004)

11. Kanzaki, A., Hara, T., Nishio, S.: On TDMA Slot Assignment Protocol Considering the Existence of Unidirectional Wireless Links in Ad Hoc Sensor Networks. In: Proc. of Int'l Workshop on Future Mobile and Ubiquitous Information Technologies, pp. 195–198 (2006)
12. Namerikawa, T., Yoshioka, C.: Consensus Control of Observer-based Multi-Agent System with Communication Delay. In: Proc. of the SICE Annual Conference, pp. 2414–2419 (2008)
13. Shi, L., Capponi, A., Johansson, K.H., Murray, R.M.: Sensor Network Lifetime Maximization Via Sensor Trees Construction and Scheduling. In: FeBID 2008 (2008)
14. Iino, Y., Hatanaka, T., Fujita, M.: Wireless Sensor Network Based Control System Considering Communication Cost. In: Proc. of the 17th IFAC World Congress, pp. 14992–14997 (2008)
15. Carli, R., Chiuso, A., Schenato, L., Zampieri, S.: Distributed Kalman Filtering using consensus strategies. In: Proc. of IEEE Confrence Decision & Control, pp. 5486–5491 (2007)
16. Song, E., Zhu, Y., Zhou, J., You, Z.: Optimal Kalman Filtering Fusion with Cross-Correlated Sensor Noises. Automatica 43(8), 1450–1456 (2007)
17. Sun, S.L.: Multi-sensor optimal information fusion Kalman filter. Automatica 40(8), 1447–1453 (2006)
18. Althaus, E., Funke, S., Har-Peled, S., Könemann, J., Ramos, E.A., Skutella, M.: Ramos and M. Skutella: Approximating k-Hop Minimum-Spanning trees. Operations Research Letters 33(2), 115–120 (2005)

Optimal Local Map Registration for Wireless Sensor Network Localization Problems

Yifeng Zhou[1] and Louise Lamont[2]

[1] Communications Research Centre, Canada
yifeng.zhou@crc.ca
[2] Communications Research Centre, Canada
louise.lamont@crc.ca

Abstract. In this chapter, we present an optimal local map registration algorithm for constructing global maps from local relative maps for wireless network localization applications. A wireless network is partitioned into sub-networks with overlapping or common nodes that shared by different sub-networks. Local maps are built for each sub-network, which consist of the relative coordinates of nodes in each network. The local maps are then transformed into a global map using a set of affine transforms with each consisting of a rotation, a reflection and a translation for each individual local map. The optimal transform is found by minimizing the discrepancies, in the global map, of the common sensor nodes shared by different local maps. A computationally efficient gradient projection algorithm is developed for finding the optimal transforms. The local map registration approach can solve many of the problems encountered by pairwise map merging based techniques and is able to achieve global optimal performance. More importantly, the technique provides a systematic approach for constructing global maps from local maps. Computer simulations are used to demonstrate the performance and effectiveness of the proposed algorithm.

1 Introduction

The application of wireless *ad hoc* sensor networks has recently attracted much interest from both academics and industries. A wireless sensor network consists of a large number of inexpensive sensors densely deployed in a certain area. Each sensor node consists of a low-power processor, a modest amount of memory, a wireless transceiver, and a sensor board. They are connected by wireless links and cooperate to self-organize into a network without the requirement for any infrastructure such as access points or base stations. Each sensor node is capable of sensing, processing data at a small scale. Wireless sensor networks provide a new means for a wide range of applications including environment monitoring [2][5][20], health care [1][25], battle-field surveillance and tracking [30][6], and environment observation and forecasting [28]. Research and development on wireless sensor networks still faces many formidable challenges due to the strict limitations on energy consumption, the simplicity of the processing power of nodes, and possibly high environmental dynamics.

In wireless sensor network applications, sensor localization capability is a highly desirable characteristic. Sensor localization refers to the process of estimating the locations of sensor nodes by using the knowledge of the absolute positions of anchor nodes and certain types of measurements such as internode distance and bearing measurements. Anchor nodes are sensor nodes that know their absolute or global coordinates. Typically, this knowledge can be acquired through some additional localization hardware such as a Global Positioning System (GPS) receiver. Anchor nodes are required for transforming local sensor coordinates into their absolute coordinates. In some approaches such as multilateration localization, anchor nodes are directly involved in the localization process. The information of sensor locations in a network is fundamental for a number of reasons. One of the key attributes of sensor networks is to self-form into an efficient network for gathering data in a useful and efficient manner. Gathering meaningful data obviously requires that the sensor locations be known. Secondly, knowing the sensor locations helps to uncover and heal, through mobility, coverage holes in a sensor network. In addition, basic services such as geographic-aware routing and multicasting often rely on sensor location information. Thirdly, knowledge of sensor locations can be used to establish efficient and low-power communications paths between nodes, *e.g.*, redundant sensors in the neighborhood of a sensor can be powered down to save energy. Finally, sensor localization is a completely automatic process and does not require human experimenters. It can help reduce sensor deployment cost and time for applications especially where a large number of sensor nodes are deployed. Although GPS can be used outdoors for positioning with a median error of meters, it is in general considered impractical to be installed on each sensor node for location discovery due to its significant power consumption, cost and line-of-sight condition requirement. Besides, GPS receivers are susceptible to jamming and attenuations.

Recently, many sensor localization techniques have been developed [29][21]. Among them, the approach that uses local map merging or patching has received considerable interest recently due to its advantages over centralized approaches [22][26][18][17][33]. In this approach, a network is divided into many small subnetworks that share common nodes, and local maps are computed for each subnetwork. The local maps are then merged into a single global map based on the common sensor nodes shared by different local maps. The coordinates of the global map are later mapped into their absolute coordinates based on the use of anchors. The local map approach has a number of advantages over a centralized approach. First, a local map approach can be implemented in a distributed fashion, *i.e.*, individual local maps are computed simultaneously and in parallel, which makes it appropriate for large-scale networks. In addition, local maps allow the use of sophisticated localization algorithms which may be too computationally expensive at the global level. In some types of networks, computing local maps can also avoid using distance information between far away nodes, which, in practice, may not be available due to the power limitation of the sensor nodes and terrain irregularities of the networks. Some localization algorithms, such as the centralized MDS method [27], often depend on estimates such as the shortest path for unavailable distance measurements. For irregularly shaped sensor networks, the shortest paths usually do not correlate well with their Euclidean distances, leading to severely deteriorated localization performance. In [26][27], a method for merging local maps into a single global map is proposed based on common sensor nodes shared by the local maps. The global map is constructed by randomly selecting a local map as the core map and sequentially merging the rest to it. The core map finally grows into a global map for the whole network when all the local maps are merged. We call this type of algorithms a pairwise map merging approach because each time the merging is between a pair of maps and based on their shared common nodes. The merging of two maps is similar to the problem of coordinate system registration in Procrustes analysis [12], where a rigid transformation is used to maps points in one coordinate system to a different one. Approaches based on pairwise map merging have several drawbacks and shortcomings. First, the pairwise map merging is a locally optimal process because it only explores the commonalities of the shared sensor nodes in two maps. In practice, the common sensor nodes are often shared by more than two local maps. Secondly, the performance of the global map depends on the order by which the local maps are merged. The problem of finding the best order is a complex scheduling problem and is difficult to solve. Currently, most approaches employ *ad hoc* methods for sequentially merging the local maps. Thirdly, in some cases, adjacent local maps may not have a sufficient number of common nodes. It is known that a minimum of three common nodes are required to define a unique merge of two local maps on a plane. Finally, sequential pairwise merging process can lead to error propagation and perhaps unacceptable errors as the network grows [22]. In a more recent paper [17], Kwon and Song explored the possibility of incorporating addition distance information in stitching two local maps. They formulated the problem as a quadratic optimization problem over a unit circle, and a suboptimal heuristic solution was proposed. A number of incremental stitching strategies were also proposed for the construction of a global map. Using simulation results, the authors showed that their method was able to prevent flip errors and provide significantly improved localization performance over the existing methods. However, the method is still a pairwise map merging approach in nature and thus cannot alleviate all the problems that are inherent to it.

In this paper, we propose an optimal local map registration algorithm for merging local relative maps for wireless sensor localization problems. Instead of using pairwise merging procedures, local map merging is considered at the global level. A set of affine transforms are used to map the local maps to a global map with each affine transform corresponding to one individual local map. Each affine transform consists of a rotation, a reflection and a translation transformation. The optimal transforms are found by minimizing the discrepancies of the locations of common sensor nodes in the global map shared by all local maps that are available. The discrepancy is represented by the sum of the squared distances of all nodes to their respective geometric centers in the global map. Since the proposed local map registration algorithm minimizes the overall discrepancies of the locations of the common nodes, it is able to counter the problems associated with approaches based on pairwise map merging and achieve the global optimal performance. A computationally efficient gradient projection algorithm is developed for finding the optimal transforms. The approach can be considered a generalized local map merging technique. It is able to provide a systematic way for constructing a global maps from local maps.

This chapter is organized as follows. In Section 2, some related work in the area of local map registration is discussed including the popular MDS-MAP and MDS-MAP(P) methods. Section 3 is devoted to the development of the optimal local map registration algorithm. The gradient projection algorithm for finding the optimal transform is also developed in this section. In Section 4, computer simulations are used to demonstrate the performance and effectiveness of the proposed algorithm. The results are compared with those of the MDS-MAP(P) and MDS-MAP methods.

2 Related Work

The construction of a global map is based on local relative maps. There are a number of techniques for building local maps for wireless sensor network applications [21][29]. For computing local relative maps, we are more interested in anchor-free localization approaches, *i.e.*, approaches without the need of using anchor nodes, and in particular approaches that are based on internode distances. In [22], Meertens and Fitzpatrick used one hop neighbor nodes to construct local maps. A local map is computed by choosing three nodes to define a relative coordinate system and using multilateration to iteratively add additional nodes to the map. Moore *et al.* [23] formulated the localization problem as a two-dimensional graph realization problem, and computed the local maps for a node based on noisy distance measurements to each of its neighbors. They also used the probabilistic notion of robust quadrilaterals as a way to avoid flip ambiguities that otherwise would corrupt localization computations. The application of the multidimensional scaling (MDS) technique for sensor localization is considered an elegant solution to wireless sensor network localization problems [27][16]. MDS has its origins in psychometrics and psychophysics. It is a data analysis technique that can be used to represent a set of data as a configuration of points in some Euclidean spaces based on their similarity measures. The distances of the resulting configuration of points resemble the original similarities. There are many types of MDS techniques, including metric MDS and nonmetric MDS, replicated MDS, weighted MDS, deterministic and probabilistic MDS [32][4]. The classical MDS method is more attractive than the others because it has analytical solutions that can be obtained via eigendecomposition of a transform of the Euclidean distance matrix. MDS is closely related to principal component analysis and factor analysis. The technique has found many applications such as cluster analysis, machine learning and computational chemistry[4]. The MDS-MAP method by Shang *et al.* [26][27] is a localization method based on the well-known classic MDS technique. The MDS-MAP uses connectivity information or distance measurements between neighbor nodes. In the approach, the shortest path is used for out-of-range nodes to construct the distance matrix. The MDS technique is relatively resilient to distance errors due to the over-determined nature of the solution. In [16], an iterative MDS algorithm was proposed based on a multivariate optimization for sensor location calculation. The iterative MDS is similar to the least squares refinement step in [27]. Unlike MDS-MAP, the iterative MDS method does not need to estimate the unavailable distances between far away sensor nodes. However, the iterative MDS approach introduces complex computations and often suffers from global convergence problems.

We summarize the procedures of the MDS-MAP localization method in the following.

1. Obtain the Euclidean distance matrix D for sensors in a network, where the ijth element is the distance between the ith and the jth sensor node. If the distance between a pair of sensor nodes is not available, it will be approximated by the shortest path by using either a single source or all-pairs shortest-path algorithms, such as Dijkstra's [10] or Floyd's algorithm [31].
2. Transform the distance matrix D into its corresponding Gram matrix form by assuming that the geometric center of the sensor coordinates is at the origin. The Gram matrix E can be computed by [8]

$$E = -\frac{1}{2} L D^2 L^T \text{ and } L = I - \frac{1}{N} \underline{1}\underline{1}^T, \tag{1}$$

where $\underline{1}$ is an all one vector of length N, and I denotes an identity matrix of proper dimensions.
3. Perform eigendecomposition on the Gram matrix and retain the first two (2-dimensional) or three (3-dimensional) columns of the eigenvectors corresponding to the largest eigenvalues. The sensor coordinates are given by the eigenvectors weighted by square roots of their corresponding eigenvalues. The resulting sensor locations are relative maps and are indeterminant up to an arbitrary translation and rotation or reflection in the origin.
4. If required, the relative sensor coordinates are aligned to provide their global coordinates with the use of location information of the anchor nodes. Given a sufficient number of anchor nodes, e.g., three or more for 2-dimensional networks, the relative coordinates of the anchor nodes are mapped to their absolute coordinates through an affine transformation that consists of a rotation/reflection and a translation. The best transformation is obtained by minimizing the conformation between the absolute and relative positions. The optimal transformation is then applied to the rest of the relative sensor coordinates to provide the global coordinates.

Once local relative maps are computed, the global map can be constructed based on the common nodes shared by the local maps. The MDS-MAP(P) method by Shang et al. [26] is MDS-MAP using patches of relative local maps. The main idea of MDS-MAP(P) is to divide a network into many sub-networks that share common sensor nodes, and compute local maps using the MDS-MAP method for each sub-network. The local maps are then patched together and merged into a global map based on the common sensor nodes shared by the local maps. The use of MDS-MAP(P) can avoid using the distance estimation between remote notes and performs well on irregular networks. Individual nodes simultaneously compute their local maps based on their local information. For each node, its neighbor nodes within

k-hops are involved in computing the local map, where k decides the size of the local map. When all the local maps are computed at each node, they are sent to a central node for merging. MDS-MAP(P) uses an incremental greedy algorithm to merge the local maps in a sequential manner. A local map is randomly selected as the core map. Each time a neighboring local map with the maximum number of common nodes with the core map is selected for merging. The shared common nodes in the local map are mapped to the core map using an affine transform that consists of scaling, rotation/reflection and translation. The best transform is found to be the one that minimizes the conformation errors of the locations of the common nodes in the core map. The local map is then mapped to and combined with the core map using the best transform. The procedure is repeated until the core map covers the whole network. In [17], the authors proposed a local map stitching method to utilize additional distance information between the sensor nodes. The incorporation of additional distance information was claimed to be able to effectively eliminate flip errors [23] and provide improved performance in map stitching. The method is applicable even when there are less than three common nodes in two local maps. The method is a two-stage approach. First, a simple heuristic method is used to solve for the translation. Next, the formulation is reduced to a quadratic optimization problem over a unit circle, which is well studies and admits an efficient solution. A number of incremental stitching strategies were also proposed.

3 Optimal Local Map Registration

The objective of local map registration is to transform the coordinates of the sensor nodes in each local map to a global system to form a global map. The transform may include rotation, reflection, translation and scaling for each local map. The optimal transform is the one that minimizes the discrepancies in the coordinates of the common nodes in the global map.

In a network, each sensor node has a certain maximum range for distance measurement due to the power limitation of the node and terrain irregularities of the network. The distance between a pair of sensor nodes can be measured if they are located within the maximum range; otherwise, the distance is not available. A neighbor node is said to be within one-hop if it is within the maximum measurement range of a node. Assume that a network consists of N sensor nodes. We want to compute a local map for each individual sensor node based on its neighbor nodes. For the ith sensor node, a local map is built based on its neighbors within k-hops. The list of neighbor nodes can be described by a neighboring vector $\underline{c}_i$ of length N, where the components of $\underline{c}_i$ have the value one if their corresponding nodes are within k-hops; otherwise, they are zeros. Define a neighboring matrix $C \in R^{N \times N}$ as

$$C = [\underline{c}_1, \underline{c}_2, \ldots, \underline{c}_N], \qquad (2)$$

where a non-zero element in the ijth location means that the ith and jth sensor nodes are connected with a known distance measurement; otherwise, the pair are disconnected. In the case of one-hop, i.e., $k = 1$, the neighboring matrix C is the adjacency matrix of the network with values of one on its diagonal. For the ith local map, define an orthogonal matrix $U_i \in \mathscr{R}^{2\times 2}$ and a row vector $T_i \in \mathscr{R}^{1\times 2}$ to represent rotation/reflection (or combination of two) and translation, respectively, for the local map. Define $U \in \mathscr{R}^{2N\times 2}$ and $T \in \mathscr{R}^{N\times 2}$ as

$$U = [U_1; U_2; \ldots; U_N] \text{ and } T = [T_1; T_2; \ldots; T_N], \tag{3}$$

respectively. Let $\mathbf{z}_{ij} \in \mathscr{R}^{1\times 2}$ denote the local coordinates of the ith sensor node in the jth local map. If the ith sensor node is not in the jth local map, then $\mathbf{z}_{ij} = \mathbf{0}$. Define a data matrix $Z_{ij} \in \mathscr{R}^{N\times 2}$, where the jth row of Z_{ij} is $\mathbf{z}_{ij}$. If the ith node is not in the jth local map, then, Z_{ij} is an all-zero matrix. Let $C_i = diag(\underline{c}_i)$ be a diagonal matrix of $N \times N$, where $diag$ puts the elements of $\underline{c}_i$ on its diagonal. For the ith local map, we construct a data matrix X_i by concatenating Z_{ij} for $j = 1, 2, \ldots, N$

$$X_i = [Z_{i1}, Z_{i2} \ldots, Z_{iN}]. \tag{4}$$

Let Y_i denote an affine transform of X_i given by

$$Y_i = X_i U + C_i T. \tag{5}$$

It can be verified that the jth row of Y_i is $\mathbf{z}_{ij}$ by an affine transform determined by U_i and T_i. If the ith sensor node is not in the jth local map, the jth row of Y_i is zeros due to $\mathbf{z}_{ij} = \mathbf{0}$; otherwise, the jth row is the coordinates of the ith sensor node in the new coordinate system. The new coordinate system is referred to as the *global coordinate system*, and the global coordinates of the sensor nodes form the *global map*. The geometric center the ith sensor node in the global map is given by

$$\bar{Y}_i = \frac{1}{N_i} \underline{c}_i^T Y_i, \tag{6}$$

where N_i denotes the number of sensor nodes in the ith local map. The difference between the non-zero components of Y_i and their geometric center can be written as

$$\Delta Y_i = Y_i - \underline{c}_i \bar{Y}_i = P_i^\perp X_i U + P_i^\perp C_i T, \tag{7}$$

where

$$P_i = \frac{1}{N_i} \underline{c}_i \underline{c}_i^T \text{ and } P_i^\perp = I - P_i, \tag{8}$$

denote the orthogonal projections onto the one-dimensional subspace spanned by $\underline{c}_i$ and its orthogonal complement, respectively. Without loss of generality, we select the first local map as a reference, i.e, $T_1 = \mathbf{0}$, meaning that there is no translation for the reference local map. Then, ΔY_i can be re-written as

$$\Delta Y_i = P_i^\perp X_i U + P_i^\perp \tilde{C}_i \tilde{T}, \tag{9}$$

where $\tilde{C}_i \in \mathscr{R}^{N\times N-1}$ is C_i with its first column being removed, and $\tilde{T} = [T_2; \ldots, T_N]$.

Since the non-zero rows of Y_i represent the coordinates of the ith sensor node in the global map, their discrepancies should be made as small as possible. One way to do this is to minimize the sum of the squared distances of all pairs of points. This can be shown to be equivalent to minimizing the Frobenius norm of ΔY_i, or the sum of the distances of all the points to their geometric center in the global map. For all the sensor nodes, the optimal U and T can be found by minimizing the sum of the Frobenius norms of ΔY_i for $i = 1, 2, \ldots, N$

$$\min_{U,\tilde{T}} J, \quad \text{where } J = \sum_i \|P_i^\perp X_i U + P_i^\perp \tilde{C}_i \tilde{T}\|_F^2, \tag{10}$$

where $\|\cdot\|_F$ denotes the Frobenius norm. The minimization of J provides the "best" tightest cluster for all of the common sensor nodes in the global map. With some matrix manipulations, we can write J as

$$J = \|AU + B\tilde{T}\|_F^2, \tag{11}$$

where

$$A = \begin{bmatrix} P_1^\perp X_1 \\ P_2^\perp X_2 \\ \vdots \\ P_N^\perp X_N \end{bmatrix} \quad \text{and } B = \begin{bmatrix} P_1^\perp \tilde{C}_1 \\ P_2^\perp \tilde{C}_2 \\ \vdots \\ P_N^\perp \tilde{C}_N \end{bmatrix}. \tag{12}$$

To find the minimizing U and $\tilde{T}$, we first fix U and minimize J with respect to $\tilde{T}$ and obtain

$$\tilde{T} = -B_s^{-1} A_s U, \tag{13}$$

where

$$B_s = B^T B = \sum_i \tilde{C}_i^T P_i^\perp \tilde{C}_i \tag{14}$$

$$A_s = B^T A = \sum_i \tilde{C}_i^T P_i^\perp X_i. \tag{15}$$

Substituting $\tilde{T}$ back into J yields

$$J = tr\{U^T \Sigma U\}, \tag{16}$$

where tr denotes the trace of a square matrix, and Σ is given by

$$\Sigma = \sum_i X_i^T P_i^\perp X_i - A_s^T B_s^{-1} A_s. \tag{17}$$

The unknown rotation matrix U can be obtained from the following optimization problem

$$\min_U J, \tag{18}$$

subject to the constraint that U_i is an orthogonal matrix for $i = 1, 2, \ldots, N$. We refer to J in (16) as the concentrated criterion since it only contains U instead of U and $\tilde{T}$ originally.

3.1 The Optimal Rotation Matrix

In (18), the optimal rotation matrix U can be obtained by minimizing the simplified criterion J. The rotation matrix $U \in \mathscr{R}^{2N \times 2}$ is structured with each $\{U_i\}$, for $i = 1, 2, \ldots, N$, being an orthogonal matrix. First, we can show that the two columns of U are orthogonal with a length of $\sqrt{N}$, i.e., $U^T U = NI$. The column orthogonality implies that $\frac{1}{\sqrt{N}} U$ must be on the Stiefel manifold $v_2(\mathscr{R}^{2N})$ [19], where in mathematics, the Stiefel manifold $v_2(\mathscr{R}^{2N})$ is defined as the set of all $2N \times 2$ orthogonal matrices. In addition, since U is constrained to contain orthogonal sub-matrices, $\frac{1}{\sqrt{N}} U$ belongs to a subset on the Stiefel manifold. We denote this subset by $\mathscr{M}$, which is the manifold consisting of all $U = [U_1, U_2, \ldots, U_N] \in \mathscr{R}^{2N \times 2}$ with each U_i being an orthogonal matrix of 2×2. Thus, finding the optimal U in (18) can be formulated as a constrain optimization problem that minimizes the criterion J with respect to U in $\mathscr{M}$.

The constrained optimization problem (18) is not trivial due to the highly nonlinear nature of J in U and the constraint conditions on U. An analytic solution does not exist and we need to resort to numerical techniques for solving the constrained optimization problem (18). Although there are many general numerical optimization techniques [9][3] for nonlinear programming, the idea of the gradient projection (GP) method by Jennrich [14][15] is adopted because of its computational efficiency and fast convergence. The gradient projection method was originally developed to solve the simple orthogonal rotation problems. The GP method is an iterative algorithm. The general idea of the GP algorithm is simple. At each step, the GP algorithm finds a next descent point in the direction of the negative gradient of the criterion function at a current point. In general, the next point will not be in $\mathscr{M}$. To deal with this, the GP method is to project the point onto $\mathscr{M}$ and obtain the next desired estimate there.

Let Q be an arbitrary matrix in $\mathscr{R}^{2N \times 2}$, and $\rho_{\mathscr{M}}(Q)$ be the projection of Q onto $\mathscr{M}$. From the definition of orthogonal projection [11], $\rho_{\mathscr{M}}(Q)$ in $\mathscr{M}$ is the closest to Q in terms the Frobenius norm on $\mathscr{M}$, i.e.,

$$\rho_{\mathscr{M}}(Q) = \arg \min_{\xi \in \mathscr{M}} \|\xi - Q\|_F. \tag{19}$$

Let $Q = [Q_1; Q_2; \ldots; Q_N]$, where $Q_i \in \mathscr{R}^{2 \times 2}$ for $i = 1, 2, \ldots, N$. Then, it can be shown that $\rho_{\mathscr{M}}(Q)$ can be obtained by finding the closest orthogonal matrix to each Q_i in the Frobenius norm sense. The problem of finding the closest orthogonal matrix is a special case of the *orthogonal Procrustes problem* [11]. Denote $\rho(Q_i)$ as the closest orthogonal matrix to Q_i

$$\rho(Q_i) = \arg \min_{\eta^T \eta = I} \|\eta - Q_i\|_F. \tag{20}$$

$\rho(Q_i)$ can be obtained by performing a singular value decomposition (SVD) on Q_i. Let $Q_i = V_L \Lambda V_R^T$ be the SVD of Q_i. The closest orthogonal matrix $\rho(Q_i)$ is given by $\rho(Q_i) = V_L V_R^T$ (see [11], Chapter 14). Since

$$\|\rho_{\mathscr{M}}(Q) - Q\|_F^2 = \sum_{i=1}^{N} \|\rho(Q_i) - Q_i\|_F^2, \tag{21}$$

it follows that the projection of Q onto $\mathcal{M}$ is given by

$$\rho_{\mathcal{M}}(Q) = [\rho(Q_1); \rho(Q_2); \ldots; \rho(Q_N)]. \tag{22}$$

The gradient projection method is an iterative algorithm, which requires a stopping rule for terminating the iteration. In the following, we discuss the problem based on the application of the Lagrange multiplier method to (18). The Lagrange function can be written as [24]

$$L = J + \sum_{i=1}^{N} \text{tr}\{\Phi_i(U_i^T U_i - I)\}, \tag{23}$$

where the matrix Φ_i, for $i = 1, 2, \ldots, N$, is the Lagrange multiplier associated with the orthogonal constraint on U_i, and is symmetric. If U is a stationary point of J on $\mathcal{M}$, a necessary condition is that there exists a Φ_i such that the partial derivative of L with respect to U_i is zero, i.e.,

$$G_i + 2U_i\Phi_i = 0, \text{ for } i = 1, 2, \ldots, N, \tag{24}$$

where G_i is the gradient of J with respect to U_i. Multiplying both sides of (24) by U_i^T yields

$$\frac{1}{2}U_i^T G_i = \Phi_i, \tag{25}$$

which implies that $U_i^T G_i$ for $i = 1, 2, \ldots, N$ is symmetric. It is known that the skew-symmetric component of a symmetric matrix is zero, which suggests that the skew-symmetry of $U_i^T G_i$, for $i = 1, 2, \ldots, N$, can be used an indication of whether or not U is a stationary point of J on $\mathcal{M}$. Define

$$\sigma = \sum_{i=1}^{N} \|skm(U_i^T G_i)\|^2, \tag{26}$$

where $skm(\cdot)$ denotes the skew-symmetric component of a square matrix. From the necessary condition, we know that $\sigma = 0$ if U is a stationary point of J on $\mathcal{M}$. This property allows the use of σ for terminating the algorithm, i.e., the algorithm stops when σ is smaller than a pre-defined threshold ε.

The gradient projection algorithm can be summarized as follows.

1. Let $k = 0$. Select an initial estimate $U^{(k)}$ and a positive step size α_0.
2. Evaluate σ at $U^{(k)}$. If $\sigma < \varepsilon$, the algorithm stops and the optimal estimate $\hat{U}_{opt}$ is given by $U^{(k)}$; otherwise, the algorithm proceeds.
3. Compute the gradient $G^{(k)}$ of J with respect to U at $U^{(k)}$. Let $\alpha = \alpha_0$ and compute a new point on $\mathcal{M}$, $\tilde{U}^{(k+1)} = \rho_{\mathcal{M}}(U^{(k)} - \alpha G^{(k)})$.
4. If $J(\tilde{U}^{(k+1)}) > J(U^{(k)})$, let $\alpha = \alpha/2$ and compute another $\tilde{U}^{(k+1)}$. Repeat until $J(\tilde{U}^{(k+1)}) \leq J(U^{(k)})$.
5. Let $U^{(k+1)} = \tilde{U}^{(k+1)}$ and $k = k + 1$. Go to Step 2 and repeat.

The initial estimate should be chosen to be sufficiently close to the global optimum in order for the algorithm to converge to the global optimum point. Consider the problem of minimizing J subject to the constrain $U^T U = N \cdot I$, i.e., U is on the Stiefel manifold scaled by a constant $\sqrt{N}$. Or in other words, U is on a manifold that consists of all matrices of $2N \times 2$ with orthogonal columns of a norm N. Denote this manifold by $\mathcal{M}_s$. From matrix theory [11], it is known that the minimizing U of J on $\mathcal{M}_s$ has a analytic solution, which is formed by the two eigenvectors of Σ, with a scaling factor $\sqrt{N}$, corresponding to the two smallest eigenvalues. Since $\mathcal{M}$ is a subset of $\mathcal{M}_s$, if the local maps contain moderate errors, it is likely that the minimizing U on $\mathcal{M}_s$ and its projection $\mathcal{M}$ will be close to the minimizing U on $\mathcal{M}$. Consequently, the minimizing U on $\mathcal{M}_s$ when projected onto $\mathcal{M}$ can be considered as an initial estimate for starting the gradient projection algorithm. An alternative is to use an initial U with $U_i = I$ for $i = 1, 2, \ldots, N$. This selection can avoid the calculation of the eigenvectors of Σ which may have large dimensions in some applications.

The convergence of the gradient projection algorithm also depends on the selection of step length α. To deal with this problem, we first introduce the following lemma.

Lemma. Let $U \in \mathcal{M}$. If U is not a stationary point of J, there exists an α such that

$$J(\rho_{\mathcal{M}}(U - \alpha G)) \leq J(U). \tag{27}$$

where G denotes the gradient of J with respect to U.

Proof. See Appendix. The proof has followed the ideas in [14]. □

From the lemma, a small step length α can be found such that the value of J decreases or remains unchanged, i.e., $J(U^{(k+1)}) \leq J(U^k)$, at each iteration. In other words, the value of J monotonically decreases after each iteration

$$J(U^{(1)}) \geq J(U^{(2)}) \geq J(U^{(3)}) \geq \cdots. \tag{28}$$

Since $J \geq 0$, the inequality (28) implies that the sequence, $\{J(U^{(k)}); k = 1, 2, 3, \ldots\}$, is convergent. That means that the gradient project algorithm will always converge to at least a local minimum after a finite number of iterations. While not guaranteed, simulations show that the algorithm converges to the desired optimal estimates almost every time. However, the algorithm may not necessarily converge to the global minimum. The global convergence depends on the selection of the initial estimates. The algorithm will converge to the global optimum if the initial estimate is in the region of convergence. However, the region of convergence is not defined rigorously. Generally speaking, the initial estimate should be close enough to the optimum to ensure the convergence. When the initial estimate is not sufficiently close to the optimal estimate, the algorithm may suffer from slow convergence and converge to a local minimum instead of the desired global one.

The gradient projection algorithm is computationally efficient. Although, each iteration of the algorithm requires the projection of a set of matrices onto $\mathcal{M}$, the implementation is relatively simple because they only involve SVD of 2×2 matrices, which has closed-form solutions.

3.2 Global Map Construction

When the optimal $\hat{U}_{opt}$ is obtained, the estimate of the translation matrix can be obtained by using (13) as

$$\hat{T}_{opt} = [\mathbf{0}; -A_s B_s \hat{U}_{opt}]. \tag{29}$$

The global map can then be constructed by estimating the global coordinates of each sensor node as

$$\hat{Y}_i = \frac{1}{N_i} \underline{c}_i^T (X_i \hat{U}_{opt} + C_i \hat{T}_{opt}), \tag{30}$$

where $i = 1, 2, \ldots, N$. Finally, since the resulting global map is a relative map, the coordinates of the sensor nodes in the global map need to be transformed into their absolute coordinates based on the absolute coordinates of the anchor nodes. The optimal transform can be found by minimizing the discrepancies between the absolute coordinates and global coordinates of the anchor nodes in the least squares sense [13]. The optimal transform is then applied to the relative sensor coordinates in the global map to provide their absolute coordinates.

4 Performance Analysis

In this section, we use computer simulations to demonstrate the performance of the proposed local map registration technique. Two type of network topologies were used in the simulation: square (regular) and C-shaped (irregular) networks. The C-shaped network has a square shape with a rectangular concave inside. As in [26], the following four sensor networks were considered.

1. C-shaped grid network: 76 nodes were placed on a grid in a square of $10r \times 10r$ with a rectangular concave of $6r \times 4r$ inside, where $r = 5$m.
2. C-shaped random network: 120 nodes were placed in a square of $10r \times 10r$ with a rectangular concave of $6r \times 4r$ inside, where $r = 1$m.
3. Square grid network: 100 nodes were placed on a grid in a square of $10r \times 10r$, where $r = 5$m.
4. Square random network: 160 nodes were randomly placed in a square of $10r \times 10r$, where $r = 1$m.

In the simulation of grid sensor networks, the locations of the sensor nodes were not a perfect grid. A uniformly distributed placement error in $[-0.2r, 0.2r]$ was added to each sensor location in both x and y directions. Figures 1 to 4 are examples of the distributions of the sensor nodes of the four simulated networks described above. The four networks each have connectivity levels of 7.58, 8.98 6.86 and 9.66, respectively. The connectivity level is the averaged number of one-hop neighbor sensors for all the sensor nodes in a network. In the figures, the large circles denote the anchor nodes.

In simulating the networks, the algebraic connectivity of each network was first checked to ensure that the networks were connected, i.e., there were no unconnected or isolated nodes. The algebraic connectivity of a network is given by the

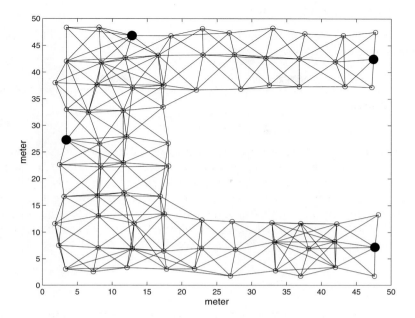

Fig. 1. C-shaped grid network with connectivity level of 7.58

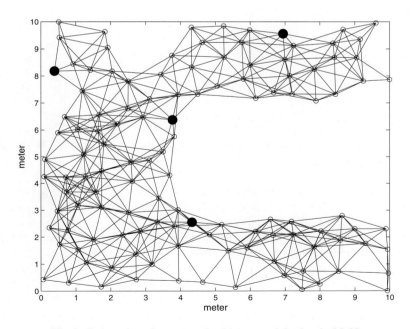

Fig. 2. C-shaped random network with connectivity level of 8.98

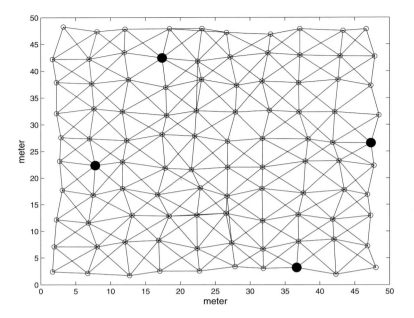

Fig. 3. Square grid network with connectivity level of 6.86

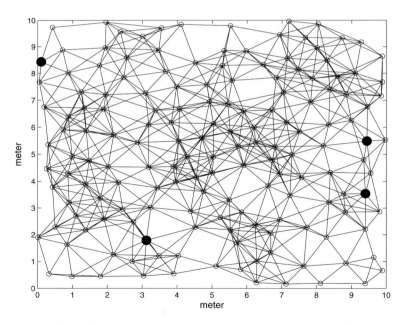

Fig. 4. Square random network with connectivity level of 9.66

second smallest eigenvalue of the Laplacian matrix of the network[7]. This eigenvalue is called the algebraic connectivity, which reflects how well the sensor nodes in a network are connected. The algebraic connectivity is non-zero if and only if all the nodes of a network are connected. In the simulations, the averaged shortest internode distance r_s was first computed and then used as an initial value for the maximum range for distance measurement. The algebraic connectivity was computed and compared with a pre-defined threshold. If the algebraic connectivity was smaller than the threshold, the maximum range for distance measurement was increased by $0.05r_s$. The process was repeated until the algebraic connectivity was equal to or larger than the threshold. For each sensor network, four anchor nodes were selected. The anchor nodes were selected by dividing the sensor deployment area into sub-regions and randomly choosing a sub-region and a node in there.

The performance of the proposed local map registration algorithm was evaluated for the four networks under different connectivity conditions. The distance measurement error was simulated as an additive Gaussian noise with zero mean and a standard deviation that was 4 percent of the actual distance. The connectivity levels were varied by changing the maximum range for distance measurement from r_s to $3.0r$ with an increment of $0.25r$. For each sensor node, a local map was built using the MDS method based on distance measurements of its one-hop neighbor nodes. The MDS-MAP(P) and MDS-MAP methods were applied for comparison purposes. The root mean square (RMS) of the absolute coordinates of sensor nodes in the global map were used as a performance measure. The RMS is the square root of the mean of the squares of the distances between the estimated and the true sensor node locations. Dijkstra's algorithm [10] was used to compute the shortest paths to approximate the distances between nodes that are more than one-hop away. For the gradient projection algorithm, the step size and the threshold were chosen to be $\alpha_0 = 0.2$ and $\varepsilon = 10^{-4}$, respectively. Figures 5 to 6 show the RMS of the sensor location estimates versus the variation of connectivity levels for the proposed local map registration algorithm and the MDS-MAP(P) method for the four simulated networks, respectively. For all the networks, the proposed technique outperformed MDS-MAP(P), especially for low connectivity levels. For the networks with random sensor deployment (both C-shaped and square), the improvement was significant considering the small averaged shortest internode distances of the networks. Table 1 to 4 list the RMS of the sensor location estimates by the centralized MDS method (MDS-MAP) for the four networks. It can be seen that, for the C-shaped networks (Tables 1 and 2), MDS-MAP failed to produce any meaningful sensor location estimates. This is mainly due to the fact that the shortest path distances (used for approximating the distances between sensors that are more than one hop away) between nodes do not correlate well with their Euclidean distances due to the shape irregularity of the networks [26]. For the square shaped networks (Table 3 and 4), MDS-MAP performed reasonably well. In these cases, the performance of MDS-MAP was comparable to those of the proposed algorithm and MDS-MAP(P) although it was still outperformed most of the time.

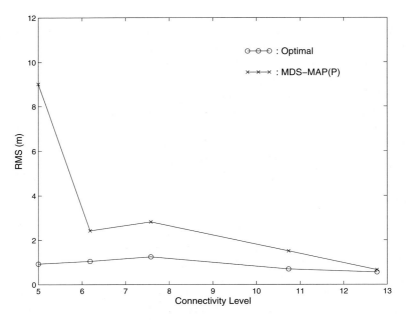

Fig. 5. RMS of sensor location estimates versus connectivity levels for the C-shaped grid sensor network. $r_s = 4.184$m.

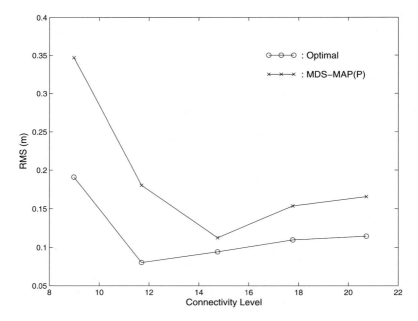

Fig. 6. RMS of sensor location estimates versus connectivity levels for the C-shaped random network. $r_s = 0.610$m.

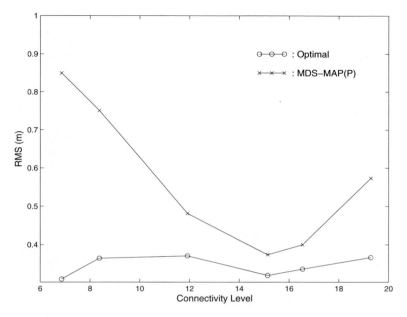

Fig. 7. RMS of sensor location estimates versus connectivity levels for the square grid network. $r_s = 4.250$m.

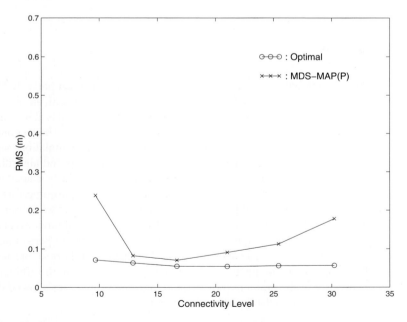

Fig. 8. RMS of sensor location estimates versus connectivity levels for the square random network. $r_s = 0.614$m.

Table 1. RMS of MDS-MAP via Connectivity Levels: C-Shaped Grid Network

Connectivity Level	5.00	6.18	7.58	10.74	12.76
RMS (m)	13.013	12.696	12.5548	10.990	10.995

Table 2. RMS of MDS-MAP via Connectivity Levels: C-Shaped Random Network

Connectivity Level	8.98	11.70	14.75	17.77	20.72
RMS (m)	2.473	2.435	2.488	2.3119	2.330

Table 3. RMS of MDS-MAP via Connectivity Levels: Square Grid Network

Connectivity Level	6.86	8.38	11.92	15.14	16.54	19.28
RMS (m)	0.774	0.488	0.580	0.629	0.579	0.693

Table 4. RMS of MDS-MAP via Connectivity Levels: Square Random Network

Connectivity Level	9.66	12.89	16.68	21.00	25.43	30.23
RMS (m)	0.145	0.089	0.103	0.134	0.157	0.167

5 Conclusions

In this chapter, an optimal local map registration technique has been presented for building a global map from relative local maps for wireless sensor localization problems. The local map registration is implemented at the global level and is able to deal with the difficulties encountered in a pairwise map merging process. The technique can be considered a generalized approach for local map merging. Simulations were used to demonstrate the performance and effectiveness of the proposed algorithm. Four sensor networks were used including both regularly and irregularly shaped networks. It was shown that the proposed local map registration technique outperformed the MDS-MAP(P) method in all of the four cases. More improvement was observed when the connectivity levels of the networks were low. In the cases of random sensor deployment, the proposed technique was able to provide a significant performance improvement over MDS-MAP(P) considering the short averaged shortest internode distances. The proposed technique is computationally efficient due to the efficiency and fast convergence of the gradient projection algorithm. Simulations showed that the algorithm often converged after only a few iterations.

Acknowledgement. The work described herein was funded by Defence Research and Development Canada (DRDC).

References

1. Akyildiz, I.F., Su, W., Sankarasubramaniam, Y., Cayirci, E.: A survey on sensor networks. IEEE Communications Magazine 40(8), 102–114 (2002)
2. Biagioni, E., Bridges, K.: The application of remote sensor technology to assist the recovery of rare and endangered species. International Journal of High Performance Computing Applications 16(3), 315–324 (2002)
3. Bertsekas, D.P.: Nonlinear programming, 2nd edn. Athena Scientific (1999)
4. Borg, I., Groenen, P.: Modern Multidimensional Scaling, Theory and Applications. Springer, New York (1997)
5. Cerpa, A., Elson, J., Estrin, D., Girod, L., Hamilton, M., Zhao, J.: Habitat monitoring: application driver for wireless communications technology. In: Proc. the First ACM SIGCOMM Workshop on Data Communications in Latin America and the Caribbean, San Jose, Costa Rica, April 2001, pp. 20–41 (2001)
6. Chong, C.Y., Kumar, S.: Sensor networks: evolution, opportunities, and challenges. Proc. IEEE 91(8), 1247–1256 (2003)
7. Chung, F.R.K.: Spectra Graph Theory. American Mathematical Society, Providence (1997)
8. Dattorro, J.: Convex Optimization and Euclidean Distance Geometry. Meboo Publishing (2005)
9. Dennisond, J.E., Schnabel, R.B.: Numerical methods for unconstrained optimization and nonlinear equations. Prentice-Hall, Englewood Cliffs (1983)
10. Dijkstra, E.W.: A note on two problems in connection with graphs. Numerische Mathematik 1, 269–271 (1959)
11. Golub, G.H., Van Loan, C.F.: Matrix Computation, 2nd edn. The Johns Hopkins University Press, London (1990)
12. Gower, J.C., Dijksterhuis, G.B.: Procrustes Problems. Oxford University Press, Oxford (2004)
13. Horn, B.K.P., Hilden, H., Negahdaripour, S.: Closed-form solution of absolute orientation using orthonormal matrices. Journal of the Optical Society of America A 5(7), 1127–1135 (1988)
14. Jennrich, R.I.: A simple general procedure for orthogonal rotation. Psychometrika 66(2), 289–306 (2001)
15. Jennrich, R.I.: A simple general method for oblique rotation. Psychometrika 67(1), 7–19 (2002)
16. Ji, X., Zha, H.: Sensor positioning in wireless ad hoc networks using multidimensional scaling. In: Proc. IEEE 23rd Annual Joint Conference of the IEEE Computer and Communications Societies (INFOCOM 2004), Hong Kong, China, March 2004, vol. 4, pp. 2652–2661 (2004)
17. Kwon, O.H., Song, H.J.: Localization through map stitching in wireless sensor networks. IEEE Trans. Parallel Distrib. Syst. 19(1), 93–105 (2008)
18. Li, L., Kunz, T.: Cooperative node localization for tactical wireless sensor networks. In: Proc. 2007 Military Communications Conference (Milcom 2007), Orlando, FL, USA, (October 2007)
19. Luenberger, D.G.: Introduction to linear and nonlinear programming, Optimization techniques. Addison-Wesley, Reading (1973)
20. Mainwaring, A., Polastre, J., Szewczyk, R., Culler, D., Anderson, J.: Wireless sensor networks for habitat monitoring. In: Proc. the 1st ACM International Workshop on Wireless Sensor Networks and Applications, Atlanta, Georgia, US, September 2002, pp. 88–97 (2002)

21. Mao, G., Fidan, B., Anderson, B.D.O.: Wireless sensor network localization techniques. Computer Networks 51(10), 2529–2553 (2007)
22. Meertens, L., Fitzpatrick, S.: The distributed construction of a global coordinate system in a network of static computational nodes from inter-node distances. Kestrel Institute Technical Report KES.U.04.04, Kestrel Institute, Palo Alto, California (2004)
23. Moore, D., Leonard, J., Rus, D., Teller, S.: Robust distributed network localization with noisy range measurements. In: Proc. the 2nd International Conference on Embedded Networked Sensor Systems (SenSys 2004), Baltimore, Maryland, November 2004, pp. 50–61 (2004)
24. Schönemann, P.H.: A generalized solution of the orthogonal Procrustes problem. Psychometrika 31(1), 1–10 (1966)
25. Schwiebert, L., Gupta, S., Weinmann, J.: Research challenges in wireless networks of biomedical sensors. In: Proc. the 7th annual International Conference on Mobile Computing and Networking, Rome, Italy, July 2001, pp. 151–165 (2001)
26. Shang, Y., Ruml, W.: Improved MDS-based localization. In: Proc. IEEE the 23rd Annu. Joint Conf. of the IEEE Computer and Communications Societies (INFOCOM 2004), Hong Kong, China, March 2004, vol. 4, pp. 2640–2651 (2004)
27. Shang, Y., Ruml, W., Zhang, Y., Fromherz, M.: Localization from connectivity in sensor networks. IEEE Trans. Parallel Distri. Sys. 15(11), 961–974 (2004)
28. Steere, D.C., Baptista, A., McNamee, D., Pu, C., Walpole, J.: Research challenges in environmental observation and forecasting systems. In: Proc. the Sixth Annual International Conference on Mobile Computing and Networking, Boston, Massachusetts, US, August 2000, pp. 292–299 (2000)
29. Stojmenović, I. (ed.): Handbook of Sensor Networks: Algorithms and Architecture. John Wiley & Sons, Chichester (September 2005)
30. D. Waller, I. Chapman and M. Michaud-Shields, "Concept of operations for the self-healing autonomous sensor network". Defence R& D Canada Technical Memorandum, TM 2008-052, July 2009
31. Warshall, S.: A theorem on Boolean matrices. Journal of the ACM 9(1), 11–12 (1962)
32. Young, F.W., Hamer, R.M.: Multidimensional Scaling: History, Theory and Applications. Erlbaum, New York (1987)
33. Zhou, Y., Lamont, L.: An optimal local map registration technique for wireless sensor network localization problems. In: Proc. 11th International Conference on Information Fusion, Cologne, Germany, July 2008, pp. 1–8 (2008)

Appendix

Appendix: Descent Property of the Gradient Projection Method

Consider the nonlinear constrained optimization problem (18), where the criterion function to be minimized is given by

$$J = \operatorname{tr}(U^T A U). \tag{31}$$

Unlike the general orthogonal rotation problem in [14], where the optimal orthogonal rotation matrix is on the Stiefel manifold, matrix U considered here lies on a sub-set or -manifold of the Stiefel manifold, i.e., U is column orthogonal but in the mean time consists of orthogonal sub-matrices.

Let $\chi = [\chi_1; \chi_2; \ldots, \chi_N]$, where $\chi_i \in \mathcal{R}^{2 \times 2}$. The closest orthogonal U_i to χ_i in terms of the Frobenius norm can be obtained by minimizing

$$f(U_i) = \|\chi_i - U_i\|_F^2, \quad \text{for } i = 1, 2, \ldots, N, \tag{32}$$

subject to the orthogonal constraint on U_i. Using the Lagrangian multiplier method, we can write the Lagrangian function as

$$F(U_i, \tilde{S}_i) = \|\chi_i - U_i\|_F^2 + \operatorname{tr}\{\tilde{S}_i(U_i U_i^T - I)\}, \tag{33}$$

where I denotes an identity matrix, and $\tilde{S}_i$ is the Lagrange multiplier associated with the orthogonal constraint on U_i. Note that $\tilde{S}_i$ is symmetric. A necessary condition [3] that U_i corresponds to a local minimum of f is that the partial derivative of the Lagrangian function with respect to U_i is zero, or equivalently

$$\chi_i = S_i U_i, \tag{34}$$

where $S_i = I - \tilde{S}_i$ is symmetric. From (34), we can obtain the following differential

$$d\chi_i = dS_i U_i + S_i dU_i, \tag{35}$$

where dS_i is symmetric. At $\chi_i = U_i$, we have $S_i = I$ (see (34)), and

$$d\chi_i U_i^T = dU_i U_i^T + dS_i. \tag{36}$$

From the constraint $U_i U_i^T = I$,

$$U_i dU_i^T + dU_i U_i^T = 0, \tag{37}$$

indicating that $dU_i U_i^T$ is skew-symmetric. It follows from (36) that

$$dU_i = \operatorname{skm}(d\chi_i U_i^T) U_i. \tag{38}$$

Thus, the differential dU in terms of dX can be obtained as

$$dU = \begin{bmatrix} \operatorname{skm}(d\chi_1 U_1^T) U_1 \\ \operatorname{skm}(d\chi_2 U_2^T) U_2 \\ \vdots \\ \operatorname{skm}(d\chi_N U_N^T) U_N \end{bmatrix}. \tag{39}$$

Let $G = [G_1; G_2; \ldots, G_N]$ be the gradient of J with respect to U. Define

$$h(\alpha) = J(\rho_{\mathcal{M}}(U - \alpha G)), \qquad (40)$$

where $\rho_{\mathcal{M}}(\cdot)$ denotes the projection onto $\mathcal{M}$. Then $h'(0)$ can be obtained as

$$h'(0) = - <G, dU(G)> = -\sum_i \|\text{skm}(G_i U_i^T)\|_F^2 \leq 0, \qquad (41)$$

In deriving (41), the property of skew-symmetric matrices is used that, if Φ is skew-symmetric, then for any real vector x, $x^T \Phi x = 0$. The inequality (41) implies that, if U is not a stationary point, then $h'(0)$ is negative and therefore, we can always find a sufficiently small α such that

$$J(\rho_{\mathcal{M}}(U - \alpha G)) \leq J(U). \qquad (42)$$

Wireless Sensor Network: Application to Vehicular Traffic

Jatuporn Chinrungrueng, Saowaluck Kaewkamnerd,
Ronachai Pongthornseri, Songphon Dumnin, Udomporn Sunantachaikul,
Somphong Kittipiyakul, Supat Samphanyuth, Apichart Intarapanich,
Sarot Charoenkul, and Phakphoom Boonyanant

National Electronics and Computer Technology Center,
National Science and Technology Development Agency,
Pathumthani, Thailand
jatuporn.chinrungrueng@nectec.or.th

Abstract. In this paper we are reporting our current development of wireless sensor network to effectively monitor vehicular traffic. A simple star configuration that consists of a server node communicating with a number of sensor nodes is proposed because of its low complexity, and easy and quick deployment, maintenance and relocation. Our system consists of the sensor, processor, and RF transceiver. We choose the magneto-resistive sensor to detect vehicles as it yields high accuracy with small size. The sensor yields important vehicle informations such as vehicle count, speed, and classification. The network topology is a simple star network. Two Medium Access Communication Protocols (MAC) are analyzed and can be automatically switched based on two different traffic scenarios. An antenna design is shown to fit with a small sensor node. Experiments show that the proposed system yields good data processing results. The classification of vehicles is very promising for major types of vehicles: motorcycle, small vehicle, and bus. RF communications is employed that cable installation can be avoided. Protocol frame formats are provided for both RF communications and RS232. This protocol is very simple and can be easily extended when new sensors or new data types are available.

1 Introduction

Traffic congestion is all big city's major concern. It hinders substantial economic and social growth. Work have been proposed with the common goal of alleviating traffic congestion. It is widely agreed that efficient traffic planning and management often reduce the congestion to a certain degree. Let's take Bangkok for instance. During rush hours, the city often manages the traffic by resorting to the traffic police as a conventional and common practice. Police officers are dispatched to major streets and junctions to help direct and control traffic flow. The traffic police manually switched traffic lights at these junctions based on real-time traffic conditions being observed and communicated among them over their trunked-radio. The lack of traffic data or wrong traffic information will worsen the traffic situation. This particular situation clearly shows that traffic data collection is very important for an effective real-time traffic management.

It is thus important that an efficient management of traffic requires data collection process in the first step. A number of work have studied on various vehicle sensors, their accuracy in collecting data, and their operation and functionality. These sensors include inductive loop [1,8,5], optical sensor [13,8], ultrasonic [6,8], and magnetic sensor [7,8]. Applications of these sensors to traffic data collection and processing have been numerously proposed [11,13,6,12,2,14,5,9]. Most of the traffic data collecting devices require signal and power cables. Recently, wireless sensor network has been applied to traffic monitoring systems as it yields several advantages including quick deployment and maintenance, less cables involved, and small size [9,4]. Therefore, applications of wireless sensor network to traffic data collection are numerous.

In this paper, we report on our development of a real-time traffic data collection system based on wireless sensor networks. We select magneto-resistive sensor as our vehicle sensor due to its reliability, versatile, and small size [3]. The system architecture provides that installation and maintenance are simple and quick. This system is therefore appropriate for those already existing road that may require temporary installation of the device to monitor traffic. We provide some backgrounds on research in this area in Section 2. This includes a review of vehicle sensors and wireless sensor network. A design of small-sized antenna at 433 MHz is shown. We describe our system and network architecture in Section 3. Section 4 explains two MAC protocols employed in the wireless part. The communications protocol frames between different communication entities are also provided. Section 5 shows data processing to obtain vehicle data including vehicle detection, length, speed, and a number of features for classifications. A number of experiments and results are shown in Section 6. Finally we conclude our work in Section 7.

2 Background

Intelligent Transportation System (ITS) is the application of the computing, communications, and sensor technologies to transportation networks. When integrated into infrastructure of the transportation system, these technologies aim to effectively and efficiently manage and control the traffic flow. In order to achieve this goal, the first important step is to monitor and collect traffic data. The data will then be analyzed, distributed, and employed to control the traffic flow. These real time traffic data can also be disseminated to travelers so that they can make proper decisions upon which route to take. It is thus an important requirement that accurate traffic data be obtained such that it represents real traffic situations. The data may include number of cars, average speed, throughput, classification and occupancy. In general, data monitoring system requires that vehicle sensors are employed. We review a number of existing traffic sensors in this section. A traffic collection system using wireless sensor networks is then described.

2.1 Vehicle Sensor Review

There are various sensor technologies in use today for monitoring vehicular traffic. We review some of common sensors here. We describe how each technology works and provide if it can obtain the following vital and fundamental data for traffic management: vehicle count, speed, and classification. Other information such as occupancy and traffic volume can often be derived from the above data.

Pneumatic Tube

Pneumatic tube is one of the first detector that is still common today. This technology uses rubber air tube for vehicle detection. It is installed across the street to be monitored. Vehicles drive over the tube will generate air pressure that is detected by the device controller. It can count vehicles based on number of axles. Conversion to vehicle count must be made resulting in its inaccuracy as different types of vehicles have different number of axles. Vehicle speed can be detected using speed trap by installing two pneumatic tubes separated by a fixed distance. Classification may be obtained according to the axle separation. The pneumatic tube wears easily as contact between cars and the tube is required for detection. It is often employed for temporarily traffic monitoring.

Inductive Loop

Inductive loop is another common vehicle detector. The technology is well developed and yields very accurate traffic data. It consists of a set of electrical coil either installed permanently under the road surface or temporarily above the road surface. Electric current in the loop will generate a strong magnetic field. When a vehicle passes over the coil, its ferrous body changes the magnetic field over the loop. This change of the magnetic field can be detected by measuring the change of electric current frequency or coil inductance. Inductive loop yields a very accurate vehicle count. Similar to the pneumatic tube, two loops are required to obtain speed. Classification can be achieved by analyzing the electric current signature in the coil but the results are not very favorable as it is very sensitive to the loop size. The installation and maintenance process requires strenuous work and needs lane closure for several days or even weeks. Road maintenance often damages the inductive loop and requires new installation.

Video Camera

Video camera is newer technology employed to obtain traffic data. It become popular because it yields live pictures of traffic. This characteristic is very advantageous as the pictures yield a real sense of traffic information as if users were observing the traffic themselves at the location. Vehicle count, speed, and classification can all be extracted with using software. The accuracy of data depends largely on the extraction software and video quality. High-quality camera systems are therefore very expensive. A support post is required to install video cameras. Another main drawback is that traffic data accuracy depends largely on light conditions.

Optoelectronic Sensor

Optoelectronic sensor is a less common vehicle detector. The sensor uses semiconductor technology that can detect light of particular frequency bands. Infrared light is common as visible light does not interfere. There are two common setups. In the first setup a light transmitter and receiver are placed opposing to one another. When light path is blocked, a vehicle is assumed. In the second setup, the light transmitter and receiver are in the same housing. It sends out light and reflects back off a vehicle body. When it detects that light reflected back, a vehicle is assumed. Optoelectronic sensors accurately gives vehicle count as it provides a fast response and can easily be focused to a small area. Two sensors are required to obtain speed data in a speed trap configuration. Classification can be obtained by measuring to the vehicle length or height. It usually has a small size and low price. However, its applications are often limited by its installation that requires spaces on the sides of each lane to be monitored. Dust and rain are often reduce its detection performance.

Magnetic Sensor

Magnetic sensor detects magnetic fields that disturbed by ferrous-body vehicles just similar to the inductive loop. There are many types of existing magnetic sensors. The magnetic sensor that becomes more popular to applications of traffic monitoring is magneto-resistive sensor. The magneto-resistive sensor is a semiconductor type whose electrical resistance varies according to a magnetic field. Its high sensitivity means that it can passively measure a magnetic field as small as the Earth magnetic field. When a vehicle passes over or in proximity to the sensor, it affects the Earth magnetic field at the sensor. The Earth magnetic field change can be detected by measuring its resistance. It's small size and low price are advantageous over the inductive loop. It accurately yields vehicle counts similar to inductive loops. Two sensors are required to obtain speed data in a speed trap configuration. Classification can be achieved by analyzing it magnetic field signature.

2.2 Wireless Sensor Network

A Wireless Sensor Network (WSN), known for its aptness of smart environment monitoring, has gain more popularity among ITS research community. This is mainly due to its capability to communicate between a sensor node and a server node (data collection point) via radio frequency. Its wireless feature and small size make the installation process quick and easy. The installation can be finished at night time to avoid traffic congestion due to lane closure. Temporary placement of WSN allows traffic data collection for a few days. Our view is that the WSN can emulate the manner the traffic police handle rush-hour traffic: observing and analyzing traffic on the spot, disseminating the information over their radio, and utilizing the information to achieve traffic flow as much as possible. As an ideal model, we automatically collect, analyze and disseminate traffic data for an effective traffic management.

3 System and Network Architecture

In this section, we describe our wireless sensor network system and network architecture for traffic monitoring. We choose the magneto-resistive sensor as the traffic sensor in our system as it yields a small size of device. The magneto-resistive sensor also yields vehicle counts, speed, and classification. A special design of antenna is also proposed such that it fits into our small size system. We simply refer to the magneto-resistive sensor as magnetic sensor for simplicity.

3.1 System Architecture

Figure 1 shows the magnetic sensor board components and its interfaces. The sensor node consists of two magnetic sensors placed at 1 inch apart, a MSP430 microprocessor, a Chipcon CC1000 RF transceiver, and an antenna. The MSP430 microprocessor is a ultra low-power 16-bit RISC microprocessor that controls the signal sampling, processing and RF communications with the server node. The Chipcon CC1000 is a low-power RF tranceiver for 433

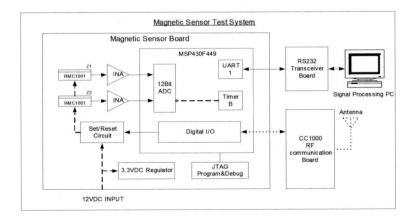

Fig. 1. Magnetic Sensor Architecture

MHz frequency band. We place the two magnetic sensors to detect changes of magnetic field in the direction of vehicle traversing. The delay between the two sensors yields information on speed and length. Since the signal obtained with the magnetic sensors is very small, signal conditioning circuit with op-amps are necessary. The microprocessor conducts the magnetic field sampling of both magnetic sensors with its 12-bit A/D converters. A real-time processing of these samples must be performed such that required traffic information is obtained. Once made, information will be transmitted to the server node such that overall traffic conditions can be processed. The sensor node can interface with a PC via RS232 for testing and parameter settings. The 12 Vdc can be supplied with batteries or power cable. Wireless communications at 433 MHz is achieved via an RF transceiver and a small-size antenna.

3.2 Network Topology

Our network has a simple star topology as shown in Figure 2. A traffic monitoring network consists of a server node and a number of sensor nodes. The server node serves as the central point for collecting traffic data from all those sensor nodes installed at various monitored points within its communications range. Each sensor node directly communicates to the server node. The sensor node is equipped with magneto-resistive sensors to detect passing vehicles. The server node can also be connected to a data server that collects traffic data. The data server can connect to several server nodes.

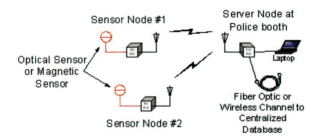

Fig. 2. WSN with a star-based topology

In monitoring vehicular traffic applications, an area being monitored is usually small as the traffic data collected in that particular small area is generally a good representative of traffic conditions for at least several hundred meters or even several kilometers. Figure 3 shows two placement configurations of WSN: at a junction and on a highway. At the junction, 8 sensor nodes are required to monitor 8 traffic lanes in all directions. On the highway, 6 sensor nodes are needed. Therefore, a WSN consisting of 10 nodes is usually practical for traffic monitoring applications. A server node (access point) can be placed at the median strip or on the side in order to collect data from these sensor nodes.

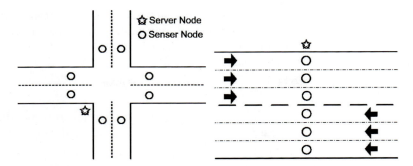

Fig. 3. Examples of sensor placement at (left) junction and (right) 6-lane highway. Eight sensor nodes are required at the junction, while six sensor nodes at the highway. In either case, the server node is in proximity to the sensor node.

3.3 Antenna

In transmitting data over radio frequency at 433 MHz, an antenna is required. We have several restrictions that our sensor node must be flat. It must also be installed easily on the road surface or even be buried under. The antenna must therefore be flat and small to fit the small size of the sensor node.

The design of our antenna structure is shown in Figure 4. It consists of the main T-strip of the width $w_1 = 3$ mm and two rectangular printed spiral loops. Each side of the rectangular printed spiral antenna has the width $w_2 = 2$ mm and the gap $G = 1$ mm. At the feed point, a U-strip is attached. This U-strip and spiral loops are used for increasing the electrical length as well as impedance matching. The result is a small size antenna with flat structure. The antenna is made on the FR-4 substrate with the dielectric constant of 4.9. The antenna size is $50x50\mathbf{mm^2}$. The antenna is 1.6 mm thick. It is noted that there is no ground plane for this antenna. The antenna's

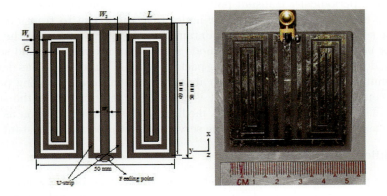

Fig. 4. Antenna structure for 433 MHz RF communications

simulated and measured return losses of -16.2 dB and -12.435dB at 433 MHz are obtained. The simulated and measured gain is -3.17 dBi and -5 dBi at 433MHz [10].

4 Protocols

Data collected at each sensor node must be transmitted to its corresponding server node and then to the PC. The communications path is divided into 2 sections:

1. the wireless section between the sensor nodes and the server nodes. This part of communications must be provided according to the star-topology network as described earlier.
2. the wired-line section between the server node and the PC. The PC is employed as database of traffic information and is connected directly to the server node.

4.1 Mac Protocol

In the wireless section, we propose two types of MAC protocols: polling and modified slotted Aloha as each protocol is suited to two different scenarios that mostly occur in traffic data collection. However, in the wired-line section, we connect our server node to the PC via RS232/RS485. This section provides exclusive connection between the server node and the PC. Therefore MAC protocol is not necessary in this part.

Polling Protocol

Polling protocol is a simple MAC protocol that fits networks in which most sensor nodes send data to the server node all the time. In this situation, the sensor node will be equally assigned time slots to send its data. The server node will periodically transmit polling packets that contain sensor nodes ID that is allowed to transmit. Sensor nodes take their turn to transmit data to the server according to the ID contained in the polling. Most of the time sensor nodes are put in the sleep mode when there are no data, and sensor nodes do not respond to the polling packet. When a sensor node has data to transmit, it wakes up and waits for its turn by monitoring a polling packet that contains its ID. Only when it is polled, it can transmit to the server node. This polling scheme provides medium access with no collision as only one sensor node can transmit to the server node at a particular instant.

Modified Slotted Aloha

Polling protocol is not efficient in situations where most of the sensor nodes are idle, then most time slots are wasted because no sensor node, except the one assigned, can access. In other words, a few busy sensor nodes cannot access those idle time slots. To cope with this situation, we propose another simple MAC protocol based on a modification of slotted Aloha. Slotted Aloha protocol allows sensor nodes to transmit at any fixed intervals (time slots). The server node synchronizes with each sensor node by sending a polling packet at the beginning of each time slot. Unlike polling protocol, this polling packet now does not contain sensor nodes ID. Any sensor node that wants to send data can send after receiving a polling packet. The main drawback is that the slotted Aloha causes collision when two or more sensor nodes transmit in the same time slot. To avoid this collision, we simply modify the protocol such that each sensor node that listens to a polling packet will have to randomly delay its start of data transmission to the server. This allows that data packet overlaps in each polling time slot and avoid collision.

4.2 Frame Format

We design a light-weight communications protocol that combine sufficient functions from data-link layer and MAC layer in our work. Our protocol is to provide medium access function and also ensure packet integrity and packet delivery acknowledgement. In this section we explain our communications frame formats for our protocol. The protocol is divided into two sections: (1) Wireless section which is communications between sensor nodes and the server node. (2) Wired-line section which is communications between the server node and the PC.

Frame Format between Sensor Node and Server Node

The frame format between the sensor node and server node is shown in Table 1. The purpose of this frame is to provide synchronization between sensor nodes and their server node. It provides addressing, acknowledgement, and error control. The frame consists of the following fields:

Preamble field (variable length) is the byte pattern that a receiver uses to synchronize its clock with the senders clock. A possible value is 0xAA as it gives the alternating pattern of 0s and 1s. The length is variable depending on environment and the receivers capability of synchronizing with the senders clock.

Sync Byte field (1 byte) is the byte pattern of 5Ah and follows the Preamble field. The Sync Byte is used to specify the start of the frame.

Destination field (1 byte) specifies the destination address for which this frame is destined.

Origin field (1 byte) specifies the sender address from which this frame is originated.

Table 1. Data frame between sensor node and server node

Header							Data Field	Tail
Preamble	Sync Byte	Destination	Origin	Length	Function	CRC8	Data	CRC16
1 byte AAh	1 byte (5Ah)	1 byte	1 byte	1 byte (data)	1 byte	1 byte	10 bytes	2 bytes compute from byte in Data

Table 2. Definitions of data field in Table 1

ID 2 bytes		Flag 1 byte	Toggle 1 byte	Info 1 byte	Info 1 byte	Info 1 byte	Info 1 byte	Info 1 byte	Info 1 byte	Description
NID	SID	00h	0/1	P	x	x	x	x	x	Polling Packet, P=01h uses Aloha; P=02h uses Polling
NID	SID	01h	0/1	x-axis	x-axis	y-axis	y-axis	z-axis	z-axis	Data Packet, Raw Magnetic Data
NID	SID	02h	0/1	x	x	in	out	speed	x	Data Packet, Processed Magnetic Data (small)
NID	SID	03h	0/1	in	in	out	out	speed	speed	Data Packet, Processed Magnetic Data (large)
NID	SID	04h	0/1	in	out	speed	speed	Class	Class	Data Packet, Processed Magnetic Data with Classification
NID	SID	05h-0Fh	0/1	ndef	ndef	ndef	ndef	ndef	ndef	Data Packets, Reserved for Magnetic Sensor
NID	SID	10h	0/1	data1	data1	data2	data2	data3	data3	Data Packet, Raw Optical Data
NID	SID	11h	0/1	x	x	in	out	speed	x	Data Packet, Processed Optical Data (small)
NID	SID	12h	0/1	in	in	out	out	speed	speed	Data Packet, Processed Optical Data (large)
NID	SID	13h-1Fh	0/1	ndef	ndef	ndef	ndef	ndef	ndef	Data Packets, Reserved for Optical Sensor
NID	SID	20h-FEh	0/1	ndef	ndef	ndef	ndef	ndef	ndef	Reserved for Additional Sensor Types
NID	SID	FFh	0/1	C	F	F	F	F	F	Command from server to node. C = Command Code; F = parameters

Length field (1 byte) specifies the length of the Data field that follows.

Flag field (1 byte) specifies optional data-link layer functions. For example: Acknowledgment request, Data encrypted request, and Sequencing bit.

CRC8 field (1 byte) specifies the cyclic redundancy check code that is computed with Destination field, Origin field, Length field and Flag field. This CRC8 is employed to verify the integrity of the above fields.

Data field (10 bytes) contains data that transmits to the receiver. A number of definitions of Data field are shown in Table 2 and described as follows:

ID field specifies the sensor node ID and the sensor ID.

Flag field specifies the type of this Data field. Examples of several options are shown in Table 2.

Toggle field specifies toggle bit which is used to identify whether a frame is duplicated. The bit value is toggled when the data frame is received and acknowledged successfully. This allows the receiver to identify whether a duplicated frame is received.

Information fields contain data according to the Flag field and are allocated for a maximum of 6 bytes.

CRC16 field (2 bytes) specifies the cyclic redundancy check code that is computed with the Data field. Acknowledgement frame uses the same frame format with out the Data and CRC16 fields. A bit in the flag field is set to specify that this frame is the acknowledge frame. The Length field is set to 0. The Destination and Origin fields are set to the receiver and the source of the acknowledge frame respectively.

Table 3. Frame definition between server node and PC

Header					Data Description (value)	Tail	
Begin 02h	Type 1 byte	Source 1 byte	Dest 1 byte	Length 2 bytes	Data Frame Formats of Data Frame defined with the Type field in Header	CRC 1 byte	End 2 bytes
02h	ACK (00h)	S	D	L	Respond upon completion of data received from Source. (06h)	CRC	CR-LF
02h	NAK (01h)	S	D	L	Respond upon incompletion of data received from Source. (15h)	CRC	CR-LF
02h	Data (10h)	S	D	L	Data frame sent from server node to data server (frame from Table 2)	CRC	CR-LF
02h	Set Values (11h)	S	D	L	Assign values to some specific parameters in server node	CRC	CR-LF
02h	Data Request (12h)	S	D	L	Data server request data transfer from data node	CRC	CR-LF
02h	Alive Signal (20h)	S	D	L	Singal Sending, in a specific interval, from Server node to Data server indicating that the Server node which report this packet still alive. (SeverID)	CRC	CR-LF
02h	Alarm (21h)	S	D	L	Alarm information generated from server node.	CRC	CR-LF

4.3 Frame Format between Server Node and PC (Data Server)

The server nodes function is to relay message from a sensor node to the data server (PC) and vice versa. It must convert into frame format as shown in Table 3. It also provides flow control, acknowledgement, parameter setting, and status and alarm report. The frame consists of the following fields:

Begin field (1 byte) with value of 02h.

Type field (1 byte) specifies the frame type, several definitions are shown in Table 3. These frames provide acknowlegdement, flow control and data frame. Additionally, parameter settings, status and alarm reports can be provided.

Origin field (1 byte) specifies the sender ID.

Destination field (1 byte) specifies the receiver ID.

Length field (2 bytes) specifies the length of the Data field and Tail field that follow.

Data field (11 bytes) specifies the data that sender transmit to receiver. When the frame is the data frame (Type = 10h), this Data field is simply modified from the Data field format in defined in communications between sensor nodes and sensor server. The Data field is divided into 2 parts: Header part and Data part as explained below.

Header part consists of the Server ID, Node ID and Sensor ID fields respectively. All fields are of 1 byte each. The above information can be obtained from the Data field in Table 2.

Data part is the Data field in Table 2 except the ID field which consists of:

Flag (1 byte) specifies the type of information field.

Toggle (1 byte) prevents package duplication.

Information is six-byte of Information field.

CRC field specifies 1 byte of cyclic redundancy check and is employed to check for data integrity. CRC is computed with bytes from Header and Data fields.

End field (2 bytes) specifies the ending of the frame and employs the carriage return and line feed characters.

The data field for other types of frame is specified as in Table 3.

5 Data Analysis

A typical magnetic signal that obtained when a vehicle passes over the sensor node is shown in Figure 5. Two signals are shown obtained with two magnetic sensors placed at distance d apart resulting in time delay between the two signals. Two main characteristics include:

1. Signal gradually builds up when vehicle moves into the sensor and tapers off when vehicle moves away from the sensor.
2. There exists several zero crossing points due to vehicle's irregularly lump ferrous structures.

The main characteristic of signal at zero crossing points however change at a more rapid rate than that when a vehicle entering or exiting the sensor node.

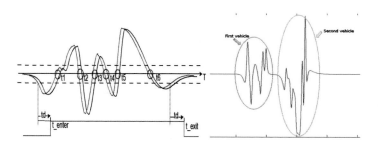

Fig. 5. Magnetic Time Series. (Left) typical time series show some notable features. (Right) real time series obtained with magnetic sensor with two vehicles passing over.

5.1 Preprocess the Data

Let $\mathbf{x}^i, \mathbf{y}^i$ and $\mathbf{z}^i$ be x-, y- and z-axis magnetic time series observed at the sensor node with magnetic sensor $\mathbf{i}$, where $\mathbf{i}$ represents one of the two magnetic sensors. When it is not significant to differiate between the two magnetic sensor in context, we will simplify our notation by dropping the superscripted index $\mathbf{i}$. Let $x_n, y_n,$ and z_n be x-, y-, and z-axis magnetic signals at time sample n respectively.

A moving average of these magnetic time series can be obtained to reduce the effect of noise as follow:

$$\overline{x}_n = \frac{x_n + x_{n-1} + ... + x_{n-S+1}}{S}, \qquad (1)$$

where $\bar{x}_n$ is x-axis moving average at time sample n. S is the number of samples that employed in the averages. The y- and z-axis moving average are defined similarly. The differential magnitude M_n and its moving average are then computed as follow:

$$M_n = \sqrt{(\bar{x}_n - \bar{x}_{n-1})^2 + (\bar{y}_n - \bar{y}_{n-1})^2 + (\bar{z}_n - \bar{z}_{n-1})^2}, \qquad (2)$$

$$\overline{M}_n = \frac{M_n + M_{n-1} + ... + M_{n-S+1}}{S}. \qquad (3)$$

5.2 Vehicle Detection

The graph in Figure 5 can represent one of the 3 axes (x-, y-, or z-axis) smoothed time series. To detect vehicle, it is enough to consider only a single axis as it also reduces complexity. We set two level-thresholds (dash lines) above and below the baseline of x-, y-, or z-axis by the value t_h. These level-thresholds can identify a vehicle as the signal will remain within this range if there is no vehicle presence around the sensor node. When a vehicle passes over the sensor node, the signal will deviate outside the thresholds. This deviation can be detected and inferred a vehicle is present. However, the signal can also oscillate about the baseline causing several zero-crossing points due to irregular mass of a vehicle body. These osillations must be distinguished from the vehicle moving away from the sensor node. Only when the vehicle moves away from the sensor node, the signal will tapers off back to the baseline. Thus, it is necessary to distinguish the zero-crossing points from vehicle exiting the sensor node. These two conditions are distinguished with a time delay t_d. The time delay threshold t_d is defined such that if magnetic signal falls within the two level thresholds t_h less than time delay t_d, it will still be considered as zero-crossing points and not the vehicle exiting the sensor node. Under the condition above, a vehicle can therefore be detected. Using definitions above, time at which a vehicle is detected t_{enter} is determined when signal is deviated outside the threshold t_h for longer than time delay t_d. Similarly time at which a vehicle moves away from the sensor node t_{exit} is determined when signal tapers back within the threshold t_h for longer than time delay t_d. Then an estimated vehicle's occupancy time, t_o, is defined as $t_{exit} - t_{enter}$.

5.3 Estimation of Vehicle's Speed and Length

An estimation of vehicle's speed and length can be obtained with the delay between signals obtained with the two magnetic sensors. Usually, an estimation can be obtained when a vehicle passes in over the sensor node. This estimate is not very accurate considering that a vehicle can change its speed resulting in speed variation over the entire occupancy time. To better represent vehicle speed in this situation, an average of several estimations obtained at different points should be obtained. We therefore propose an estimate of vehicle speed based on an average of time delays at zero-crossing points as shown in Figure 5. There are two main advantages of using these zero-crossing points over the entry point for estimating vehicle speed:

1. The signal changes more rapidly at these zero-crossing point compared to a gradual change of signal at entry point or exit point. Therefore, it is more accurate to locate these zero-crossing points than to locate the entry point or exit point. The time delay t_n between these two signals can then be obtained at each zero-crossing points. These time delays are shown in Figure 5 as $t_1, t_2, ..., t_6$.
2. The variation of vehicle speed can be taken into consideration by averaging these delay times t_n at all zero-crossing points over the entire occupancy period.

The average time delay $\bar{t}_s$ is defined as follows,

$$\bar{t}_s = \sum_{s=1}^{n} t_s. \qquad (4)$$

If magnetic signal does not possess zero-crossing points, which can happen to small vehicles, the entry point and the exit point will be used in the average instead. Once the average time delay is found, an estimate of average speed is defined as follows,

$$\bar{v} = \frac{d}{\bar{t}_s}, \qquad (5)$$

where d is the separation distance between the two magnetic sensors. Then an estimate of vehicle length is obtained with the speed estimate $\bar{v}$ and the occupancy time t_o,

$$\bar{L} = \bar{v} \cdot t_o. \qquad (6)$$

The complexity of computations above are very low and can be well implemented on a resource-limited microprocessor.

5.4 Averaged Magnetic Energy

We define averaged vehicle magnetic energy E as follow:

$$\frac{\sum_{n \in t_o} \overline{M}_n}{\bar{v} \cdot \overline{L}}, \qquad (7)$$

which is a sum of differential magnitude sampled over the occupancy time normalized to a unit speed and unit magnetic length. Averaged vehicle magnetic energy can give information about the vehicle ferrous body as a whole. It yields information on how much effect a vehicle has on the earth magnetic field at the sensor node. It is expected that vehicles with large ferrous body and engine have more effect than with small ferrous body. In addition, the closer the body to the sensor node also yields more changes of the signal.

5.5 Hill Pattern Peaks

The vehicle magnetic energy investigates the ferrous body as a whole. In order to scan the vehicle ferrous body in more detail, Hill pattern is used in [3]. It is a pattern extracted from magnitude signal based on its slope. Hill-pattern is employed as it is believed that different vehicles possess different parts of structures and different numbers of axles. This is represented as the number of peaks in the Hill-pattern. We represent positive slope with 1 and negative slope with 0. A peak is considered valid when a consecutive number of samples are larger than a threshold.

6 Experiments and Results

We perform a number of experiments to test our magnetic sensor. The following sections report on the results of our tests. Figure 6 shows our sensor node package with sensor and transceiver boards. The antenna is fit on the cover of the package. The package size is 8x8 **in²**.

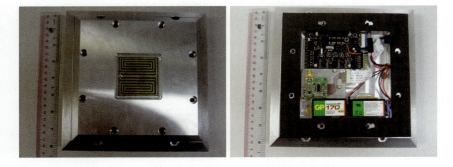

Fig. 6. (Left) Sensor node with sensor and transceiver board. (Right) Antenna fit on the cover of the package.

6.1 Detectability, Length and Speed Estimation Experiment

We first test our device for its detectabiltiy and its accuracy on estimating vehicle length and speed. We setup an experiment by placing our magnetic sensor board on the middle of the lane. This magnetic sensor board with two sensor installed at 1 inch apart works as a sensor node. We repeatedly drive a car over the sensor node for 30 times at different speed (9 times at 15 km/hr, 11 times at 20 km/hr and 10 times at 30 km/hr). The measured physical body length of the vehicle is 177 inches. We collect vehicle count, estimate its speed and length within the sensor node microprocessor and find the averages and standard deviations of speed and length estimation. The sampling rate is 4,000 samples per second. The threshold employed for detecting vehicle 48mV deviation from its baseline and time delay is 0.1 seconds.

The results show that the sensor node can detect the vehicle arrival at node with 100% accuracy at all speed. In Table 4, the average and standard deviations of speed and length estimations are shown at different speeds. The results show that the sensor can accurately estimates speeds in all test ranges at slightly lower than the actual speed. For length estimation, they are always longer than the actual values because magnetic sensor can sense the vehicle before and after physical occupation period. The actual length of vehicle is 177 inches. And our estimate yields the length of 187 inches. The different speed does not change our accuracy in estimating the length. The standard deviations also show that our results are very consistent as each estimate yields very similar result.

Table 4. The statistic of estimated velocity and length of vehicle at different speeds

Speed (km/hr)	Estimated Speed (km/hr) mean	SD	Estimated Length (inches) mean	SD
15	12.28	1.09	186.44	12.51
20	18.84	1.69	186.27	10.16
30	26.49	2.27	187.70	16.20

6.2 RF Communications Experiment

In this experiment, we test our RF communications capability. We employ the same experiment setup as in the previous experiment. We place the server node at three different locations (3, 5, and 7 meters) from the sensor node. We drive the car over the sensor node at 20 km/hr. The signal is processed and results are transmitted to the server using the protocol described in Section 4. We record the number of transmissions and the number of successful transmissions. The communications process is specified to complete when the sensor node recieves acknowledgement frame from the server.

Our experiment shows that the sensor node can successfully communicate with its server node at approximately 50 percentage as shown in Table 5. The Success rate represents percentage of successfully transmitted data over all transmitted data. Those failed communications are due to the sensor node cannot receive poll or acknowledgment from its server node. The experimental results show the low communication rates due to the characteristic of radio propagation is dominated in upward direction while server's antenna is placed on the road surface. This make the server node does not receive signal very well. If retransmission is allowed, higher successful rate of transmission is expected.

Table 5. The success rate of RF communications between sensor node and server node

Distances (meters)	Success Rate (%)
3	50.0
5	58.33
7	47.37

6.3 Classification Experiment

We setup the sensor node and collect our data at an entrance gate into Thammasart Univerity, Rangsit Campus, Thailand. At this entrance, there is a security gate where an entering vehicle must stop and take a ticket then start off driving. We place on the middle of the lane our sensor node about 5 meters behind the security booth. To obtain the ground truth, we also record visual data with a digital video camera.

With our sensor node, we obtain the earth magnetic field variations in 3 axes: x-axis, y-axis, and z-axis. X-axis is the direction perpendicular to the direction of moving vehicle pointing to the drivers left. Y-axis is the direction parallel to the moving vehicle pointing toward the vehicle. Z-axis is the direction perpendicular to the ground pointing upward. We sample the signal at 4000 samples/sec. The total number of vehicles we observed is 393. Out of this total, we visually classify into 65 motorcycles, 154 passenger cars, 98 pickup trucks, 34 vans, and 42 buses. Figure 7 shows this vehicle distribution obtained from our data collection.

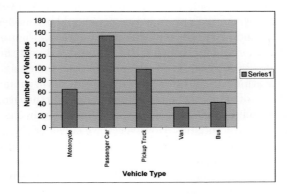

Fig. 7. Vehicle distribution

Distribution of Vehicle Magnetic Length

Figure 8 shows the distribution of relative vehicle length of all 5 types of observed vehicles. We observe that the distribution of motorcycle and bus lengths are clearly separated from the others types. The distributions of car, pickup and van lengths are very much overlapped. It means that only relative vehicle length can be employed to classify motorcycles and bus from small passenger cars. However, in order to classify car, pickup and van, additional features will be required.

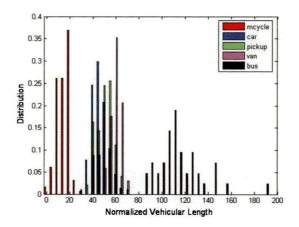

Fig. 8. Vehicle length distribution

Distribution of Averaged Energy

Figure 9 shows the distribution of averaged energy of 5 types of vehicles. Due to the size of the motorcycles, their averaged energy should be smaller than others. However the results show the distribution of motorcycles in Figure 9 is not much different from the others. This is because the sensors can detect the motorcycles only when they pass by very closer to the sensors that makes significant earth magnetic field variations. In addition, the bus energy is very much overlapped with the other types of vehicles due to the bus body is farther from the sensor. This feature is not useful for classifying motorcycle and bus from smaller cars as they have different characteristics. However, if we consider the group of small vehicles, i.e., car, pickup and van, the distributions show significantly that the van's energy is higher than the car's and the pickup's. We can therefore take advantage of this feature to classify van from car and pickup. The distributions of car and pickup are almost the same.

Distribution of Number of Hill-Pattern Peaks

Figure 10 shows that the distributions of number of Hill-pattern peaks for 5 types of vehicles. The graph shows that the motorcycle yields smallest number

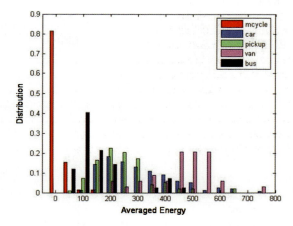

Fig. 9. Distribution of averaged energy for all types of vehicles

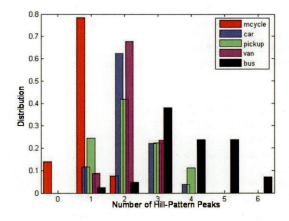

Fig. 10. Distribution of Number of Hill-Pattern Peaks

of peaks comparing to other types. The bus yields significantly larger than the other types. For the group of small vehicles, the pickup tends to yield a flat distribution of energy, while the car and the van yield a sharper peak at 3.

All means and standard deviations of the three features extracted from each type of vehicles are summarized in Table 6.

Based on these three distributions, we propose a simple classification based on hierarchical tree methodology. The classification trees are shown in Figure 11. Following the tree in Figure 11(left), we start off identifying buses using the normalized vehicle length. This is obvious as buses normalized vehicle length distribution is clearly distinguished from other types of

Table 6. Means and Standard Deviations for the three distributions

	Vehicle Magnetic Length		Averaged Energy		Hill-Pattern Peaks	
	mean	SD	mean	SD	mean	SD
Motorcycle	17.17	4.8	36.93	27.42	0.94	0.46
Passenger Car	48.19	6.58	329.5	133.1	2.18	0.68
Pickup	52.71	7.3	272.7	108.1	2.2	0.94
Van	59.43	8.21	476.7	119.9	2.15	0.96
Bus	117	20.56	178.1	103.6	3.63	1.15

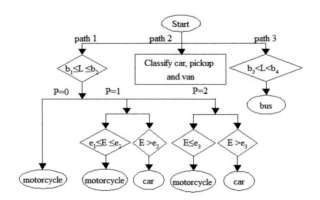

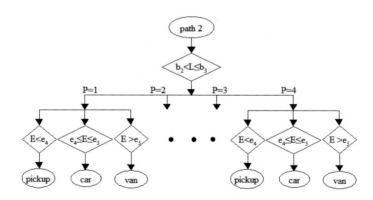

Fig. 11. Classification Trees. (Top) to classify vehicles into motorcycle, bus, and small vehicle. (bottom) to classify small vehicles further into car, van, and pickup truck.

vehicles. Similarly, the motorcycles normalized length, averaged energy, and number of Hill-pattern peaks are much lower than the other types of vehicles. These three features are employed in path 1 to mainly identify motorcycles and cars. A more thorough look into the groups of cars, pickup, and vans is then considered based on combination of the three features by following path 2 as shown in Figure 11(right).

The classification results are shown in Table 7. When vehicles are classified into 3 types: motorcycle, small vehicle, and bus, the results are very promising. However, classification of small vehicles further into car, pickup, and van results in lower accuracy. This is due to the features employed do not well distinguish these 3 types of vehicles.

Table 7. Classification results showing a high classification accuracy when 3 types of vehicles: motorcycle, small vehicle, and bus, are to obtained

Type	Classification Result (%)
Motorcycle	100
Passenger Car	77.92
Pickup	66.33
Van	76.47
Bus	100

7 Conclusion

We describe our wireless sensor network for traffic data collection in this work. The system consists of two magneto-resistive sensors, a microprocessor, and an RF transceiver. The magneto-resistive sensor measures the Earth magnetic field that can be disturbed when a vehicle passes. This signal will be processed to yield information such as vehicle count, speed, and classification. This traffic information will be sent to the server via RF communications. The experiments performed show promising results with high accuracy in obtaining vehicle counts, speed and length. The classification into 3 main types: motorcycle, small vehicle, and bus, yields a good result. Classification of the small vehicles into car, van and pickup truck yields 60-70 % classification rate. A simple star configuration that consisting of a server node communicating with a number of sensor nodes is proposed because of its low complexity, easier and quick deployment and maintenance, and also relocatable. Two Medium Access Communication Protocols (MAC) are analyzed and can be automatically switched based on two different traffic scenarios. Experiments show that the proposed system yields good data processing results. Protocol frame formats are provided for both RF communications and RS232. This protocol is very simple and is extensible when new sensors or new data types are available.

References

1. Anderson, R.L.: Electromagnetic loop vehicle detectors. IEEE Transactions on Vehicular Technology (1970)
2. Bertini, R.L., Lindgren, R.V., Helbing, D., Schonhof, M.: The german autobahn: An its test bed for examining dynamic traffic flow phenomena. In: Proceedings of the 8th International IEEE Conference on Intelligent Transportation Systems, pp. 502–507 (2005)
3. Caruso, M.J., Withanawasam, L.S.: Vehicle detection and compass applications using amr magnetic sensors. Technical report, Honeywell
4. Cheung, S.-Y., Varaiya, P.: Traffic surveillance by wireless sensor networks: Final report. Technical report, University of California, Berkeley (2007)
5. Guo, H., Jiang, G., Zhu, W.: Study on identification method for urban road traffic conditions with inductive loop data. In: IEEE International Conference on Automation and Logistics, pp. 154–156 (2007)
6. Kim, H., Lee, J.-H., Kim, S.-W., Ko, J.-I., Cho, D.: Ultrasonic vehicle detector for side-fire implementation and extensive results includiing harsh conditions. IEEE Transactions on Intelligent Transportation Systems (2001)
7. Lenz, J.E.: A review of magnetic sensors. Proceedings of the IEEE (1990)
8. Mills, M.K.: Future vehicle detection concepts. IEEE Transactions on Vehicular Technology (1971)
9. Pelczar, C., Sung, K., Kim, J., Jang, B.: Vehicle speed measurement using wireless sensor nodes. In: IEEE International Conference on Vehicular Electronics and Safety, pp. 195–198 (2008)
10. Phaebua, K., PhongCharoenpanich, C., Torrungrueng, D., Chinrungrueng, J.: Two-rectangular-printed-spiral antenna with u-strip. In: IEEE International Symposium on Antennas and Propagation (2008)
11. Reijmers, J.J.: On-line vehicle classification. IEEE Transactions on Vehicular Technology (1980)
12. Sakamoto, K., Takimoto, H.: Comparative study for performance level between two types of vehicle detector and comprehensive results. IEEE Intelligent Transportation Systems, 1008–1012 (1999)
13. Yoshida, M.: Optical vehicle detector for traffic control. In: IEEE-IEE Vehicle Navigation & Information Systems Conference, pp. 154–156 (1993)
14. Yoshida, M., Aoyama, K.: New concept for utms. In: Information Systems Conference, pp. 191–194 (1993)

Thermal Energy Harvesting for Wireless Sensor Nodes with Case Studies

C. Knight[1] and J. Davidson[2]

[1] CSIRO Energy Technology
Mayfield West, NSW, Australia
[2] James Cook University
Townsville, QLD, Australia

Abstract. Over the last decade, wireless computing and mobile devices have decreased in size and power requirements. These devices traditionally had significant power requirements that necessitated the use of batteries as their power source. However, as the power requirements are reducing, with wireless sensor nodes rarely exceeding 75mW, alternative means of power provision become available. One of these alternatives is the use of thermal energy harvesting from waste heat or environmental sources. This report discusses the field of thermal energy harvesting, with a particular focus on those thermal technologies that provide direct electricity as output. There are a number of technologies in this field. The general technologies discussed in this report are thermoelectric devices, such as devices utilising the Seebeck effect, thermocouples and, thermionics. Sources of thermal energy discussed include solar, the human body, vehicle exhaust systems and subsurface heating. Solar has an obvious advantage of being a very large source of energy. Cases studies are presented to show magnitudes and the daily variation of power output from Seebeck style thermoelectric devices. Overall, this report reveals that while present thermal energy harvesting technologies suffer from very low efficiencies, there are a number of promising technologies that are increasing in reliability and efficiency. This, along with a continuing decrease in power requirements on the demand side of the equation, marks thermal energy harvesting as a very promising field of research.

1 Introduction

A modern wireless sensor network consists of individual sensor nodes which measure various environmental variables. These variables depend on the application and can range from simple physical parameters, such as temperature or humidity, to more abstract parameters like indications of local flora and fauna through the distinctive sounds of a frog or the location of a cow in a particular paddock. This information can be stored as data at the node or relayed through the network, using wireless communications, for access by the user.

Recent advancements made in the miniaturisation of electronics have sparked a growing interest into the vast possibilities and applications of wireless sensor networks. This reduction of size, energy consumption and cost of the wireless sensor node components i.e. sensors, circuits, wireless communication; has made the vision of autonomous sensor networks deployed throughout the environment, a near reality. However, battery technology has been unable to keep pace with the exponential miniaturisation of electronics.

Battery life becomes the limitation in network deployment and the greater the potential for network deployment, the greater the liability for battery replacement. Thus alternative means to satisfy the sensor node's energy requirements are being actively explored [1]. One area which may prove useful in providing alternative energy supplies is to exploit a local temperature difference to generate power at the node.

A temperature difference existing between two locations will result in a flow of heat energy from hot to cold in an attempt to develop thermal equilibrium. The heat flow can be exploited to harness useful energy. This process is governed by the laws of thermodynamics therefore its efficiency, the ratio of the useful work extracted out, W, to the input heat, Q, is constrained by the fundamental Carnot efficiency. The Carnot efficiency applies to all heat engines and generators and can be expressed in terms of the hot, T_H, and cold, T_C, temperatures as [2].

$$\eta = \frac{W}{Q} = \frac{T_H - T_C}{T_H} \quad (1)$$

This shows that the efficiency is very low for small to modest temperature differences. As an example, a heat source, 5K above room temperature (298K), could be used to harness energy with an efficiency of 1.7%. Even if that heat source was increased to 373K the maximum efficiency would only be 25%. The Carnot efficiency is the limit or maximum theoretical efficiency; real world conversion devices however do not achieve efficiencies as high as this. As will be shown in later sections current commercial devices operate below 40% of the Carnot efficiency.

The low efficiencies of this process necessitate a large amount of heat to be transferred in order for a device to harvest useful amounts of work. The transfer of heat can occur in three different ways; via conduction, convection and/or radiation. Roundy *et al* [3] derives an analysis to demonstrate the power levels achievable from temperature gradients by assuming heat conduction through a material. At small scales and temperature differentials convection and radiation would be negligible compared to conduction and as such the amount of heat flow can be given by;

$$q = k \frac{\Delta T}{L} \quad (2)$$

Where q is the heat, k is the thermal conductivity of the material, ΔT is the temperature difference and L is the length of the material.

Although thermal power harvesting, across large temperature differentials spanning hundreds of degrees Kelvin, has been available for many years, the focus of this study are those technologies which are able to utilise the smaller ambient temperature differences available at the site of a sensor node. These temperature differences exist in many everyday situations, such as between an animal body and surrounding air, or between soil and air, and are often in the range of one to tens of degrees Kelvin.

2 Conversion Methods and Technologies

There are many types of heat engines designed to extract useful work from sources of heat, examples of which range from thermally powered wrist watches, the internal combustion engine in a car to nuclear power plants. These engines can be broadly classified into two categories; mechanical and solid state. For the wireless sensor node application of harvesting tiny amounts of power from small ambient temperature gradients, solid state devices offer the best potential as they lack moving parts thus are robust and require low maintenance. In fact, life testing of thermoelectric devices has shown their capability for over 100,000 hours of continuous operation [4]. They are compact and light, noiseless in operation, are highly reliable and eliminate power losses in extra conversion steps needed for mechanical engines. An overview of the different solid state thermal energy harvesting techniques currently available is given.

2.1 Thermoelectric

Thermoelectricity is by far the dominant solid state conversion technique so a more in depth description will be given for it over other techniques. It involves the direct conversion of heat into electricity and was discovered by Thomas John Seebeck in 1821 when he found that current would flow through two different metals joined in a loop while their ends were held at different temperatures. This phenomenon is now known as the Seebeck effect; a thermoelectric EMF will be produced across two different metals or semiconductors in the presence of a temperature difference.

Modern thermoelectric devices use n and p type bismuth telluride semiconductors instead of two different metals. Their operation can be seen in Figure 1. The two semiconductors are connected electrically in series and thermally in parallel with one end exposed to a heat source and the other to a cooler side in a configuration known as a thermocouple. The temperature difference is conducted through the two carriers, one an n-type material and the other a p-type material. A voltage difference is generated by this temperature difference at the base of the electrodes. The charge carriers in the n-type material have a negative charge producing a current from the cold to hot side whereas the carriers in the p-type material have a positive charge producing a current from the hot to cold side, with the total result of a current flowing anti-clockwise around the circuit shown.

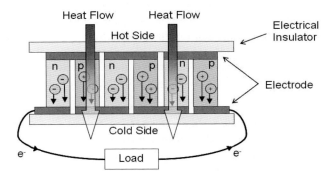

Fig. 1. Thermoelectric module concept [1]

The voltage produced across a thermoelectric device is proportional to the temperature difference across it and to the difference between the Seebeck coefficients, $S_1(T)$ and $S_2(T)$, of the two materials. The p and n type semiconductors have a positive and negative Seebeck coefficient respectively. As the Seebeck coefficients are functions of temperature, the value of the voltage across a thermocouple exposed to a temperature difference, $T_H - T_L$, can be found from the integral in equation 3. This voltage is generally quite small so many thermocouples are usually connected in series to form a 'thermopile' in order to achieve useable voltages. The power harvested by a thermocouple, or thermopile, is proportional to the square of the voltage and therefore the temperature difference.

$$V = \int_{T_L}^{T_H} (S_1(T) - S_2(T))dT \qquad (3)$$

Ferrari et al [5] investigated using a thermoelectric generator to power a wireless sensor node. The paper presents the characterization of three different commercial thermoelectric modules designed for heating/cooling applications. Their analysis included the effects of electrical load resistance, thermal conductivities of the thermoelectric and heat exchanger modules and different temperature gradients. They found that thermoelectrics could be used for their application of powering a wireless sensor node consuming 32mW, when the temperature difference exceeded 30K.

The efficiencies of thermoelectric generators have remained low and unchanged for the past 50 years. The reason for this is that in order to exploit a temperature gradient the thermoelectric device must be a good electrical conductor to allow the flow of charge but be a thermal insulator to maintain the temperature difference. This is contrary to most conventional materials as good electrical conductors are also good thermal conductors therefore a large portion of energy is transferred across the device as heat and not as electrical energy. The dimensionless thermoelectric figure of merit, ZT, is a measure of this ability and is roughly proportional to the device's efficiency[6]. ZT is given by equation 4, where σ is the electrical conductivity, λ is the thermal conductivity and S is the Seebeck coefficient. It has remained around the value of 1 for more than 50 years however modern research into thermoelectric materials is improving this by a factor of over 2 [7].

$$ZT = \frac{\sigma S^2}{\lambda}\left(\frac{T_H - T_L}{2}\right) \qquad (4)$$

Improving this ZT value is crucial for the widespread implementation of thermoelectric converters as typical commercial converters currently operate at efficiencies of less than 6%.

2.1.1 Thermoelectric Materials and Figures of Merit ZT

A whole field has evolved to improve the ZT value for thermoelectric devices involving research into nanoscale thermoelectric materials. Venkatasubramanian *et al* have quoted an improvement of ZT to a value of 2.4 by using Bi2Te3/Sb2Te3 super lattice devices which control the transport of heat carrying phonons and charge carrying electrons in the super lattice [7]. Through the fine-tuning of carrier levels they have updated this ZT value to greater than 3.5 in more recent work [8]. Joint research efforts between MIT and Boston college has increased the efficiency of bismuth antimony telluride thermoelectric devices to a ZT value of 1.4 by crushing it into a powder consisting of an average particle size 20 nanometers then sintering it into bars or disks at high temperatures. These new bars have a much finer crystalline lattice structure than the original material which consisted of millimetre size grains [9]. Many other approaches and materials are currently being investigated with more improvements to the ZT value expected in coming years.

2.2 Thermionic and Thermo-tunnelling

Another type of solid state heat engine which has been around for decades is that based on thermionic conversion. A thermionic converter is a system in which electrons are ejected via thermionic emission from a hot electrode over a potential barrier to a cooler electrode. The barrier the electrons must overcome is known as the work function of the material and is essentially the heat of vaporization of the electrons from the surface. Due to this the thermionic conversion works best with large temperature differentials. Although thermionic conversion has better efficiencies than thermoelectric devices, its reliance on high temperatures would make it unsuitable for most wireless sensor applications.

A method similar to thermionic conversion is thermo-tunnelling which narrows the potential barrier using properties of quantum physics known as quantum tunnelling. This technology seems plausible to use for small temperature gradients for lower power applications [10]. In order for this effect to be useable at the much lower temperatures envisaged for sensor networks the two surfaces need to be held at very close separations – nanometres – over large areas, with no part actually touching thus causing an electrical short circuit. Methods that allow manufacturing of this gap are being researched. One current method is to electro deposit layers of two metals such as silver and titanium. The newly created material is then thermally shocked so it breaks at the interface. This leaves an atomically rough surface that fits together very well. The gap is then adjusted using piezo devices [30].

2.3 Power Management Systems

The voltage and power output from a thermal energy harvesting device is dependent on the temperature difference across it. As this temperature difference fluctuates so too does the voltage and power output. The sensor node electronics require power at specific voltage levels with varying current and duty cycles depending on the task. The mismatch between the output power from the harvesting device and the input power required by the electronics necessitates power management systems.

A power management system interfaces the irregular output from the energy harvester to a robust source usable by the electronics. It generally does this in three stages; input power conversion, energy storage and then output power conversion. The input power conversion step needs to facilitate the flow of energy from the thermal energy harvesting device into the storage while trying to maximise the power and efficiency. This is achieved through such strategies as load matching and maximum power point tracking. Different storage devices require different charging profiles to maximise their capacity and it is the job of the power management system to implement this. Distinct required voltages are achieved through DC/DC conversion. Becker *et al* demonstrate the power management considerations employed to enable thermal harvesting in aircraft [11].

3 Heat Sources and Applications

3.1 Solar

The surface of the Earth receives about $1kW/m^2$ of peak power from the Sun. This offers enormous potential for energy harvesting. Rather than use the conventional solar conversion method of photovoltaics, a growing field is exploring converting this energy indirectly to electricity through thermal energy harvesting.

On a large scale, arrays of mirrors are used to reflect and concentrate the Sun's power to a single point where the temperature rises to hundreds of degrees Celsius. On a smaller scale suitable for wireless sensor nodes, a black surface facing the Sun will absorb the sun's rays and heat up relative to its shaded underside. The difference in temperature of the top and bottom faces can be used to generate power by sandwiching a thermal energy harvester between them.

Yu *et al* [12] investigated the use of a hybrid power system for wireless sensors which incorporated solar and thermoelectric conversion. Solar photovoltaic (PV) cells heat up when in operation, so to harness this waste heat they attached thermoelectric harvesters underneath the cells with a heat sink underneath the thermoelectric harvesters to the atmosphere.

In their experiments, for a solar irradiance of 744 W/m^2 and ambient temperature of 34°C, they found that the rear of the PV cells reached 61°C. In other research by Wang [13], it was found that the rear of solar cells reached over 70°C in stronger summer light. The advantages of harvesting this relatively large 30-40°C temperature difference are twofold. Firstly and most obviously the thermoelectric devices are harvesting and providing extra power to the sensor node. The second benefit is that the presence of the thermoelectric device increases the efficiency of the solar cell. This is due to the fact that a PV cell's efficiency drops with increasing temperature by about 0.4% of total efficiency per degree. By including the thermoelectric device to harness heat energy from the cell its temperature drops and efficiency increases. In their experiment Yu *et al* found that the rear of the cells equipped with thermoelectric harvesters were 13°C colder than those without and measured a 5.2% increase in their efficiency.

Other research by the same authors showed the operation of such a hybrid PV/thermal solar system was used to successfully power an actual sensor node [14]. They developed an intelligent power management system to control the flow of energy between the two sources (PV and thermal), the storage (ultracapacitors and lithium ion batteries) and the load (sensor node).

3.2 Ground to Ambient Air

The temperature of the air will vary across the day and night in response to changes in the incoming solar radiation. Soil temperatures vary in response to both solar radiation and conduction from the air. The temperature of the soil changes at a slower rate than that of the air due to the larger thermal mass of the soil. This creates a temperature difference between the air and the ground. This is shown in data taken from the Australian Bureau of Meteorology's East Sale Airport station in 2007, figure 2.

Figure 2 shows the temperatures of the air and ground at 10cm and 100cm as recorded at 9am over a year. The air temperature is seen to vary greatly on the daily basis compared to the soil temperature. The temperature at 10cm depth follows the fluctuations in the air more than the temperature at 100cm which is seen to remain more constant on the short term and varying sinusoidally with the changing seasons across the year. The largest temperature difference is 10.7 K with an average difference of 2.5K existing between the air and the 100cm soil temperature.

These temperature differences can be used to drive a thermal generator. Figure 2 illustrates this concept. The data shown in Figure 2 is provide by the Australian Bureau of Meteorology and is from East Sale in southern Victoria. The air side heat exchanger emits or receives thermal energy from the ambient air, depending on whether the air is cooler or warmer than the soil. Likewise the soil heat exchanger receives or emits thermal energy from the ground at a desired depth. The heat pipe then transfers this heat to/from the thermoelectric device allowing the temperature gradient across tens of centimetres of soil to be applied directly across the much thinner thermoelectric device.

Stevens first proposed this energy harvesting idea and has shown that using a maximum power approach rather than a maximum efficiency approach will lead to the greatest energy harvest output [16]. The low temperature differences encountered in this application lead to very low fundamental conversion efficiencies

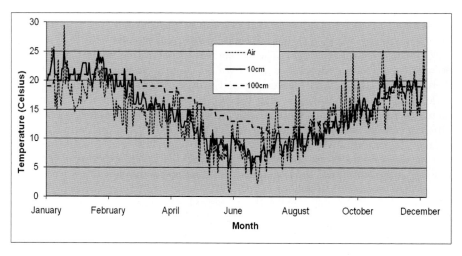

Fig. 2. Measured temperatures for air and at soil depths of 10cm and 100cm at 9am

which can only be counteracted through large amounts of heat transfer. In order to maximise the flow of heat through the system while maintaining an optimal operating temperature difference the thermal conductance of the thermoelectric device must match the sum of all the external thermal conductances. This is analogous to load matching in electrical circuits for maximum power transfer and has the effect in this application of splitting the total air/soil temperature drop evenly between thermoelectric device and the heat exchangers. In later work he showed that there is an optimal depth at which the ground side heat exchanger should be buried and which theoretically leads to a 7% increase in the temperature difference across the thermoelectric device over the difference which could be achieved by placing the heat exchanger at an infinite depth [16].

Meydbray et al [17] experimented on the effects of the thermoelectric generator module surface area when exploiting the soil to ambient air temperature difference. They ran three modules with different surface areas for 110 hours with the results shown in Table 1, indicating a strong dependence on the surface area. Additionally, during the course of this experiment, their data also showed a large decrease in power output during times of cloud cover. Lawrence and Synder [18] investigated the performance of heat sinks for this method with theoretical results also showing power outputs in the order of 0.1mW.

Table 1. Power from TE modules of differing surface areas in soil thermal gradient [17]

Set up	Total Energy (mWh)	Average Power (mW)	Area (cm^2)
1	2.5	0.0228	9
2	7.8	0.071	33
3	63.3	0.575	131

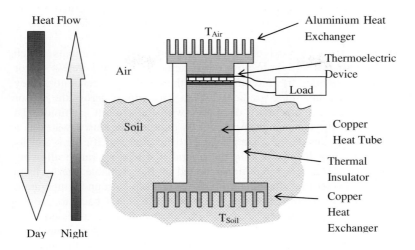

Fig. 3. Simplified diagram of temperature harvesting device [1]

3.3 Water to Ambient Air

Inspired by the concepts of ground to ambient air thermal energy harvesting outlined in section 3.2 coupled with the huge potential for wireless sensor networks in and around marine/aquatic environments, the present authors have investigated powering wireless sensor nodes by exploiting the temperature difference existing across the air/water interface [19][20]. This scheme is examined in the case study in section 4.

3.4 Transport

Wireless sensor networks offer huge capability for applications throughout the transport sector. For example as the accuracy of motor control units in automobiles increases, the need for improved real-time engine data for use in closed loop control arises. This and many other applications across the transport sector possess significant temperature gradients inherent throughout the working environment which opens many opportunities for sensors self powered by thermal energy harvesting.

For the automobile industry's use of sensors in engine bays, waste heat is available from a variety of sources, principally the engine block, exhaust and coolant flow. Bodensohn *et al* [21] report thermoelectric generators supplying autonomous sensor systems with up to 7mW of power utilizing the engine block, typically at temperatures around 450-500K, as the heat source.

Sensors being used for structural health monitoring in aircraft are poised to reduce inspection costs and increase safety. Becker *et al* [11] investigated system development for using thermal energy harvesting in aircraft applications and compared results against other feasible energy harvesters i.e. solar and vibration. Their results showed that expected power densities (W/kg) from thermal energy harvesting in aircraft to be orders of magnitude larger than for other harvesting methods.

Bailey *et al* [22] investigated using the transient thermal gradients which develop during an aircraft's flight to power structural monitoring wireless sensor nodes deployed throughout the aircraft. Using thermoelectric generators they looked to exploit the temperature gradients which may appear across the sensor node as the atmospheric temperature decreases with the aircraft's altitude. To enhance the value of the thermal gradient and its duration they used water as a thermal energy storage inside the generator due to its high specific heat and its fusion point (0°C) being located within the aircraft's operating range allowing high energy storage. To give a rough upper bound on the available energy from this method they consider that changing 1mL of water from a ground temperature of 15°C at take-off, to a cruising temperature of -60°C, then back to ground temperature at landing requires 1.3 kJ of energy. They assessed this experimentally by using a 6mm^2 thin film thermoelectric generator compressed between a 10mL tank of water and a piece of aluminium simulating the aircrafts structure. This module was then placed into an environmental test system which simulated the aircraft's temperature conditions on a typical 1 hour flight for a commercial airliner. They were able to extract 34J of energy which compared to the theoretical upper bound shows a system conversion efficiency of 0.3%.

It is recognized that the solar thermal energy absorbed by black asphalt roads is a large possible source for energy harvesting [23]. A traffic monitoring system such as that proposed by Coleri could benefit from utilising a thermal energy harvesting power system. They acknowledged that in order to extend the life of the sensor network, currently powered by a pair of AA batteries, to acceptable levels, extra relay nodes coupled with energy efficient routing would be needed [24].

3.5 Industrial Waste Heat

There are abundant sources of waste heat littered throughout industry which produce thermal gradients in machinery, process piping or vents etc. Waste heat recovery is a topic of much research and although this source is very site specific it is logical to power sensors developed to monitor such industrial equipment from the rich energy source they are attached to.

Draney [25] presented an autonomous "smart bearing" which is a combined sensor and bearing used for monitoring bearing health in harsh environments (e.g. turbine engines). The sensor provided temperature and vibration data via wireless transmission and scavenged thermal energy from the environment as thermoelectric generators can operate in high temperatures whereas batteries cannot.

Meydbray *et al* [26] investigated exploiting the temperature difference between solid structures and the ambient air. They reported an average power density on the order of 5mW/m^2 from their air-solid structure setup with temperature differentials of ±4K. Although the setup detailed is very simple and no attempt at improving it appears to have been made, the power output indicates that further work may be useful in improving the overall efficiency and power available from this sort of process.

3.6 The Human Body

The human body self regulates its temperature at a constant 37°C. Harnessing this against the ambient air temperature offers a source for thermal harvesting for sensor nodes applied on the human body. Many companies producing body worn products, for example watches, have already developed devices which utilise the small difference between our body heat and the ambient temperature, generating power on the order of microwatts demonstrating that similar techniques can be applied to sensor nodes [27]. In 1998 Seiko released the first wrist watch driven by body heat, the Seiko Thermic. It uses the small thermal gradient between the wearer's arm and the ambient air to generate microwatts of power which run the movement of its mechanical clock.

Thermal energy harvesting is not only applicable on the outer surface of the human body but can also be utilised internally to power implantable medical devices and sensors. Watkins *et al* [8] investigated using their advanced thin film superlattice thermoelectric technology [7] to harness electrical energy from the temperature differences which can exist between the inner surface of the skin and the core body temperature. These temperature differences are typically as low as 0.3 – 1.5 K but they showed that the power levels needed for pacemakers and other similar implantable medical devices are easily achieved with current thermoelectric technology.

These ideas can be extended to cases involving other warm blooded animals. CSIRO recently investigated providing virtual livestock fencing to limit the location of cattle, separate bulls from each other and protect environmentally sensitive areas, used GPS information in a wireless sensor network. This system was powered by solar panels but could equally be powered by thermal devices [28].

4 Case Study: The Use of Environmental Heat

As can be seen in this chapter, energy from the differences in temperature between two objects can be used to generate useable electrical energy. Many devices can be used to generate power from heat transfer such as heat engines and thermoelectric devices. Solid-state thermoelectric devices are reliable and robust as they have no moving parts. The efficiency of such devices has remained constant for around 50 years but recent advances in the composition of the devices have shown efficiency gains approaching twice those previously achieved [7]. This has sparked renewed interest in the use and development of thermoelectric technology.

Modern thermoelectric devices commonly use bismuth telluride (BiTe) doped to n- and p-type semiconductors. These are connected electrically in series and thermally in parallel. For this case study the hot side is provided by connection to a black disk illuminated by the sun, while the cold side is connected to a heat exchanger immersed in water. Water has a high thermal mass and also good convective heat transfer properties, which makes it ideal for use as a heat sink. The water temperature of large bodies of water remains relatively constant during a day, regardless of large air temperature fluctuations. The difference in temperature between the water and ambient air temperature could be used as a natural thermal gradient.

Individual nodes within a network will have power requirements that are dependant on aspects like transmission and reception cycle and data sampling frequency. For example the Fleck ™ series of nodes operates at 3.3V and consumes 30-40mA when transmitting, 15-20mA when receiving data and 1-2mA when idle. With a well designed duty cycle a node will consume much less than 50mW. This power consumption does not include the power required by the sensors or other equipment attached to the node.

Because of this low power requirement thermoelectric devices may be suitable where a temperature difference of a few degrees Kelvin can be found. For example, Ferrari *et al* [5] and Knight and Collins [19] investigated using a thermoelectric generator to provide sufficient power for a wireless sensor node. Ferrari *et al* presents the characterization of three different commercial thermoelectric modules designed for heating/cooling applications. Their analysis included the effects of electrical load resistance, thermal conductivities of the thermoelectric and heat exchanger modules and different temperature gradients. They found that thermoelectrics could be used for their application of powering a wireless sensor node consuming 32mW, when the temperature difference exceeded 30K.

Figure 4 shows the result of some of their work, displaying the maximum power density generated by the three different thermoelectric generators vs. temperature difference.

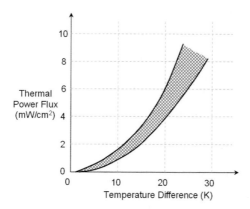

Fig. 4. Power flux generated by example Thermoelectric Generators. [5]

4.1 Experimental Details

4.1.1 Collector Design

The basic design of the module is shown in Figure 5. The design uses a simple 40mm x 40mm square aluminium block to sink the heat into the water on the cold side. The thermoelectric device is sandwiched between a circular disk, which has been painted matt black, and the aluminium heat sink. Heat transfer paste is used to facilitate good heat flow at each of these junctions. A foam collar was placed below the disk and that assembly slid through a hole in a circular acrylic disk. This acrylic disk has a number of fasteners around its periphery which mate with the acrylic dome which covers the device. This dome assembly floats by virtue of a foam raft.

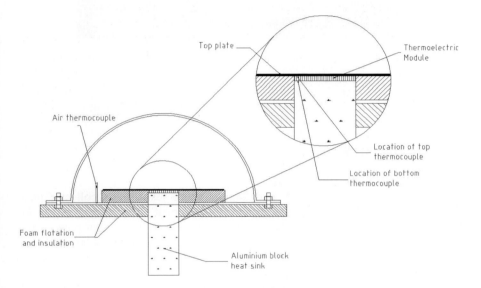

Fig. 5. Collector design [19]

4.2 Experimental Variations

Many aspects of the module design can be varied to examine the effect on output power levels; in this paper we will vary just two features. The first series of tests measured the effect of changing the collector disk size. Small, medium and large disks were used to provide heat to the TEM. The small disk had a diameter of 160mm, the medium disk had a diameter of 200mm and the largest disk had a diameter of 240mm.

The second series of tests varied the number of thermo electric modules, connected in series. One, two and four TEMs were connected in series each with a medium sized collector disk. Although it may appear that connecting four devices in series should quadruple the power output, a secondary effect is introduced: By

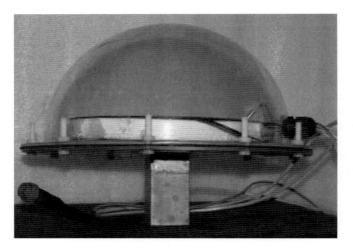

Fig. 6. Experimental module showing the acrylic dome and interior foam

having 4 devices there is four times the conduction area for any heat difference. Thus the heat flow will be altered and the dome with four TEMs will have a lower temperature difference between the hot side and cold side.

4.3 Experimental Results

4.3.1 Variation of Collector Size

Three collector sizes, 160, 200 and 240mm, were tested. The collector plate consists of a round aluminium plate, painted with a matt black paint. The data for this was collected over a three day period in mid winter. The output power from each of the collector areas is shown in Figure 7 along with the solar insolation data. The insolation data was collected from an on-site weather station located approximately 50m away.

The slight dip that occurs in the power output at 12:45 is caused by a shadow cast from a nearby wind turbine tower, which did not shade the Insolation data collection location.

Insolation data is on the secondary axis and shows that the day was cloud free until early afternoon. The solar insolation data shows a peak of more than 635W/m^2 which is a typical mid-winter value for the Newcastle latitude. The power output tracks the insolation data well and shows that the collector with the largest diameter provides the largest output. The peak output from the largest collector is 47mW, the medium size collector has a peak output of 31mW and the smallest collector has an output of 22mW.

The fact that the larger collector has the largest power output should not be particularly surprising. Of more interest is the normalised power output for each size disk. This is achieved by normalising power output against disk area and gives a value of power flux for each disk and an indication of the efficiency of the process. The results for this calculation are shown in Figure 8.

Thermal Energy Harvesting for Wireless Sensor Nodes with Case Studies 235

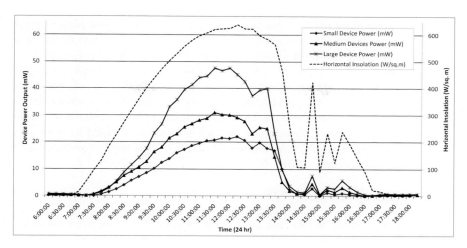

Fig. 7. Power output across three different sized collectors

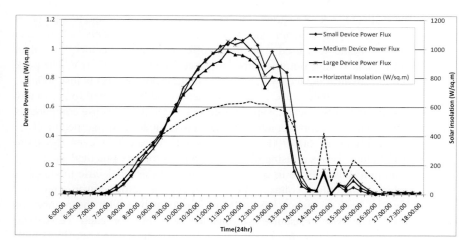

Fig. 8. Solar Insolation and Device Power Flux

This data shows some important results. Firstly the power output from the devices is quite low and peak efficiencies are between 1% and 2% for all devices. Although low compared to solar photovoltaic technology, this is quite high for thermoelectric devices and is driven by the design of the device: Each collector is covered by a plastic dome. This has the effect of increasing the temperature inside the dome like a miniature greenhouse. The domes are made of an acrylic which allows visible light to pass but is opaque to infrared light. The black plates reflect infrared but this is blocked by the dome and increases the temperature inside the dome. The use of the dome has effectively boosted the power output from a mid-winter sun to a typical mid summer sun value.

The next important result is that the small device has the largest power output flux. This indicates that the ratio between collector area and the TEM device area is closer to optimal. The TEM to collector area is 8% for the small collector, 5.1% for the medium collector and 3.5% for the largest collector. Although the largest collector will gather the largest amount of energy from the sun it will not use it optimally and in fact requires a TEM device larger than the one 40 x 40mm device used. In order to get an 8% ratio the TEM device needs to be 60 x 60mm or two to three of the 40 x 40mm devices could be used (two devices would have an area ratio of 7.1% and 3 would have a ratio of 10.6%). This result correlates with previous work published in [20].

4.3.2 Variation of Number of Thermoelectric Modules

Another variable tested in this case study is the number of thermoelectric modules. Although the previous section indicates quite useable levels of power coming out of the domes, the reality is that this power is only practical if it is capable of recharging a battery or being converted to a voltage sufficient to power a wireless device. The CSIRO Fleck ™ requires a voltage range between 3.3 and 8V. Figure 9 shows the output power from the first range of experiments where one thermoelectric module was used with three different sized collectors. This data shows that the peak voltage output for the largest collector is just 530mV. The peak output from the smallest device (which had the highest power flux) is just 360mV. At such small levels a multi-stage voltage doubler would be required to get to any practical battery recharging level. Typically these voltage doublers use components that have potential drops corresponding to that across diodes and microcontrollers. This is in the region of 500-700mV. This would indicate that even the large collector would have issues utilising the harvested power in any practical way.

One solution to this may be to connect a number of thermoelectric modules in series to increase the output voltage. In order to keep the output power conditions comparable, the load resistance has to be increased in line with the number of devices used. The base resistance for this test was 6Ω for each TEM. The described test involved three different devices. The first device had one TEM with a 6Ω load, the second had two TEMs with a 12Ω load, and the third device had four TEMs in series and was loaded with 24Ω.

The results for this series of tests are shown in Figure 10. The solar insolation data indicates that for this particular day the maximum solar resource was 620W/m^2. The results indicate that the voltage output has increased in line with the number of modules connected in series. The peak output voltage from four thermoelectric modules connected in series is 935mV. This would be useable in a standard voltage doubler circuit. The peak voltage out of just one module was 320mV and the device with two modules in series peaked at 550mV. These would not be useful in any practical sense due to predicted voltage requirements of any power conditioning circuitry.

Thermal Energy Harvesting for Wireless Sensor Nodes with Case Studies 237

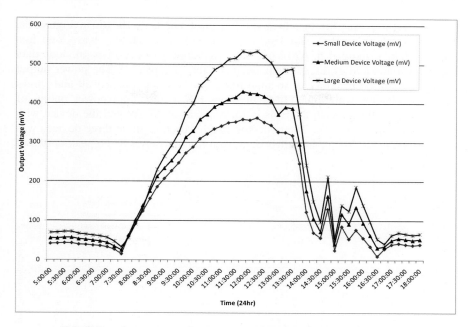

Fig. 9. Output voltage from three different sized collectors and one TEM

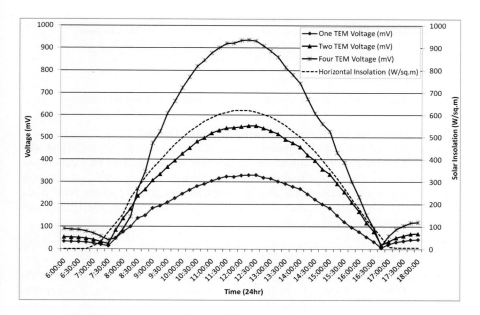

Fig. 10. Output voltage from multiple thermoelectric modules in series

The output power for the three different test points is shown in Figure 11. Insolation data is on the secondary axis and shows that the day was cloud free with a peak of more than 620W/m² which is slightly lower than previous data (635W/m²) and allows direct comparison of other output data. The power output for each variation of collector module tracks the insolation data well and shows that the collector with four TEMs in series has a maximum power output of 36mW. The collector with 2 TEMs in series peaks at 25mW and the device with one TEM peaks at 18mW. The collector size for each of these three devices was the medium disk. The medium disk data from the first series had a peak output of 30mW from one TEM.

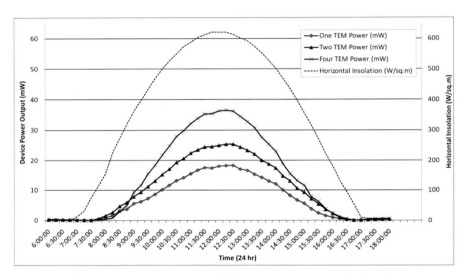

Fig. 11. Output power from multiple thermoelectric modules in series

4.3.3 Comparison of Ideal and Measured Power Output

A comparison of the ideal and measured output for one of the devices is provided in Figure 12. The determination of the ideal power output is based on a calculation using the power factor of the thermoelectric device. This is a relatively simple calculation and the equation for this is shown in equation 1, where KPF is the power factor for the particular TEM, a is the area and ΔT is the temperature difference across the Seebeck device.

$$P = K_{PF} a \Delta T^2 \quad (5)$$

For this particular TEM, the power factor is $0.09\mu W/(mm^2 K^2)$ and the area of the device is $1600mm^2$. Figure 12 indicates that the ideal power output is larger than the measured power output by about three times. This difference is mainly explained by the different load resistance and the experimental temperature. The ideal power output is calculated based on using a matched load, which is one where the load resistance is the same as the internal resistances of the TEG, to

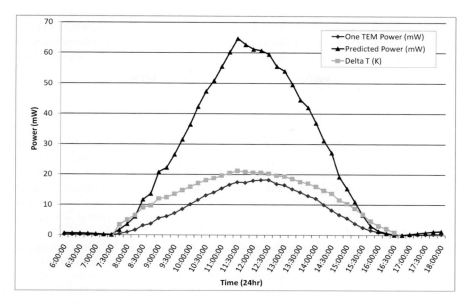

Fig. 12. Experimental and predicted power output from a single TEM

ensure maximum power transfer. For this style of TEG the matched load is approximately 10Ω, with the variation dependent on the actual temperature difference across the device and variations in manufacturing methods and materials. The detailed experiments used a load resistance of 6Ω. In addition the ideal power output is calculated based on much larger operating temperatures, where the hot side is approximately 350K. The hot side for these experiments peaked at 310K and was usually much lower.

A second series of tests was run where the load resistance of the clear dome with heat sink module had been reduced from 100Ω to 9Ω by the addition of a 10Ω resistor in parallel with the 100Ω resistor. This second series of tests was performed at a farm dam at Armidale in northern NSW. As the volume of the dam was small compared to the initial deployment, the water temperature showed a larger variation. Due to space constraints for this paper the results will not be repeated for this second deployment. However, Figure 12 shows that this simple variation in load resistance has a marked effect on the power output. Recall from above that the power output of the three variations of clear dome was similar with a load resistance of 100Ω. When the load resistance of module 3 was dropped to 9Ω the peak power output climbed to 85mW, which is 58% of the ideal output of 146mW at the same conditions. In this case, the output from the two remaining clear domes was approximately 10mW. This is similar to the Newcastle deployment and indicates that the deployments were somewhat similar. These results are in line with [29]. Further work is required in this area to optimise these results.

The power output produced in these preliminary tests is enough to run a range of wireless sensors including the CSIRO wireless node. With improvements in load matching electronics and thermoelectric materials, it will become possible to use similar devices to power sensor nodes over much smaller gradients, even without solar assistance.

5 Conclusion

The continuing decrease in both size and power consumption of personal electronics and wireless nodes has driven a search for power sources that can power these devices indefinitely. While technologies such as solar or wind can be applied to particular locations, they are of little use in some locations, such as mine shafts or under dense tree cover. In these locations a more basic device has been considered. These devices generate power directly from thermal differences in the surrounding area.

The technologies considered for this energy harvesting include thermoelectric devices based on the Seebeck effect, thermionic and thermo-tunnelling devices. The sources of energy for these technologies include solar energy and the temperature difference between ambient air and local soil or water. One excellent resource of relatively high temperature energy is from vehicle exhausts and industrial waste heat. The forced cooling of the Seebeck cold side is also relatively easy to implement in these situations. Finally a reasonable source of heat is that of the body heat of mammals, including humans.

A case study using commercially available Seebeck devices across an air-water interface is discussed in detail. The output from this study peaks around 50mW for mid-winter sun, which is sufficient to power small wireless nodes. The issue of the quality of this power is discussed in terms of producing power at a useable voltage. One way to achieve sufficient quality power is by connecting multiple devices in series.

This report reveals that while present thermal energy harvesting technologies have low efficiencies, there are a number of promising technologies that are increasing in reliability and efficiency via increased ZT values. This, along with continued decreases in power requirements on the portable electronic side, marks thermal energy harvesting as a very promising field of research.

References

[1] Knight, C., Davidson, J., Behrens, S.: Energy Options for Wireless Sensor Nodes. Sensors 8(12), 8037–8066 (2008)
[2] Halliday, D., Resnick, R., Walker, J.: Fundamentals of Physics, 6th edn. John Wiley and Sons, Chichester (2001)
[3] Roundy, S., Wright, P.K., Rabaey, J.M.: Energy Scavenging for Wireless Sensor Networks. Kluwer Academic Publishers, Boston (2003)
[4] Riffat, S.B., Ma, X.: Thermoelectrics: a review of present and potential applications. Applied Thermal Engineering (2003)

[5] Ferrari, M., Ferrari, V., Guizetti, M., Marioli, D., Taroni, A.: Characterization of Thermoelectric Modules for Powering Autonomous Sensors. In: Instrument and Measurement Technology Conference, Warsaw Poland (2007)
[6] Weiling, L., Shantung, T.: Recent Developments of thermoelectric power generation. Chinese Science Bulletin 49(12) (2004)
[7] Venkatasubramanian, R., Siivola, E., Colpitts, T., O'Quinn, B.: Thin-film thermoelectric devices with high room-temperature figures of merit. Nature (2001)
[8] Watkins, C., Shen, B., Venkatasubramanian, R.: Low-grade-heat energy harvesting using superlattice thermoelectric for applications in implantable medical devices and sensors. In: IEEE 24th International Conference on Thermoelectrics (2005)
[9] Poudel, B., Hao, Q., Ma, Y., Lan, Y., Minnich, A., Yu, B., Yan, X., Wang, D., Muto, A., Vashaee, D., Chen, X., Liu, J., Dresselhaus, M.S., Chen, G., Ren, Z.: High-Thermoelectric Performance of Nanostructured Bismuth Antimony Telluride Bulk Alloys. Science (March 2008)
[10] Hishinuma, Y., Geballe, T.H., Moyzhes, B.Y., Kenny, T.W.: Refrigeration by Combined Tunneling and Thermionic Emission in Vacuum: Use of Nanometer Scale Design. Applied Physics Letters 78(17) (2001)
[11] Becker, T., Kluge, M., Schalk, J., Otterpohl, T., Hilleringmann, U.: Power Management for thermal energy harvesting in aircrafts. IEEE Sensors (2008)
[12] Yu, H., Li, Y., Shang, Y., Su, B.: Design and Investigation of Photovoltaic and Thermoelectric Hybrid Power Source for Wireless Sensor Networks. In: Proceedings of the 3rd IEEE Int. Conf. On Nano/Micro Engineered and Molecular Systems, Sanya, China (2008)
[13] Wang, J.: The temperature influence on solar PV generation. Journal of Qinghai Normal University (Natural Science) (1), 28–30 (2005)
[14] Li, Y., Yu, H., Su, B., Shang, Y.: Hybrid Micropower Source for Wireless Sensor Network. IEEE Sensors Journal 8(6) (June 2008)
[15] Stevens, J.: Optimal Design of small ΔT thermoelectric generation systems. Energy Conversion and Management (2001)
[16] Stevens, J.: Optimal placement depth for air–ground heat transfer systems. Applied Thermal Engineering 24 (2004)
[17] Meydbray, Y., Singh, R., Shakouri, A.: Thermoelectric Module Construction for Low Temperature Gradient Power Generation. In: IEEE 24th International Conference on Thermoelectrics (2005)
[18] Lawrence, E.E., Snyder, G.J.: A study of heat sink performance in air and soil for use in a thermoelectric energy harvesting device. In: IEEE Proceedings ICT 2002, Twenty-First International Conference on Thermoelectrics (2002)
[19] Knight, C., Collins, M.: Results of a water based thermoelectric energy harvesting device for powering wireless sensor nodes. In: Active and Passive Smart Structures and Integrated Systems 2009, SPIE, San Diego (2009)
[20] Davidson, J., Collins, M., Behrens, S.: Thermal energy harvesting between the air/water interface for powering wireless sensor nodes. In: SPIE Smart Structures & Materials/NDE 2009, San Diego, California USA, March 8 - 12 (2009)
[21] Bodensohn, A., Falsett, R., Haueis, M., Pulvermüller, M.: Autonomous Sensor Systems for Car Applications. In: Advanced Microsystems for Automotive Applications, Springer, Heidelberg (2004)

[22] Bailey, N., Dilhac, J., Escriba, C., Vanhecke, C., Mauran, N., Bafleur, M.: Energy Scavenging based on transient thermal gradients: Applications to structural health monitoring of aircrafts. In: 8th International Workshop on Micro and Nanotechnology for Power Generation and Energy Conversion Applications (PowerMEMS 2008), Sendai Japan (2008)
[23] Mallick, R., Chen, B., Bhowmick, S.: Harvesting energy from asphalt pavements and reducing the heat island effect. International Journal of Sustainable Engineering 2(3) (2009)
[24] Coleri, S., Cheung, S., Varaiya, P.: Sensor Networks for Monitoring Traffic. In: Allerton Conference on Communication, Control and Computing (2004)
[25] Draney, R.: High Temperature Sensor for Bearing Health Monitoring. In: IEEE Aerospace Conference (2008)
[26] Meydbray, Y., Singh, R., Nguyen, T., Christofferson, J., Shakouri, A.: Feasibility Study of Thermoelectric Power Generation for Stand Alone Outdoor Applications. In: IEEE 23rd International Conference on Thermoelectrics (2004)
[27] Paradiso, J.A., Starner, T.: Energy Scavenging for Mobile and Wireless Electronics. In: Energy Harvesting and Conservation, IEEE (2005)
[28] Wark, T., Swain, D., Crossman, C., Valencia, P., Bishop-Hurley, G., Handcock, R.: Sensor and Actuator Networks: Protecting Environmentally Sensitive Areas. IEEE Pervasive Computing 8(1), 30–36 (2009)
[29] Dalola, S., Ferrari, M., Ferrari, V., Guizzetti, M., Marioli, D., Taroni, A.: Characterization of Thermoelectric Modules for Powering Autonomous Sensors. IEEE Transactions on Instrumentation and Measurement 58(1), 99–107 (2009)
[30] Cox, I., Tavkhelidze, A.: Power Chips for Efficient Energy Conversion. In: El-Genk, M.S. (ed.) Space Technology and Applications International Forum-STAIF (2004)

IEEE 1451.5 Standard-Based Wireless Sensor Networks

Eugene Y. Song and Kang B. Lee

National Institute of Standards and Technology
Gaithersburg, Maryland USA
ysong@nist.gov, kang.lee@nist.gov

Abstract. Recent advances in wireless communications, networking, electronics, and microprocessors have enabled the development of low-cost, low-power wireless sensor nodes. Wireless sensor networks have become more widely used to monitor the condition of the environment, machining systems, and other applications. Sensor data exchange, sharing, and interoperability are major challenges when using wireless sensor networks for condition monitoring. Standardized sensor data formats and communication protocols can help to solve these problems. This paper mainly focuses on wireless sensor networks based on the Institute of Electrical and Electronics Engineers (IEEE) 1451.5 standard, including the architecture, network interfaces, wireless transducer physical interfaces, a prototype system, and case studies.

1 Introduction

Sensors are ubiquitous. They are used in a variety of applications that touch people's daily lives, ranging from smart homes to industrial automation to condition monitoring and control to homeland defense [1]. A sensor generates an electrical signal related to a physical, biological, or chemical parameter. A smart sensor consists of sensing elements, a data processing unit, and a communication interface. Wireless technologies not only facilitate the application of sensors in remote sensing systems, they also help to reduce system installation time and cost by eliminating cabling. Recent advances in wireless communications, networking, electronics, micro-electro-mechanical-system technology, and microprocessors have enabled the development of low-cost, low-power wireless sensor nodes that are small in size and can communicate over short distances [2]. Table 1 shows a list of wireless sensor nodes related to wireless technologies [3]. BTNode[1], IMote, and

---

[1] Commercial equipment and software, many of which are either registered or trademarked, are identified in order to adequately specify certain procedures. In no case does such identification imply recommendation or endorsement by the National Institute of Standards and Technology, nor does it imply that the materials or equipment identified are necessarily the best available for the purpose.

Table 1. Wireless sensor nodes related to wireless technologies

Wireless Sensor Node Product Name	Wireless Protocol	Microcontroller	Programming	Operating system
BTNode	Bluetooth (2.4 GHz)	Atmel ATmega 128L	C	BTnut and TinyOS
IMote	Bluetooth (2.4 GHz)	ARM core 12 MHz	C	TinyOS
Mulle	Bluetooth (2.4 GHz)	Renesas M16C	NesC, C	TinyOS
Iris	802.15.4/ZigBee(2.4 GHz)	ATmega1281	NesC	TinyOS, MoteWorks
Redbee	802.15.4 (2.4 GHz)	MC13224V	C	Contiki, Standalone
SenseNode	802.15.4 (2.4 GHz)	MSP430F1611	C and NesC	GenOS and TinyOS
Ember	802.15.4/ZigBee(2.4 GHz)	ARM Cortex-M3	C	
SunSPOT	802.15.4 (2.4 GHz)	ARM 920T	Java	Squawk J2ME

Mulle are Bluetooth wireless sensor nodes and the others, such as Iris, Ember, and SunSPOT, are the Institute of Electrical and Electronics Engineers (IEEE) 802.15.4-compliant wireless sensor nodes.

A wireless sensor network (WSN) shown in Figure 1 consists of one gateway node and a number of wireless sensor nodes deployed to monitor physical or environmental conditions, such as temperature, sound, vibration, pressure, or motion at different locations. Users can remotely access the sensor data from the sensor nodes of a wireless sensor network through the gateway node. User remote access can be done via wired or wireless means. Wireless sensor networks have become popular in the condition monitoring of the environment, machining systems, and other applications.

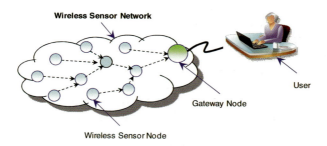

Fig. 1. Wireless sensor network

With the availability of low-cost wireless technologies, sensor manufacturers can use them to build wireless sensors very economically. Thus, various wireless communication protocols have emerged in recent years in the marketplace. Each wireless technology has its own strengths and weaknesses for a specific application. But it is too costly for sensor manufacturers to make unique sensors for each network in the market. Interfacing the sensors to all of these wireless networks and supporting the wide variety of protocols require very significant efforts and are costly to sensor manufacturers [4]. Sensor interfaces, sensor data exchange, and interoperability are major issues. How can these issues be solved? Standardized sensor data formats and protocols can help to enhance sensor interoperability. Standardizing wireless sensor interfaces can solve the problem mentioned above, and is serving to shape the next-generation wireless monitoring landscape [5]. In fact, the IEEE 1451 transducer interface standard was developed specifically to address these issues.

As defined in the IEEE 1451 smart transducer interface standards, a smart transducer provides functions beyond those necessary for generating a correct representation of a sensed or controlled quantity. This functionality typically simplifies the integration of the transducer into applications in a networked environment [6]. Figure 2 shows an IEEE 1451 smart transducer, which consists of a Network Capable Application Processor (NCAP), a network interface between the user network and NCAP, a Transducer Interface Module (TIM), and a transducer physical interface between the NCAP and TIM [1]. An NCAP consists of hardware and software and provides a gateway function between the TIMs and the user network or host processor. The NCAP is an application processor capable of network communications. A TIM is a module that contains the interface, signal conditioning, analog-to-digital and/or digital-to-analog conversion, Transducer Electronic Data Sheets (TEDS), and the transducer(s). The TIM may range in complexity from a single sensor or actuator to units containing many transducers (sensors and actuators). The NCAP communicates with a TIM via a physical interface, either wired or wireless. The IEEE 1451 family of standards defines a set of common communication interfaces for connecting smart transducers to microprocessor-based systems, instruments, or networks in a network-independent environment [7-8]. As shown in Figure 2, three user network interfaces are used to access IEEE 1451 smart transducers: the IEEE 1451.1 communication protocols [9], IEEE 1451.0 Hypertext Transfer Protocol (HTTP), and the proposed Smart Transducer Web Services (STWS) developed at the National Institute of Standards and Technology (NIST) based on the IEEE 1451.0 standard [10, 11]. The STWS allows Internet access of IEEE 1451 transducers via Web Services. The STWS source code is available at the Open1451 project in SourceForge [12]. The IEEE 1451.0 standard defines a common set of functions, commands, TEDS formats, and communication protocols for the IEEE 1451 family of standards. The connectivity between the NCAP and TIM is defined by the IEEE 1451.X physical interfaces, which include the point-to-point interface specified in IEEE Std. 1451.2-1997 [13], the distributed multi-drop interface specified in IEEE Std. 1451.3-2003 [14], and the wireless interface specified in IEEE Std. 1451.5-2007 [15]. Normally analog and digital transducers are connected to a TIM directly. However, IEEE 1451.4 transducers can be connected to a TIM through the IEEE 1451.4 Mix-Mode Interface [16]. Likewise, IEEE 1451.7 transducers can be connected to a TIM through radio frequency (RF) interfaces [17].

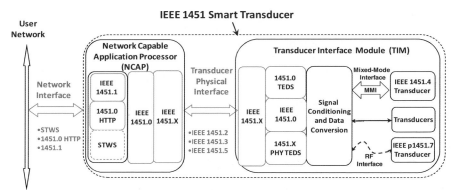

Fig. 2. IEEE 1451 smart transducer

This paper mainly focuses on the discussion on the IEEE 1451.5 standard-based wireless sensor networks. Section 2 describes related work. Section 3 describes the architecture of the IEEE 1451.5 standards-based wireless sensor network. A service-oriented and IEEE 1451.5-802.11 standard-based wireless sensor network, and a few case studies are presented in section 4. The conclusion and summary are given in section 5.

2 Related Work

In this section we overview the existing wireless communication protocols, some of them adopted in the IEEE 1451.5 standard, illustrate the history of the IEEE 1451.5 standard, and then introduce work related to the IEEE 1451.5 standard. Existing wireless sensor communication protocols shown in Table 2 include IEEE 802.11 (WiFi), IEEE 802.15.1 (Bluetooth), IEEE 802.15.4 (ZigBee, 6LoWPAN, WirelessHART, ISA-100.11a), and IEEE 802.16 (WiMAX).

Table 2. Wireless Protocols

Standards	IEEE 802.15.4 (Low Power-Wireless Personal Area Network) (LP-WPAN)				IEEE 802.15.1 (WPAN)	IEEE 802.11 (WLAN)	IEEE 802.16 (WWAN)
	ZigBee	6LoWPAN	WirelessHART	ISA100.11a	Bluetooth	WiFi (802.11a,b,g,n,y)	WiMAX (802.16d,16e)
Range	100 m	50 m			100 m	5 km	15 km
Data rate	250-500 Kbps	250 Kbps	250 Kbps	250 Kbps	1 Kbps – 3Kbps	1 Mbps-450 Mbps	75 Mbps
Frequencies (Bandwidth)	2.4 GHz	2.4 GHz	2.4 GHz	2.4 GHz	2.4 GHz	2.4, 3.7, and 5 GHz	2.3, 2.5, and 3.5 GHz
Network Topology	Star, mesh, cluster tree	Star, mesh, cluster tree IPv6	Star, mesh, cluster tree	Star, mesh, cluster tree	Star	Star, Tree, P2P	Star, Tree, P2P
Applications	Wireless sensors (monitoring and control)	wireless internet, automation and entertainment	Wirelessly process monitoring and control applications	Wireless Systems for Automation	Wireless sensors (monitoring and control)	PC-based Data Acquisition Mobile Internet	Mobile Internet

- IEEE 802.15.1 (Bluetooth)

Bluetooth is a short-range wireless communication technology intended to replace the cables connecting portable and/or fixed devices. The key features of Bluetooth technology are robustness, low power, and low cost [18]. The basic framework of Bluetooth conceives a 1-meter to 100-meter communication area with a transfer rate of 1 Mbps to 3 Mbps at 2.4 GHz. Bluetooth is mainly used for personal area sensor network and short-range portable personal devices. The IEEE 802.15.1-2002 standard is based on the Bluetooth specifications.

- IEEE 802.15.4

The IEEE 802.15.4 standard investigates a low data rate solution with multi-month to multi-year battery life and very low complexity [19]. It specifies the physical layer and media access control for a low-rate wireless personal area network (LR-WPAN). The basic framework of the IEEE 802.15.4 conceives a 10-meter communication area with a transfer rate of 250 Kbps at 2.4 GHz. Its potential applications are sensors, interactive toys, smart badges, remote controls, and home automation. IEEE 802.15.4 devices are characterized by short range, low bit rate, low power, and low cost. This standard is the basis for ZigBee, WirelessHART, 6LoWPAN, and ISA-100, each of which further attempts to offer a complete networking solution by developing the upper communication layer.

- ZigBee

The ZigBee Alliance is an association of companies working together to enable reliable, cost-effective, low-power, wirelessly networked, monitoring and control products based on the IEEE 802.15.4 standard for Wireless Personal Area Networks (WPANs) [20]. ZigBee is intended for use in embedded applications requiring low data rates and low power consumption, such as smart homes, building automation, health care, and industrial plants.

- 6LoWPAN

The Internet Protocol version 6 (IPv6) over Low power Wireless Personal Area Networks (6LoWPAN) allows IPv6 packets to be sent and received over IEEE 802.15.4 based networks through encapsulation and header compression mechanisms [21]. 6LoWPAN brings IP to the smallest of devices, such as sensors and controllers. The 6LoWPAN protocol is targeted at wireless IP networking applications in the home, office, and factory environments.

- WirelessHART

WirelessHART is a wireless networking technology developed by the Highway Addressable Remote Transducer (HART) Communication Foundation. The protocol utilizes a time synchronized, self-organizing, and self-healing mesh architecture. WirelessHART enables users to quickly and easily gain the benefits of wireless technology while maintaining compatibility with existing devices, tools, and systems. The WirelessHART technology aims to provide a robust wireless protocol for the full range of process measurement, control, and asset management applications [22].

- ISA-100.11a

ISA-100.11a-2009 is a standard for wireless systems for industrial automation. This standard is intended for process control and related applications to provide reliable and secure wireless operations for non-critical monitoring, alerting, supervisory control, open-loop control, and closed-loop control applications. This standard defines a protocol suite, system management, gateway, and security specifications for low-data-rate wireless connectivity with fixed, portable, and moving devices supporting very limited power consumption requirements [23]. The standard also addresses coexistence with other wireless devices anticipated in the industrial workspace, such as cell phones and devices based on IEEE 802.11x, IEEE 802.15.X, IEEE 802.16.X, and other relevant standards.

- IEEE 802.11 (WiFi)

IEEE 802.11 is a set of standards for Wireless Local Area Network (WLAN) communications in the 2.4 GHz, 3.6 GHz, and 5 GHz frequency bands [24]. The Wi-Fi name is a trademark of the Wi-Fi Alliance for certified products based on the IEEE 802.11 standards. There are a variety of protocols currently in use for wireless networking. The most prevalent is 802.11b (2.4 GHz at 11 Mbps). The newer 802.11g (2.4 GHz at 54 Mbps) standard improves on 802.11b. The 802.11a is also capable of transmission speeds up to 54 Mbps, like the 802.11g standard, but in a different frequency (5 GHz) range. IEEE 802.11n uses multiple input, multiple output antennas for higher throughput improvements (2.4 GHz or 5 GHz at 50 Mbps to 600 Mbps). IEEE 802.11 is widely used for residential, business, and industrial applications for general LAN access in work areas.

- IEEE 802.16 (WiMAX)

IEEE 802.16 is a series of wireless broadband standards for the Metropolitan Area Network (MAN) [25]. The standard provides a transfer rate of up to 100 Mbps within 10 GHz to 66 GHz. The Worldwide Interoperability for Microwave Access (WiMAX) system is a wireless communication system that allows computers and workstations to connect to high-speed data networks (such as the Internet) using radio waves as the transmission medium with data transmission rates that can exceed 120 Mbps for each radio channel [26]. IEEE 802.16 is used for wide area network or Metropolitan Area Networks (MANs) applications.

Many of the standard wireless communication protocols mentioned above are not specifically designed for sensor applications, even though they have been used for wireless sensor networks. Each wireless sensor network defines its own specific sensor data formats. The sensor data cannot be exchanged or shared among different wireless sensor networks. Factory automation applications require a common interface for sensors and other devices to collect process data from the shop floor, a standard network protocol to communicate and exchange information for system control and diagnosis, and an open system architecture to facilitate the integration of components from variety of sources [27]. How can sensor data be shared or exchanged among different wireless sensor networks? There are several approaches that contribute to making these systems more interoperable.

Standardizing sensor interfaces and sensor data formats helps to facilitate sensor device and sensor data interoperability and thus shorten the development cycle and reduce the costs of wireless sensor network deployment. The IEEE 1451 family of standards defines a set of common commands and functions to access sensors in a wireless sensor network. These standards provide a useful framework for these types of sensor applications [28-29].

On June 4, 2001, the first Wireless Sensing Workshop was held at the Sensors Expo in Chicago, IL, to determine the interest and requirements for wireless interfaces for sensor-based networks. In this workshop, the various wireless technologies, such as the IEEE 802.11x and Bluetooth, were presented and discussed. The medium access and physical layer definitions from the IEEE 802 family address some of the needs of a wireless IEEE 1451 network. In addition, the general structure of the IEEE 802 family can also be used as a basis for the structure of the IEEE 1451 family. In order to greatly reduce the time to develop the IEEE 1451 wireless standard, the IEEE 1451 wireless standard adopted parts of these wireless standards [29]. The second workshop was held on October 4, 2001, in Philadelphia, PA, and examined alternative wireless communication technologies for sensors. The workshop focused on wireless features necessary for inclusion in the IEEE 1451 wireless sensor interface standard [29]. An IEEE 1451.5 study group was formed to further explore this idea, along with wireless sensor interface requirements at the 2002 Sensors Expo/Conference held on September 23-26 in Boston, MA. Later an IEEE 1451.5 working group was formed to develop a wireless sensor standard.

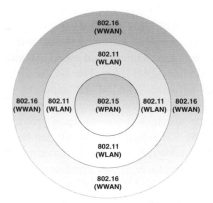

Fig. 3. Relative coverage area of wireless networks

Wireless networking technologies are targeted at PANs, LANs, WANs, and even larger networks. Therefore, the IEEE 1451.5 wireless sensor network standards should cover the wireless networks mentioned above. Figure 3 shows the relative coverage area of wireless networks. IEEE 802.15 supports a Wireless Personal Area network (WPAN) with a range of 10 m to 100 m. IEEE 802.11 supports Wireless Local Area Network (WLAN) with a range of 20 m to 5 km. IEEE 802.16 supports

Wireless Wide Area Network (WWAN) with a range of 5 km to 15 km. As of today, IEEE 1451.5 has adopted WPAN and WLAN. In the future, it could also adopt WWAN. The first draft of the IEEE 1451.5 standard was proposed at the 2004 Sensors Expo/Conference. After much effort during three workshops, the IEEE 1451.5 standard was finally developed and published in 2007. This standard was developed to be compatible with the IEEE 1451.0 standard. It also adopts multiple wireless communication protocols, including the IEEE 802.11(WiFi), Bluetooth, ZigBee, and 6LoWPAN. A modular approach to the IEEE 1451.5 wireless sensor development based on Bluetooth is provided in the reference [30-31]. An implementation of IEEE 1451.0 and 1451.5 based on IEEE 802.11 is provided in the reference [32]. A wireless environmental monitoring system based on the IEEE 1451.1 and 802.11 standards is described in the reference [2]. A Smart Transducer Web Service, as a unified Web service for IEEE 1451 smart transducers, is proposed and described in the reference [10, 11]. Integration of IEEE 1451 smart transducers with the Open Geospatial Consortium – Sensor Web Enablement (OGC-SWE) using STWS is described in the reference [33].

3 IEEE 1451.5 Standard-Based Wireless Sensor Networks

3.1 Architecture of IEEE 1451.5 Standard-Based Wireless Sensor Networks

Figure 4 depicts an architecture of IEEE 1451.5 standard-based wireless sensor networks. Each oval shows one of the IEEE 1451.5 standard-based wireless sensor networks. Figure 4(a) shows a view of an IEEE 1451.5-802.11 wireless sensor network, whereas Figures 4(b), 4(c), and 4(d) show the IEEE 1451.5-Bluetooth, -ZigBee, and -6LoWPAN wireless sensor network, respectively. Each wireless sensor network shown in Figure 4 consists of an NCAP and a number of Wireless Transducer Interface Modules (WTIMs). The NCAP is a gateway between the WTIMs and the user network. The NCAP supports the required functional specifications, the message structures, the required commands of the IEEE 1451.0 standard, and one of the wireless communication protocols and physical media defined by the IEEE 1451.5 standard. The NCAP can be connected or associated with a number of WTIMs using the same wireless medium. The WTIM is a module that contains a wireless interface, signal conditioning, analog-to-digital and/or digital-to-analog conversion capability, TEDS, and transducers (sensors and actuators). It also supports the required functional specifications, message structures, commands, TEDS format of the IEEE 1451.0 standard, and one of the communication protocols and physical media defined by the IEEE 1451.5 standard. Thus the NCAP can communicate with each WTIM via one of the four wireless protocols using IEEE 1451.0 messages and it may also communicate with a host via an external network through one of the IEEE 1451 network interfaces.

The IEEE 1451.0 standard defines a set of common functions and commands for the entire family of IEEE 1451.X smart transducer standards. It operates under either a client-server or request-response protocol, which is independent of the physical

IEEE 1451.5 Standard-Based Wireless Sensor Networks

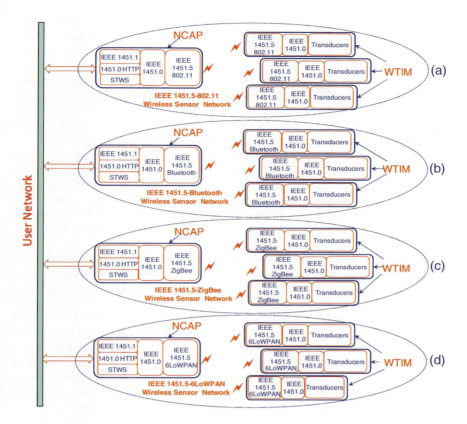

Fig. 4. Architecture of IEEE 1451.5 Standard-based Wireless Sensor Networks

communication media. Basic functions include reading from and writing to transducer channels, reading and writing TEDS, and sending configuration, control, and operation commands. The IEEE 1451.0 was designed to facilitate interoperability and compatibility among the family of IEEE 1451 standards when multiple wired and wireless sensor networks are connected to form a system of networks.

The communications between the NCAP and WTIM are based on IEEE 1451.0 messages through one of the IEEE 1451.5 media. In this architecture, two kinds of interfaces exist for the IEEE 1451.5 wireless sensor networks, 1) network interface and 2) wireless physical interface. The network interface is an interface between the NCAP and user network. The wireless physical interface is an interface between the NCAP and WTIMs.

3.1.1 Network Interface

The IEEE 1451.5 wireless sensor networks have a network interface for an NCAP to communicate with high-level sensor applications through a user network, such as Ethernet/Internet or any control network. The network interface can be based on the IEEE 1451.1-1997 standard, IEEE 1451.0 HTTP protocol, and the proposed Smart

Transducer Web Service. Choosing a network interface for an IEEE 1451.5 wireless sensor network highly depends on the sensor applications as discussed below.

3.1.1.1 IEEE 1451.1

IEEE 1451.1 is a standard for a Smart Transducer Interface for Sensors and Actuators -Network Capable Application Processor Information Model [9]. It defines a common object model and interface specification for the components of a networked smart transducer. The IEEE 1451.1 standard specifies a software architecture, which includes three models: data model, object model, and communication model.

The data model specifies the type and form of information communicated across the IEEE 1451.1 object interfaces for both local and remote communications. The IEEE 1451.1 data types can be classified into primitive data type and structured data type (or derived data type). The primitive data type includes Boolean, Integers (8, 16, 32, 64 bits and unsigned), Floats (32, 64 bits), and Octet (an 8 bit unsigned char). The structured or derived data types include data types such as Arrays, Structs, Unions, and Enumerations.

The object model specifies the software component types used to design and implement application systems. The object model provides software building blocks for the application systems. Figure 5 shows the hierarchical class diagram of IEEE 1451.1 Objects [8]. The top class is IEEE1451_Root that shall be the root for the class hierarchy of all IEEE 1451.1 objects. The subclass of IEEE1451_Root class is IEEE1451_Entity that shall be the root for the hierarchy of all objects that may be visible over the network. The IEEE 1451.1 application system, such as a remote monitoring system or a measurement and control system, consists of application objects, which inherit from the IEEE 1451.1 standard object classes. Application developers can use aggregation and composition relations to combine the application object classes to construct and design their specific applications.

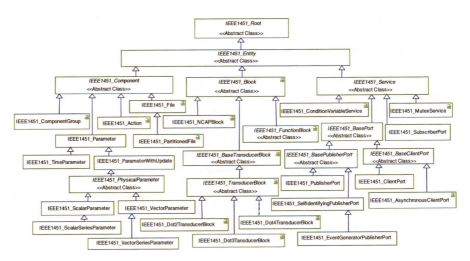

Fig. 5. Hierarchical class diagram of IEEE 1451.1 Objects

IEEE 1451.5 Standard-Based Wireless Sensor Networks

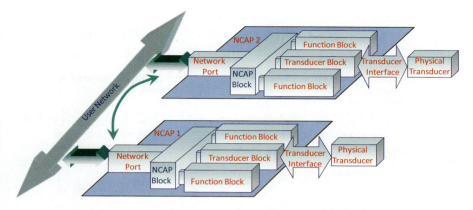

Fig. 6. Communication between two IEEE 1451.1 NCAPs

As shown in Figure 6, each NCAP consists of a NCAP Block, Function Block, Transducer Block, Network Port, and transducer interface. The IEEE 1451.1 standard mainly focuses on the communication among NCAPs through a user network [2]. The IEEE 1451.1 communication model including the client-server model and publish-subscribe model defines the syntax and the semantics of the software interfaces between a communication network and the application.

The client-server model is a tightly coupled, point-to-point communication model. The client-server communication is a request-response process. A client initializes a communication or sends a request to a server, which waits passively for and then responds to the client. The client-server communication model of IEEE 1451.1 is supported by two complementary application-level operations: Execute() that operates on client-side ClientPort Objects, and Perform() that operates on all network visible, server-side objects. Figure 7 shows the client-server communication between client object and server object through IEEE1451_ClientPort. They work together to provide a remote-object-operation-invocation message service.

The IEEE 1451.1 publish-subscribe model is a loosely coupled communication model for one-to-many and many-to-many communications. The publish-subscribe communication is a communication pattern where one or more publisher objects communicate information on a specific topic to one or more subscriber objects interested in that topic without the necessity of any of these objects knowing the identity of any other objects. In an IEEE 1451.1 system, the pattern is established via the Publication Topic, Publication Key, and Publication Domain of the publication. The publish-subscribe communication model shown in Figure 8 is supported by two operations: Publish() on PublisherPort objects, and AddSubscriber() with an associated callback operation on the SubscriberPort objects. Each IEEE 1451.1 publication carries with a publication domain, a publication key, and a publication topic. A SubscriberPort object has the attributes: subscription domain, subscription key, and subscription qualifier. The combination of domains, keys, and topics/qualifiers allows only those publications of interest to a subscriber object to be selected from publications received by a subscriber port.

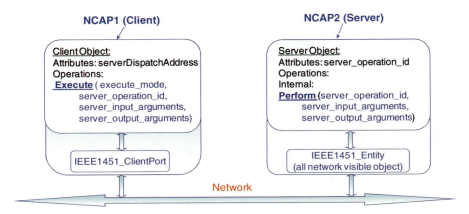

Fig. 7. Client-Server communication model

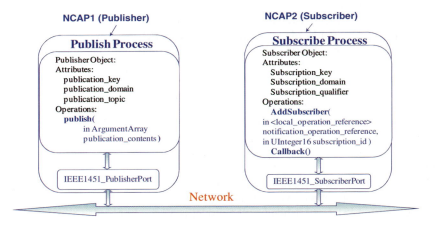

Fig. 8. Publish-Subscribe communication model

The IEEE 1451.1 is suitable for a wide range of application prospects, such as remote monitoring and distributed measurement and control (DMC) applications.

3.1.1.2 IEEE 1451.0 HTTP Protocol

The IEEE 1451.0 HTTP protocol focuses mainly on Web access of transducer data and TEDS through the HTTP 1.1 protocol [6]. Users or clients can send a HTTP request to the HTTP server of the NCAP. The server processes the request and then returns a response to the client. The response can be in eXtensible Markup Language (XML), Hyper Text Markup Language (HTML), or text format. Thus sensors and actuators can be remotely accessed anywhere using the IEEE 1451.0 HTTP Protocol. The request – response process shown in Figure 9 can be described as follows:

- A HTTP client sends a HTTP request to the HTTP server on the NCAP.
- The HTTP server on the NCAP receives a HTTP request, processes it, and then calls the corresponding IEEE 1451.0 transducer service.
- The IEEE 1451.0 transducer service calls on the IEEE 1451.X communication module to communicate with the IEEE 1451.X communication module in TIM and get the results from the TIM.
- The HTTP server on the NCAP gets the results from the IEEE 1451.0 transducer service, and then returns the HTTP response to the client.

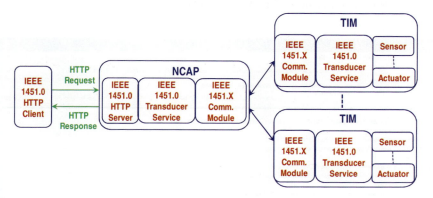

Fig. 9. IEEE 1451.0 HTTP Protocol

3.1.1.3 Smart Transducer Web Service

Smart Transducer Web Service (STWS) consists of a set of web services for IEEE 1451 smart transducers, such as TIMDiscovery, TransducerDiscovery, ReadTransducerData, ReadGeolocationTEDS, and ReadTIMMetaIDTEDS services. The Web Services Description Language (WSDL) is an XML format for describing network services as a set of endpoints operating on messages. The operations and messages are abstractly described and then bound to a concrete network protocol and message format to define an endpoint.

Figure 10 shows the STWS deployed, which is defined in WSDL based on the IEEE 1451.0 transducer services. The WSDL specification of the STWS is divided into six major elements: definitions, types, messages, portType, binding, and service [10-11]. The definition element defines the name of the Web service, declares multiple namespaces used, and contains all of the service elements. The type element describes all the data types used in the communications between the service provider and the consumer. All data types of the STWS include the simple and complex data types, TEDS of the IEEE 1451.0 standard, and the types used in the request and response messages. The message element defines the name of the message and contains one or more message part elements, which can be referred to in message parameters or message return values. The portType element describes a named set of abstract operations and the abstract messages involved. The portType of the STWS defines multiple operations or services, which are based on IEEE 1451.0 transducer services. The binding element describes how the services can be implemented on the network. The WSDL specifies the style of the binding as either Remote Procedure Call or Document. For the Document, the content of <soap:body> is specified by the XML schema defined in the WSDL-type section. It does not need to follow specific Simple Object Access Protocol (SOAP) conventions. We use the SOAP Document/Literal style. The service element defines the address to invoke the specified service. Most commonly, this includes a Uniform Resource Locator to invoke the SOAP service.

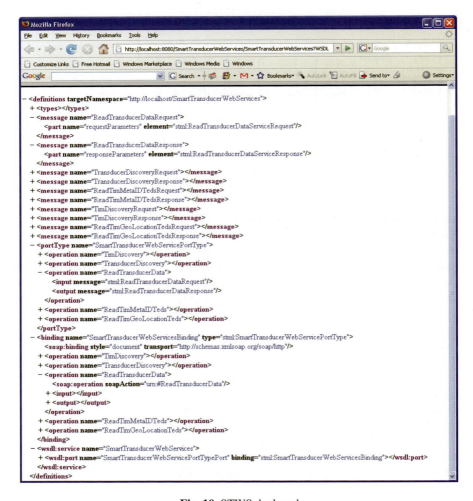

Fig. 10. STWS deployed

IEEE 1451.5 Standard-Based Wireless Sensor Networks

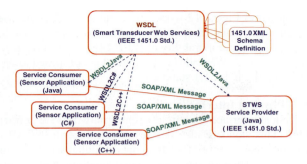

Fig.11. WSDL-based STWS Interoperability

Figure 11 shows an example of WSDL-based STWS interoperability. The STWS service provider can be generated from the WSDL file using web service development tools. The web services of the STWS can be implemented, then built and deployed. On the other hand, the STWS web service consumers can also be generated from the web reference of the STWS. The STWS consumers can be bounded to the STWS provider with SOAP, and invoke the web services of the STWS through a SOAP/XML message. The STWS service consumers implemented in Java, C#, or C++ can be interoperable with the STWS service provider implemented in Java. The STWS consumer could be any sensor application, such as a sensor alert or machine monitoring for condition-based maintenance. Therefore, these sensor applications can interoperate with the IEEE 1451 smart transducers through the STWS based on the WSDL of STWS. The STWS allows access to IEEE 1451 smart transducers through SOAP/XML messages over the Internet.

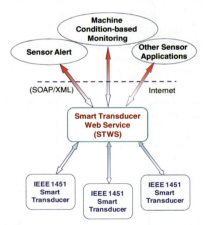

Fig. 12. STWS – a bridge between IEEE 1451 smart transducers and sensor applications

The STWS shown in Figure 12 works as a bridge or integrator between IEEE 1451 smart transducers and the sensor application. The STWS provides a standardized method for interoperability of IEEE 1451 smart transducers with other sensor applications. The STWS and HTTP protocol are proposed to be included in a future version of the IEEE 1451.1 standard.

3.1.2 Wireless Physical Interfaces

The wireless physical interfaces of wireless sensor networks are defined in the IEEE 1451.5 standard. This standard also defines a transducer-to-NCAP interface and TEDS for wireless transducers [15]. It adopts popular wireless communication protocols. The radio range of each associated medium highly depends on the antenna design and transmitter gain. The IEEE 1451.5 wireless media transfer IEEE 1451.0 messages in an octet array format. Thus sensor information from the IEEE 1451.0 system is bundled together into a payload encoded as an octet array. The IEEE 1451.5 layer transports the payload without needing to parse it. It is expected to package the octet array into a series of appropriate IEEE 1451.5 wireless protocol packets. Also, the IEEE 1451.5 layer is responsible for segmenting the octet array into appropriately sized network packets and reassembling them back into the octet array. Similarly, all issues with encryption, authentication, and compression are the responsibilities of IEEE 1451.5 and they are handled by the IEEE 1451.5 wireless network layer.

3.1.2.1 IEEE 1451.5-802.11

The IEEE 1451.5-802.11 wireless interface adapts IEEE 802.11 a/b/g radios and is compatible with IEEE 802.11e and IEEE 802.11i security constructs. The IEEE 1451.5-802.11 stack shown in Figure 13 is based on the Open System Interconnection (OSI) model. The IEEE 1451.5-802.11 uses the IEEE 802.11 Physical (PHY) and Medium Access Control (MAC). Its network layer supports Internet Protocol (IP). The transport layer uses either Transmission Control Protocol (TCP) or User Datagram Protocol (UDP). The application layer consists of IEEE 1451.5-802.11, IEEE 1451.0, and either NCAP application or WTIM application. The NCAP can communicate with the WTIM using IEEE 1451.0 messages through IEEE 802.11 protocol.

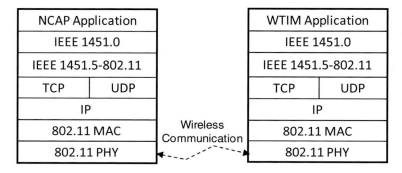

Fig. 13. IEEE 1451.5-802.11 Stack

3.1.2.2 IEEE 1451.5-Bluetooth

The IEEE 1451.5-Bluetooth interface specifies an IEEE 1451.5-Bluetooth profile acting as an application layer on top of a Bluetooth stack. The IEEE 1451.5-Bluetooth stack shown in Figure 14 conforms to the OSI model. The IEEE 1451.5-Bluetooth supports the Link Controller (PHY layer) and Link Manager (MAC layer). The network layer uses either the Logical Link Control and Adaptation Layer Protocol (L2CAP) or Service Discovery Protocol (SDP). The L2CAP provides connection-oriented and connectionless data services. The SDP discovers peer devices. The transport layer supports either the Bluetooth Network Encapsulation Protocol (BNEP) or Serial Link Internet Protocol (SLIP) over radio frequency communication (RFCOMM). The BNEP provides networking capabilities for Bluetooth devices over L2CAP. The SLIP translates between a byte stream and a packet stream. The RFCOMM provides emulation of serial ports over L2CAP as a byte stream. The application layer consists of IEEE 1451.5-BlueTooth, IEEE 1451.0, and either an NCAP application or WTIM application. The NCAP can communicate with the WTIM through the Bluetooth protocol with IEEE 1451.0 messages.

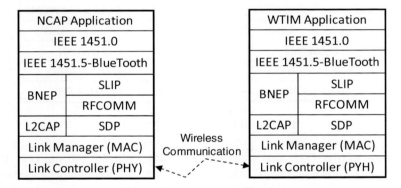

Fig. 14. IEEE 1451.5-BlueTooth Stack

3.1.2.3 IEEE 1451.5-ZigBee

The IEEE 1451.5-ZigBee interface enables wireless communication via the ZigBee network protocol between the NCAP and WTIM. The IEEE 1451.5-ZigBee stack shown in Figure 15 is based on the OSI model. The IEEE 1451.5-ZigBee standard supports the PHY and MAC of IEEE 802.15.4-2003. The ZigBee protocol provides the Network Layer (NWK) and Application Support and ZigBee Device Object (ZDO). The application layer consists of IEEE 1451.5-ZigBee, IEEE 1451.0, and either the NCAP application or WTIM application. The NCAP can communicate with the WTIM through the ZigBee protocol using IEEE 1451.0 messages.

3.1.2.4 IEEE 1451.5-6LoWPAN

This IEEE 1451.5-6LoWPAN interface specifies the functions, protocols, and interfaces required by the 6LoWPAN communication module interconnecting the WTIM and NCAP. The IEEE 1451.5-6LoWPAN stack shown in Figure 16 is

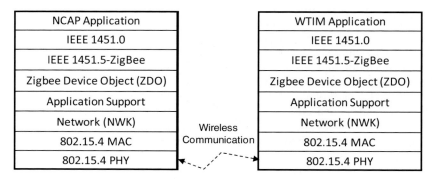

Fig. 15. IEEE 1451.5-ZigBee Stack

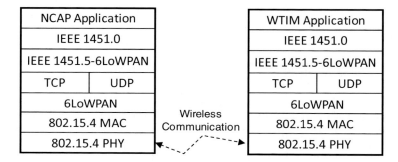

Fig. 16. IEEE 1451.5-6LoWPAN Stack

according to the OSI model. The IEEE 1451.5-6LoWPAN supports IEEE 802.15.4-2003 PHY and MAC. The network layer uses the Internet Engineering Task Force (IETF) 6LoWPAN protocol. The transport layer supports either TCP or UDP. The application layer consists of IEEE 1451.5-6LoWPAN, IEEE 1451.0, and either the NCAP application or WTIM application. The NCAP can communicate with the WTIM through the 6LoWPAN protocol using IEEE 1451.0 messages.

3.2 IEEE 1451.5 Standard-Based Wireless Sensor Networks

3.2.1 IEEE 1451.5-802.11 Wireless Sensor Network

Figure 17 shows an IEEE 1451.5-802.11 wireless sensor network, which supports a star network configuration. The NCAP, a gateway, is a device that contains one IEEE 802.11 wireless radio and can wirelessly communicate with one or more IEEE 802.11 WTIM. Each WTIM contains one IEEE 802.11 wireless radio, signal conditioning, required TEDS, analog-to-digital and/or digital-to-analog conversion unit, and the transducers. The NCAP can wirelessly communicate with each WTIM using an IEEE 802.11 wireless protocol and may also be connected to an external network. One NCAP can communicate with a number of WTIMs.

IEEE 1451.5 Standard-Based Wireless Sensor Networks 261

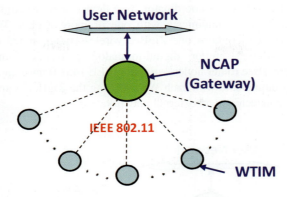

Fig. 17. IEEE 1451.5-802.11 WSN

3.2.2 IEEE 1451.5-Bluetooth Wireless Sensor Network

A Bluetooth wireless network uses a master-slave network architecture that consists of one master and one or more (up to seven) active slaves. Figure 18 shows an IEEE 1451.5-Bluetooth wireless sensor network. The NCAP, a master, is a device that contains one Bluetooth wireless radio and can communicate with one or more Bluetooth WTIMs. Each WTIM, a slave, contains one Bluetooth wireless radio, signal conditioning, TEDS, analog-to-digital and/or digital-to-analog conversion unit, and the transducers. The NCAP can wirelessly communicate with each WTIM using a Bluetooth wireless protocol, and may also be connected to an external network.

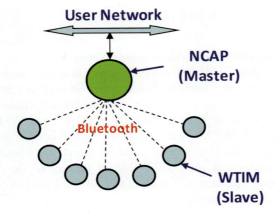

Fig.18. IEEE 1451.5-Bluetooth WSN

3.2.3 IEEE 1451.5-ZigBee Wireless Sensor Network

Figure 19 shows an IEEE 1451.5-ZigBee wireless sensor network, which supports a mesh network. The NCAP, a coordinator, is a device that contains one ZigBee wireless radio and can communicate with one or more ZigBee WTIMs. Each WTIM, an end device, contains one ZigBee wireless radio, signal conditioning, TEDS, analog-to-digital and/or digital-to-analog conversion unit, and the transducers. The ZigBee router nodes can route IEEE 1451.0 messages. The NCAP can wirelessly communicate with each WTIM using the ZigBee wireless protocol and may also be connected to an external network.

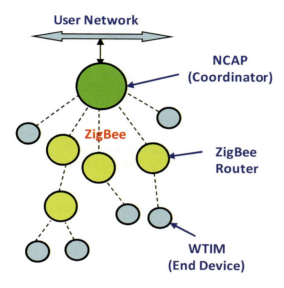

Fig. 19. IEEE 1451.5-ZigBee WSN

3.2.4 IEEE 1451.5-6LoWPAN Wireless Sensor Network

Figure 20 shows an IEEE 1451.5-6LoWPAN wireless sensor network, which supports a mesh network and Internet Protocol version 6 (IPv6) network. The NCAP, a coordinator, is a device that contains one 6LoWPAN wireless radio and can communicate with one or more 6LoWPAN WTIMs. Each WTIM, an end device, contains one 6LoWPAN wireless radio, signal conditioning, TEDS, analog-to-digital and/or digital-to-analog conversion and the transducers. The 6LoWPAN Router nodes can route IEEE 1451.0 messages. The NCAP can wirelessly communicate with each WTIM using a 6LoWPAN wireless protocol and may also be connected to an external network.

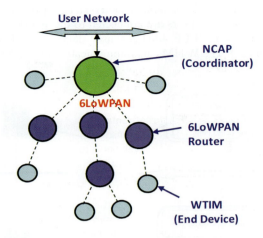

Fig. 20. IEEE 1451.5-6LoWPAN WSN

4 Service-Oriented and IEEE 1451.5-802.11 Standard-Based Wireless Sensor Network

4.1 Service-Oriented and IEEE 1451.5-802.11 Standard-Based Wireless Sensor Network

Figure 21(a) shows a service-oriented and IEEE 1451.5-802.11 standard-based wireless sensor network. The network interface uses the proposed STWS, and the wireless physical interface uses IEEE 802.11. Figure 21(b) shows a prototype IEEE 1451.5-802.11 wireless sensor network, which consists of a sensor application (STWS service consumer), an IEEE 1451 NCAP (STWS service provider), and an IEEE 1451.5-802.11 WTIM (wireless sensor node).

- Sensor Application (STWS consumer)

The sensor application is a STWS consumer including a STWS client running on a laptop. The service consumer can find the web services of the STWS, and then invoke these web services on the NCAP using Simple Object Access Protocol (SOAP)/XML messages through the STWS client.

- IEEE 1451 NCAP (STWS provider)

An IEEE 1451 NCAP, a STWS provider, consists of the STWS, IEEE 1451.0, and IEEE 802.11 communication module. The NCAP can communicate with a sensor application through the STWS interface. It can also communicate with a WTIM through the IEEE 802.11 interface.

- WTIM (wireless sensor module)

An IEEE 1451.5-802.11 WTIM is implemented on a laptop computer with an IEEE 802.11 radio. IEEE 1451.2 sensors are connected to the WTIM through a RS-232 serial interface. Accessing IEEE 1451.2 sensor data and TEDS is through the RS232 connection [23]. The WTIM can communicate with the NCAP through the IEEE 802.11 interface.

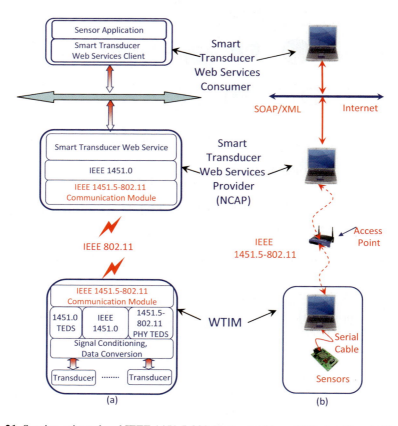

Fig. 21. Service-oriented and IEEE 1451.5-802.11 standard-based Wireless Sensor Network

The STWS consumer and provider can communicate with each other using SOAP/XML messages. The client-server and publish-subscribe communications between the NCAP and WTIM have been implemented using TCP/IP and TCP/UDP, respectively. In Figure 22, the communication processes among the sensor application, NCAP, and WTIM can be described as follows:

- The sensor application, through a STWS client of the STWS consumer, sends a request to the NCAP (service provider) using SOAP/XML messages.

- In the STWS provider (NCAP), the NCAP receives a request from the sensor application through the STWS, processes it, and then calls the corresponding transducer service of IEEE 1451.0. The transducer service in turns calls the IEEE 1451.5-802.11 communication module on the NCAP.
- Then the communication module in the NCAP communicates the request to the corresponding communication module in the WTIM through the IEEE 802.11 wireless interface.
- In the WTIM, the IEEE 1451.5-802.11 communication module obtains a response from the request, for example, sensor reading from the TIM 1451.0.
- This sensor reading is sent to the sensor application through a series of reverse processes.

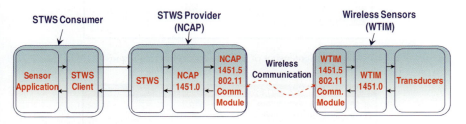

Fig. 22. Communications among sensor application, NCAP, and WTIM

4.2 Case Studies

A few case studies, such as TIMDiscovery, TransducerDiscovery, ReadTIMMetaIDTEDS, and ReadTransducerData services, are discussed in detail in the following.

4.2.1 TIM Discovery

When a TIM is connected to a network or system, it performs a TIM announcement to an NCAP. After the TIM announcement, the NCAP performs a TIM discovery. Through this discovery operation, the NCAP finds out all the TIMs connected to it, and obtains the TIM identity number (timIds) of all the TIMs. Figure 23 shows the screenshot of a TIM discovery service request by the NCAP with ncapId (1). When the parameter ncapId is input, and this request is submitted, a response of timId (11) is shown in Figure 24. Thus, through the TIM discovery operation, one TIM with an identification number (11) is found to be connected to the NCAP with identification number (1).

4.2.2 Transducer Discovery

After the discovery of TIMs, a transducer discovery service can be invoked to obtain all the transducers connected to the TIM and the transducer channel names of the specified TIM. Figure 25 shows the screenshot of the transducer discovery service request, while Figure 26 shows the response. The result shows that the TIM with a timId (11) has two transducer channels. The first transducer channel is a hall-effect sensor with a transducerIds (1) and the second transducer channel is an illumination sensor with a transducerIds (2).

Fig. 23. Screenshot of TIM discovery service request

Fig. 24. Screenshot of TIM discovery service response

Fig. 25. Screenshot of transducer discovery service request

Fig. 26. Screenshot of transducer discovery service response

4.2.3 Reading TIM MetaID TEDS

Figure 27 shows a user input interface of reading TIM MetaID TEDS service request. The user inputs the parameters of the reading TIM MetaID TEDS service request, submits it, and then obtains a response from the service provider. Figure 28 shows the result of reading MetaID TEDS of a TIM with a timId (11). The MetaID TEDS consists of the manufacturer ID, model number, version code, serial number, date code, number of channels, group name, and production description.

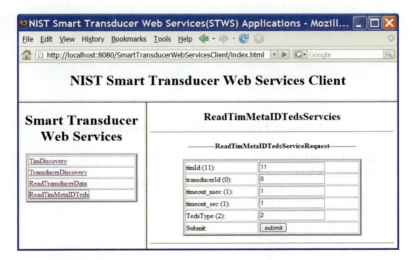

Fig. 27. Screenshot of reading TIM MetaID TEDS service request

Fig. 28. Screenshot of reading TIM MetaID TEDS service response

4.2.4 Reading Transducer Data

The reading transducer data service is used to get transducer data from a specified transducer channel of the specified TIM. Figure 29 shows the screenshot of the reading transducer data service request. The user inputs the parameters of the reading transducer data service request, submits it, and then obtains a response from the service provider. Figure 30 shows the screenshot of the reading transducer data service response, where the result of transducer reading is returned.

Fig. 29. Screenshot of reading transducer data service request

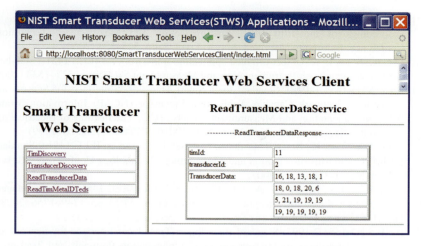

Fig. 30. Screenshot of reading transducer data service response

5 Summary

This paper describes wireless sensor networks based on the IEEE 1451.5 standards. An architecture of an IEEE 1451.5 standard-based wireless sensor network is presented. The network interfaces and transducer physical interfaces of IEEE 1451.5 wireless sensor networks are described. A prototype of a service-oriented and IEEE 1451.5-802.11 standard-based wireless sensor network is presented. A few case studies are shown to illustrate the operation and resulting data of the wireless sensor network.

References

[1] Song, E.Y., Lee, K.: Understanding IEEE 1451-Networked smart transducer interface standard. IEEE Instrumentation & Measurement Magazine, vol 11, number 2, pp. 11–17 (2008)
[2] Lee, K., Song, E.: A Wireless Environmental Monitoring System Based on the IEEE 1451.1 Standards. In: IMTC 2006 - Instrumentation and Measurement Technology Conference Sorrento, Italy, April 24-27, pp. 1931--1936 (2006)
[3] Wireless sensor nodes, http://en.wikipedia.org/wiki/List_of_wireless_sensor_nodes
[4] IEEE 1451, http://ieee1451.nist.gov/
[5] Lee, K., Song, E.Y.: Wireless Sensor Network Based on IEEE 1451.0 and IEEE 1451.5-802.11. In: The Eighth International Conference on Electronic Measurement and Instruments, ICEMI 2007, Xian, China, August 16-18, vol. 4, pp. 7–11 (2007)
[6] IEEE STD 1451.0-2007, Standard for a Smart Transducer Interface for Sensors and Actuators – Common Functions, Communication Protocols, and Transducer Electronic Data Sheet (TEDS) Formats, IEEE Instrumentation and Measurement Society, TC-9, The Institute of Electrical and Electronics Engineers, Inc., New York, N.Y. 10016, SH99684, October 5 (2007)

[7] Lee, K.: IEEE 1451: A Standard in Support of Smart Transducer Networking. In: Proceedings of the 17th IEEE Instrumentation and Measurement Technology Conference, Baltimore, Maryland, May 1-4, vol. 2, pp. 525–528 (2000)
[8] Lee, K., Song, E.Y.: Object-oriented application framework for IEEE 1451.1 standard. IEEE Transactions on Instrumentation and Measurement 54(4), pp. 1527–1533 (2005)
[9] IEEE STD 1451.1-1999, Standard for a Smart Transducer Interface for Sensors and Actuators – Network Capable Application Processor (NCAP) Information Model, IEEE Instrumentation and Measurement Society, TC-9, The Institute of Electrical and Electronics Engineers, Inc., New York, N.Y. 10016, SH94767, June 26 (1999)
[10] Song, E.Y., Lee, K.: Smart Transducer Web Service Based on IEEE 1451.0 Standard. In: IMTC 2007-Instrumentation and Measurement Technology Conference, Warsaw, Poland, May 1-3, pp. 1–6 (2007)
[11] Song, E.Y., Lee, K.: STWS: A unified Web service for IEEE 1451 smart transducers. IEEE Transaction on Instrumentation and measurement 57(8), pp. 1749–1756 (2008)
[12] Open1451, http://sourceforge.net/projects/open1451/
[13] IEEE STD 1451.2-1997, Standard for a Smart Transducer Interface for Sensors and Actuators–Transducer to Microprocessor Communication Protocols and Transducer Electronic Data Sheet (TEDS) Formats, IEEE Instrumentation, SH99685, and Measurement Society, TC-9, The Institute of Electrical and Electronics Engineers, Inc., New York, N.Y. 10016, SH94566, September 25 (1998)
[14] IEEE STD 1451.3-2003, Standard for a Smart Transducer Interface for Sensors and Actuators–Digital Communication and Transducer Electronic Data Sheet (TEDS) Formats for Distributed Multidrop Systems, IEEE Instrumentation and Measurement Society, TC-9, The Institute of Electrical and Electronics Engineers, Inc., New York, N.Y. 10016, SH95174, March 31 (2004)
[15] IEEE STD 1451.5-2007, Standard for a Smart Transducer Interface for Sensors and Actuators – Wireless Communication and Transducer Electronic Data Sheet (TEDS) Formats, IEEE Instrumentation and Measurement Society, TC-9, The Institute of Electrical and Electronics Engineers, Inc., New York, N.Y. 10016, SH99685, October 5 (2007)
[16] IEEE STD 1451.4-1994, Standard for a Smart Transducer Interface for Sensors and Actuators – Mixed-Mode Communication Protocols and Transducer Electronic Data Sheet (TEDS) Formats, IEEE Instrumentation and Measurement Society, TC-9, The Institute of Electrical and Electronics Engineers, Inc., New York, N.Y. 10016, SH95225, December 15 (2004)
[17] IEEE P1451.7 – Proposed Standard for a Smart Transducer Interface for Sensors and Actuators - Transducers to Radio Frequency Identification (RFID) Systems Communication Protocols and Transducer Electronic Data Sheet Formats, August 30 (2007),
http://standards.ieee.org/board/nes/projects/1451-7.pdf
[18] Bluetooth,
http://www.bluetooth.com/Bluetooth/Technology/Basics.htm
[19] IEEE 802.15.4, http://www.ieee802.org/15/pub/TG4.html
[20] ZigBee,
http://www.zigbee.org/About/OurMission/tabid/217/Default.aspx
[21] IPv6 over Low-Power Wireless Personal Area Networks (6LoWPANs),
http://www.ietf.org/rfc/rfc4919.txt

[22] WirelessHART,
 http://www.hartcomm.org/protocol/wihart/wireless_overview.html
[23] ISA100.11a-2009, http://www.isa.org/ISA100-11a
[24] IEEE 802.11, http://ieee802.org/11/
[25] IEEE 802.16, http://standards.ieee.org/getieee802/802.16.html
[26] WiMAX, http://www.wimax.com/education
[27] Zhuang, L.Q., Goh, K.M., Zhang, J.B.: The Wireless Sensor Networks for Factory Automation: Issues and Challenges,
 http://saturn.ee.psu.ac.th/~graduate/doc/paper2.pdf
[28] The Next Step-Wireless IEEE 1451 Smart Sensor Networks,
 http://archives.sensorsmag.com/articles/0901/35/main.shtml
[29] Gilsinn, J.D., Lee, K.: Wireless Interfaces for IEEE 1451 Sensor Networks. In: SICon 2001 sensors for industry conference, Rosemont, Illinoisa, U.S.A., November 5-7, pp. 45–50 (2001)
[30] Sweetser, D., Sweetser, V., Nemeth-Johannes, J.: A Modular Approach to IEEE-1451.5 Wireless Sensor Development. In: Proceedings of the 2006 IEEE Sensors Applications Symposium, Houston, Texas USA, February 7-9, pp. 82–87 (2006)
[31] Nemeth-Johannes, J., Sweetser, V., Sweetser, D.: Implementation of An IEEE-1451.0/1451.5 Compliant Wireless Sensor Module. In: The 42nd Annual AUTOTESTCON Conference, Baltimore, MD, USA, September 17-20, pp. 364–371 (2006)
[32] Song, E.Y., Lee, K.: An Implementation of the Proposed IEEE 1451.0 and 1451.5 Standards. In: Proceedings of the 2006 IEEE Sensors Applications Symposium, Houston, Texas, USA, February 7-9, pp. 72–77 (2006)
[33] Song, E.Y., Lee, K.: Integration of IEEE 1451 Smart Transducers and OGC-SWE Using STWS. In: Sensors Applications Symposium (SAS), February 17-19, pp. 298–303. IEEE, New Orleans (2009)

Fuzzy Based Optimized Routing Protocol for Wireless Sensor Networks

P. Manjunatha, A.K. Verma, and A. Srividya

IDP in Reliability Engineering,
Indian Institute of Technology Bombay, Mumbai, India
manjup@ee.iitbac.in, akv@ee.iitb.ac.in, asividya@ee.iitb.ac.in

Abstract. Wireless sensor networks are used in military, health monitoring, tracking and security applications. These critical applications require reliable and fault tolerant operation of the network. This paper proposes a novel reliable routing protocol for wireless sensor networks (WSN) to extend the network lifetime using fuzzy inference system (FIS). Due to limitation in communication range, sensor nodes transmit their sensed data through multiple hops. Each sensor node acts as a routing element for other sensor nodes for transmitting data.The purpose of this work is to extend the network lifetime using FIS. Each sensor node selects the optimized path for forwarding packets to the base station based on routing metrics. Simulation results shows that proposed algorithm can achieve higher network lifetime by comparison to other routing protocols.

Keywords: Wireless Sensor Network, Routing, Fuzzy Inference System, Energy Efficiency.

1 Introduction

With the rapid advancement in Micro-Electronics technology, low power digital circuitry and RF communication capabilities have enabled the development of low-cost, low power, small size sensor nodes[1]. Wireless sensor network consists of tiny sensor nodes which are able to sense the environmental conditions and communicate with each other wirelessly. These sensor nodes collect the sensed data and forward it to the base station (BS) through multi-hop communication. Due to limitation in communication capability data is sent to the BS usually by a multi-hop communication.The base station collects the data from all the sensor nodes, and then forwards the gathered information to the user. Typically, every node shares its resources and serves as a router for other nodes.

Sensor nodes are typically deployed in hostile environments and operated by battery power; once the battery power is exhausted it is difficult to recharge or replace the battery and these nodes are expected to operate for a few months to a year. Every sensor node participates in routing process by sharing its resources. Nodes are limited in resources (like battery power, communication range, processing capability). Sensor nodes expend most of their energy in communication activity than sensing and processing activity[2]. Communication links between nodes are wireless and are more easily prone to failure due to its limited transmission power and its environmental conditions such as noise, collision and multi-path fading. Hence the wireless channel itself unreliable for communication. Network topology is highly dynamic in nature due to node failure, node addition, and exhausted battery energy of a node. Sensor nodes are also more prone to failure due to its limited battery power. This provides the motivation for the energy aware reliable multi-path routing protocol for WSN.

Several routing protocols have been proposed to minimize the energy consumption in order to prolong the network lifetime. Most of the proposed routing protocols uses shortest single path for data transmission. With continuously using shortest path will deplete nodes energy at a much faster rate than the other path nodes. As a result nodes in theses path will die much faster and causes network partition[3]. Most of the sensor network applications require reliable data delivery to the user.

Basically there are two types routing protocols based on its network structure as flat based and cluster based. In cluster based routing protocol sensor nodes are grouped into clusters with one node is elected as cluster head for each cluster. Sensor nodes in the cluster are transmitting their sensed information the cluster head. The cluster head process the data and perform the data aggregation from the collected sensor nodes.

Routing protocols in wireless sensor network may be either proactive or reactive. In proactive routing protocols, routing tables are updated periodically at every node and on the other hand in reactive protocol routing is calculated when it is needed.

This paper proposes an energy efficient reliable multi-path routing protocol for WSN. A fuzzy inference system is used to optimize the routing protocol. The optimal path is selected based on the residual energy, distance from the node, and distance to the sink.

The remaining part of the paper is organized as follows. Related work and motivation for the research work is discussed in section 2. Section 3 presents the proposed energy efficient reliable routing protocol for wireless sensor network. The simulation set up and its performance analysis is presented in section 4 and in section 5 we conclude this paper.

2 Related Work

Many routing algorithms have been proposed for WSN to maximize the network lifetime by optimizing the energy usage [4–7]. Wireless sensor network requires energy efficient and reliable routing protocols. Multi-path routing algorithms [8] have been proposed to maximize the reliable transmission of an event detected by sensor nodes.

In particular, for larger network, the cluster based routing protocol is more efficient in minimizing the energy consumption. Heinzelman, et al.[4] have proposed cluster based (LEACH) routing protocol, in which sensors transmit their data to the cluster head. Cluster head collect the data from all the sensors, fused it and then transmit the data to the sink. In [3], Chang et al have proposed routing protocol in that which selects the routing path which uses link costs as communication energy and remaining energy in a node. Geographical Multipath Routing (GMR) scheme was proposed in [10] which uses location information as routing metric. In [11], the authors proposed energy and mobility aware geographical multipath routing (EM-GMR) using FIS. The remaining battery capacity, mobility, and distance to the sink are considered for next hop relay node selection. Chiang et.al [12]have analyzed routing protocol using fuzzy which considers remaining energy, distance from the shortest path, distance from the node and traffic load as input variables while selecting routing path.

Based on the above mentioned exhaustive literature survey, a next hop selected node should be nearest to the sink, nearest to the source and higher remaining energy. The proposed protocol will selects the next hop node based on these metrics by using FIS.

3 Proposed Routing Protocol

The following section describes the proposed novel routing protocol using fuzzy logic for sensor networks. The proposed protocol extends the network lifetime, equal distribution of energy consumption as well as selects the reliable path. When a sensor node detects an event and needed to send to the sink, the protocol selects the optimal path using FIS. Proposed algorithm considers the remaining battery energy, distance from the sensor and distance to the sink are considered for selecting next hop node.

Assumptions

It is assumed that sensor network comprises of a large number of sensor nodes which are aware of their locations and all nodes are stationary. Sensor node location information can be find through GPS receivers or localization algorithms [14–16]. The node batteries are neither replaceable nor rechargeable and all the nodes have equal energy levels. Assuming there is enough memory and processing power available in the sensor node and node can execute some optimization algorithm.

3.1 Protocol Operation

The proposed routing algorithm finds the suitable path as follows. Source node initiates a routing request for sending its sensed data to a sink node. When a routing request is initiated source node selects the path based on maximum remaining energy, minimum distance to the source node and minimum distance to the sink. These linguistic variables are fuzzified and are given to fuzzy inference system (FIS). The inference system processes the fuzzified values using rule base system. Defuzzified value is used to select the path.

Fuzzy Inference System

The decision network of the new algorithm is realized by a fuzzy expert system. Fig. 1 illustrates the structure of a fuzzy logic system. The model of fuzzy logic system as shown in Fig. 1 consists of fuzzification, fuzzy rules, fuzzy inference system, and defuzzification process.

Fuzziness describes event uncertainty and impreciseness of linguistic terms. Fuzzy logic fits best in applications where the variables are continuous and/or mathematical models do not exist or traditional system models become overly complex [18]. The atmospheric events are complex, ambiguous and vagueness embedded in their nature. Consequently, a fuzzy based approach is a viable option.

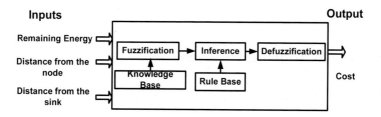

Fig. 1. Structure of a fuzzy logic system

Fuzzification

The fuzzification is the process of transforming crisp values into fuzzy linguistic variables. The membership function is used to associate a grade to each linguistic variable. Selection of the number of membership functions and their initial values is based on process knowledge and intuition. A membership function has value between 0 and 1 over an interval of crisp variable. The number of membership functions can vary to provide the resolution needed. Number of rules can grow exponentially as the number of input membership functions increases.

In our path selection algorithm the remaining energy, distance from the node, and distance form the sink are the input fuzzy variables and the cost as the output variable which determines the cost of the link between two sensor nodes. The membership functions LOW, MEDIUM and HIGH are defined on each input variable and VLOW, LOW, MEDIUM, HIGH and VHIGH are defined on output variable.

Fig. 2 shows the membership graph for input variables remaining energy, distance from the node and distance form the sink respectively. In our

Fuzzy Based Optimized Routing Protocol for Wireless Sensor Networks

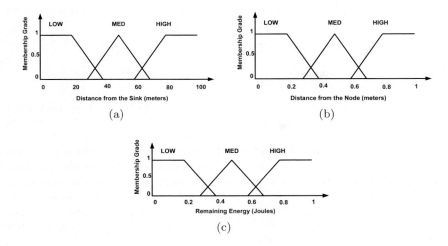

Fig. 2. Membership functions for input variables

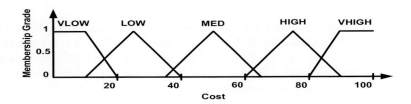

Fig. 3. Membership functions for output variables

example the fuzzy sets LOW, MEDIUM and HIGH are defined as trapezoidal, triangular, and trapezoidal membership function respectively. Fig. 3 shows the membership graph for output variable cost.

Fuzzy Inference System

The fuzzy inference system consists of fuzzy rules (IF antecedent THEN consequent) that are devised by an expert knowledge base or through system input-output learning. Gaussian, triangle, and trapezoid functions are the most commonly used membership functions. In the fuzzy rules, triangular and trapezoidal-shaped membership functions are used for the variables to simplify the computations. The core of fuzzy system is this rule base system which mimics human reasoning. The most commonly used fuzzy inference technique is Mamdani method. Fuzzy rule base drive the inference system to produce fuzzy outputs, which are defuzzified to get system outputs.

The fuzzy if-then rules in expert system are usually in the following form [17, 18]:

IF x_1 is A_{11} **and** x_2 is A_{21} ... **THEN** y is B_1 else
$\vdots$

IF x_1 is A_{1k} **and** x_2 is A_{2k} ... **THEN** y is B_k

where $x_1, x_2 \ldots$ are the fuzzy input(antecedent) variables, y is a single output(consequent) variable and $A_{11} \ldots A_{1k}$ are the fuzzy sets.

There are 3 input variables and each consists of 3 fuzzy linguistic variables. Therefore, the total $3^3 = 27$ rules are used, which are all possible combinations of the input variables. Thus some of the example rules in this rule based system is as follows:

IF remaining energy is $High$ **and** Distance is Low **and** distance to the sink is Low **THEN** Cost is is $VLOW$
$\vdots$

IF remaining energy is Low **and** Distance is $high$ **and** distance to the sink is $High$ **THEN** Cost is is $VHIGH$.

The rules are created using the Fuzzy Inference System (FIS) editor contained in the Matlab Fuzzy Toolbox [19]. Figs. 4(a) and 4(b) shows a control surface of a cost based on the above parameters. Fig 5 shows a sample fuzzy calculation of a fire probability based on the amount of temperature, humidity, light intensity and carbon monoxide.

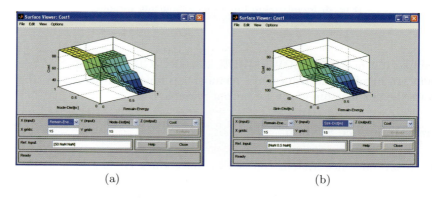

Fig. 4. Surface view of the cost with respect to Remaining Energy and distance

Defuzzification

The transformation from a fuzzy set to a crisp number is called a defuzzification. There are many kinds of defuzzification methods, usually maximum

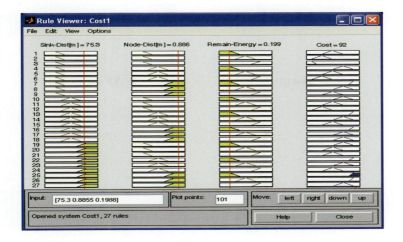

Fig. 5. Membership functions and sample fuzzy calculation of a fire probability

membership and centroid techniques are used. In practice defuzzification is done using centroid method. It is given by the following formula [17]:

$$z^* = \frac{\int \mu_A(z) * z \, dz}{\int \mu_A(z) \, dz} \quad (1)$$

where, $\mu_A(z)$ is the membership function of fuzzy set A.

4 Performance Analysis and Simulation Results

The simulation of fuzzy path selection protocol was implemented in MATLAB. In the proposed simulation, 100 sensors were deployed randomly in 100m by100m field with sink is located at the center as shown in Fig. 6. The radio propagation range was selected by 10 meters. All the sensors energy was set with 1J initially. We have used Heinzelman et al. [4] energy dissipation radio model. In this model, energy E_{TX} dissipation in the transmitter node to transmit a k bit message through a distance r with path loss exponent α is defined as:

$$\begin{aligned} E_{TX}(k,r) &= E_{Tr-elec}(k,r) + E_{Tr-amp}(k,r) \\ &= E_{Tr-elec} * k + E_{Tr-amp} * k * r^\alpha \end{aligned} \quad (2)$$

Energy E_{RX} consumed in the receiver node to receive a k bit message is defined as:

$$\begin{aligned} E_{RX}(k) &= E_{Rr-elec}(k) \\ &= E_{Rr-elec} * k \end{aligned} \quad (3)$$

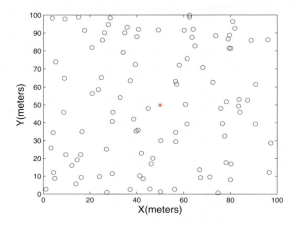

Fig. 6. Sensor node deployment

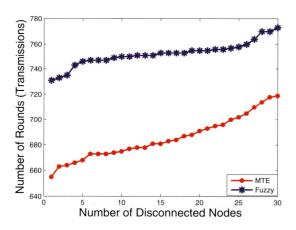

Fig. 7. Performance comparison of the proposed fuzzy algorithm with MTE

In eq. 2 and 3 $E_{Tr-elec}$ and $E_{Rr-elec}$ denotes the total energy required for transmission and reception respectively, and E_{Tr-amp} denotes the energy required for the transmitter amplifier and depends upon the acceptable bit error rate. In this paper we use first order radio model where $E_{Tr-elec} = E_{Rr-elec} = 50$ nJ/bit, and $E_{Tr-amp} = 100 pj/bit/m^2$.

The performance of the proposed protocol was compared with minimum transmission energy (MTE) protocol. Each sensor generates the packet randomly with equal distribution. As shown in Fig. 7 it clearly shows that the proposed scheme out performs the MTE, which is due to the equal participation of sensor nodes in communication activity.

5 Conclusions

This paper proposes fuzzy logic based routing protocol for wireless sensor networks to maximize the network lifetime. While most of the research works focuses on energy efficient routing algorithms by considering the minimum energy to transmit the sensed information to the sink. Proposed fuzzy path selection algorithm can select the next hop relay node based on residual energy, distance from the node and distance to the sink. The proposed protocol helps to increase the lifetime of the network compared to the other routing protocols. Simulation results show that the proposed fuzzy based multi constraint routing protocol outperforms the existing routing protocols.

References

[1] Akyildiz, I.F., Su, W., Sankarasubramaniam, Y., Cayirci, E.: Wireless sensor networks: A survey. Computer Networks (Elsevier) 38(4), 393–422 (2002)
[2] Pottie, G., Kaiser, W.: Wireless integrated network sensors. Communications of the ACM 43(5), 51–58 (2000)
[3] Leandros, J.C., Tassiulas, L.: Maximum lifetime routing in wireless sensor networks. IEEE/ACM Trans. on networking 12(4), 609–619 (2000)
[4] Heinzelman, W., Chandrakasan, A., Balakrishnan, H.: Energy-efficient communication protocol for wireless microsensor networks. In: Proceedings of the 33rd International Conference on System Sciences (HICSS 2000), Anchorage Alaska, January 2000, vol. 2, pp. 1–10 (2000)
[5] Heinzelman, W., Chandrakasan, A., Balakrishnan, H.: An Application-Specific Protocol Architecture for Wireless Microsensor Networks. IEEE Trans. on Wireless Communications 1(4), 660–670 (2002)
[6] Manjeshwar, A., Agarwal, D.P.: TEEN: A Routing Protocol for Enhanced Efficiency in Wireless Sensor Networks. In: IEEE Parallel and Distributed Processing Symposium, San Francisco, April 23-27, pp. 2009–2015 (2001)
[7] Lindsay, S., Raghavendra, C.: PEGASSIS: Power Efficient Gathering in Sensor Information Systems. In: IEEE Aerospace Conference, Big Sky, Montana, March 2002, vol. 3, pp. 1125–1130 (2002)
[8] Ganesan, D., Govindan, R., Shenker, S., Estrin, D.: Highly-resilient, energy-efficient multipath routing in wireless sensor networks. ACM SIGMOBILE Mobile Computing and Communications Review 5(4), 11–25 (2001)
[9] Chang, J.H., Tasiulas, L.: Maximum Lifetime Routing in Wireless Sensor Networks. IEEE/ACM Trans. on Networking 12(4), 609–619 (2004)
[10] Jain, R., Puri, A., Sengupta, R.: Geographical Routing using partial information for wireless sensor networks. IEEE Personal communications, 48–57 (February 2001)
[11] Liang, Q., Ren, Q.: Energy and Mobility aware Geographical Multipath Routing for Wireless Sensor Networks. In: Wireless communications and networking conference, New Orleans, LA, March 2003, vol. 3, pp. 1867–1871 (2003)
[12] Chiang, S., Wang, J.: Routing Analysis Using Fuzzy Logic Systems in Wireless Sensor Networks. In: Lovrek, I., Howlett, R.J., Jain, L.C. (eds.) KES 2008, Part II. LNCS (LNAI), vol. 5178, pp. 966–973. Springer, Heidelberg (2008)

[13] Shin, K., Chung, M.Y., Won, J., Choo, H.: Routing Based on Ad Hoc Link Reliability. In: Min, G., Di Martino, B., Yang, L.T., Guo, M., Rünger, G. (eds.) ISPA Workshops 2006. LNCS, vol. 4331, pp. 341–350. Springer, Heidelberg (2006)

[14] Albowicz, J., Chen, A., Zhang, L.: Recursive position estimation in sensor networks. In: Proc. IEEE Int. Conf. Network Protocols, November 2001, pp. 35–41 (2001)

[15] Savarese, C., Rabaey, J.M., Beutel, J.: Locationing in distributed ad-hoc wireless sensor networks. In: Proc. ICASSP, May 2001, pp. 2037–2040 (2001)

[16] Patwari, N., Hero, A., Perkins, M., Correal, N.S., O'Dea, R.J.: Relative Location Estimation in Wireless Sensor Networks. IEEE Tran. On signal processing 51(8), 2137–2148 (2003)

[17] Ross, T.J.: Fuzzy logic with engineering applications. McGraw-Hill Inc., New York (1997)

[18] Simon, Y.: A fuzzy logic approach to fire detection in aircraft dry bays and engine compartments. IEEE Tran. on industrial electronics 47(5), 1161–1171 (2000)

[19] Fuzzy Logic Toolbox user's guide, http://www.mathworks.com/

Energy Aware Sensor Group Scheduling to Minimise Estimated Error from Noisy Sensor Measurements

Siddeswara Mayura Guru[1] and Suhinthan Maheswararajah[2]

[1] CSIRO Tasmanian ICT Centre, GPO Box 1538, Hobart 7001, Australia
siddeswara.guru@csiro.au
[2] University of Melbourne, Parkville, Vic 3010, Australia
smah@unimelb.edu.au

Abstract. In wireless sensor network applications, sensor measurements are corrupted by noise resulting from harsh environmental conditions, hardware and transmission errors. Minimising the impact of noise in an energy constrained sensor network is a challenging task. We study the problem of estimating environmental phenomena (e.g., temperature, humidity, pressure) based on noisy sensor measurements to minimise the estimation error. An environmental phenomenon is modeled using linear Gaussian dynamics and the Kalman filtering technique is used for the state estimation. At each time step, a group of sensors is scheduled to transmit data to the base station to minimise the total estimated error for a given energy budget. We considered a diverse solution for scheduling sensors from heuristic-based Particle Swarm Optimisation (PSO) to dynamic programming and one-step-look-ahead methods. The simulation results show that PSO outperforms all other proposed methods and it requires less computational time than dynamic programming.

1 Introduction

Generally, it is impossible to capture error-free measurements from the environment. The measurement may be corrupted due to several reasons including environmental conditions such as temporal and spatial variation of temperature, pressure, electromagnetic noise and radiation interference [1]. This problem is rampant in wireless sensor networks due to the use of low-cost sensors and transmission system [2]. The low-cost sensors are subject to greater drift than expensive sensors because the latter have internal stabilisation mechanisms [3]. Errors associated with sensor measurements can be classified as systematic or random errors [4]. Systematic errors are predictable, caused by environmental conditions and can be corrected by calibration [5]. However, random errors are inherent and unavoidable due to hardware noise and inaccurate measurement techniques. Filtering techniques are commonly used to alleviate random error. We maintain that any error associated with an observation cannot be ignored if a critical decision needs to be made.

The work in this chapter is largely motivated by our experience with developing a wireless sensor network application for water use efficiency in irrigation [6]. The aim of this application is to develop closed-loop sensor technology to irrigate the field only when it is necessary. As decisions for irrigation will be made partly based on various sensor measurements gathered from the field, errors in measurements will lead to inaccurate actuation of the irrigation system. Therefore, we should minimise the impact of measurement errors on decision making as much as possible.

There are several instances where an environmental phenomenon is represented as a linear Gaussian stochastic model. The daily average surface temperature was modeled using a spatio-temporal linear Gaussian stochastic approach [7]. This model was validated using real data captured from different geographical areas and found to fit well. In [2], a physical phenomenon of an indoor environment and sensor measurements were again modeled using a linear Gaussian stochastic approach and validated using the publicly available Intel Lab Dataset [8]. It was found that Kalman filtering gives better state estimation results than the linear regression method. The same model and technique were also used in a target tracking application [9]. The target was modeled as a linear Gaussian system and estimation was performed using Kalman filtering. Similar model and estimation is performed to develop an energy efficient scheduling technique for individual sensors to track a target with a mobile base station [10]. In this chapter, we assume that environmental phenomena vary linearly over time with white Gaussian noise. Moreover, we assume that sensor nodes are free from drifting but sensor measurements are corrupted by white Gaussian noise and linearly related to the phenomena. It is considered that the phenomenon is measured by a group of sensors at any instance of time and transmitted to a stationary base station.

2 Chapter Overview

In this chapter, we study the problem of selecting an optimal sequence of sensor groups which minimises the total estimation error for a given energy budget in an energy-constrained wireless sensor network. The energy budget is the initial energy available in a network. Each sensor group consists of a certain number of sensor nodes and each node can have different error covariance. Selecting all sensor groups may produce better estimation but dissipate more energy due to a large volume of data transmission in the network. Therefore, it is necessary to select an optimal sensor group at each time step in order to minimise the estimation errors while considering the energy budget. The use of Kalman filtering technique allows us to calculate the estimation error before obtaining any measurements [11]. Therefore, it is possible to calculate the total estimated error associated with a given sequence of sensor groups off-line. This helps us to find the best sensor group scheduling sequence in advance. We consider this problem as a combinatorial optimisation and solve it using dynamic programming. We have considered evolutionary computing based meta-heuristic particle swarm optimisation to schedule sensor groups. The problem is formulated as a discreate optimisation in the context of PSO and constraints are relaxed using a penalty function method. We also use a sub-optimal method, one-step-look-ahead, to produce a result more quickly.

The remainder of this chapter is organised as follows: Section 3 briefly describes the problem description and formulation. Section 4 deals with the modelling of the process and sensor measurements as linear Gaussian dynamics and the Kalman filtering technique used to estimate the state of the phenomena. Section 5 presents dynamic programming, particle swarm optimisation and one-step-look-ahead methods to find the sequence of sensor groups to minimise the estimated error of the phenomena whilst satisfying the energy budget. Section 6 provides experimental setup and simulation results. The chapter is concluded in Section 7.

3 Problem Description and Formulation

In a battery-operated wireless sensor network deployed to measure the physical phenomena of an environment, sensor nodes produce measurements and transmit them to a base station (BS). The communications between the base station and sensor nodes are considered single-hop. The measurement inherently contains random noise and the base station can estimate the state (i.e., true value) of the phenomena based on noisy measurements collected from sensor nodes. The error in state estimation due to noise needs to be minimised.

In practice, noise in measurements varies with different phenomena even though the measurements are from the same sensor node. For example, a sensor node may produce a temperature measurement with lower noise but humidity measurement with higher noise. Although it is possible to get a better error estimation with all measurements from all sensor nodes in a network at a given time, it is not feasible due to the resultant high energy consumption due to transmission. Therefore, there should be a trade-off between error estimation and energy consumption in an energy-constrained wireless sensor network.

The problem we intend to address in this chapter is how to select only a subset of sensor nodes at each time step that could minimise the total estimation error while satisfying the energy budget. However, the possible number of sets of sensor nodes available at each time step depends on the size of the network. Let a sensor network have M sensor nodes with a maximum of U nodes selected to form a set at a given time. A sensor set could have $1, 2, \ldots$ or U sensor nodes and a total of $^{M}C_1 + ^{M}C_2 + \ldots\ldots ^{M}C_U$ sensor sets could be formed at each time step. It is a time consuming and complex task to identify the sensor sets and the combination of sensor nodes in a set. Therefore, without loss of generality, pre-defined set of sensor groups are considered and the selection is made from the set at each time step.

4 System Models

In this section, the models that represent environmental phenomena, sensor measurements, and energy dissipation are described. The phenomena we are observing in a network can be temperature, humidity, pressure, moisture etc. However, in this study, only temperature and humidity are considered. The state of a phenomenon varies with time (but not with space) according to a stochastic equation (1):

$$X_k = F_{k-1} X_{k-1} + w_{k-1}, \tag{1}$$

where, the column vector $X_k = [t_k h_k]'$ represents the state of the phenomena and t_k and h_k denote the true temperature and humidity of an area at time step k. The meaning of the notations used in this chapter can be found in Table 1. State and true temperature and humidity are used interchangeably in this section. w_k represents a white Gaussian process noise with covariance matrix Q_k, such that $w_k \sim N(0, Q_k)$. According to (1), X_k (i.e., true temperature and humidity at time step k) is transited from X_{k-1} by the transition

Table 1. Summary of the notations used in the system model

Notation	Description
k	time step
X_k	true state of the phenomena or process
F_k	transition matrix
w_k	process noise
Q_k	error covariance of the process noise
N	total number of sensor groups in the network
g_i	i-th sensor group
S_{i_j}	j-th sensor node in i-th sensor group
n_i	total number of sensor nodes in i-th sensor group
z_k^{ij}	observed measurement of X_k from the j-th sensor node in i-th sensor group at k-th time step
H_k^{ij}	observation matrix of the j-th sensor node in i-th sensor group at k-th time step
v_k^{ij}	measurement noise of the j-th sensor node in i-th sensor group at k-th time step
R_k^{ij}	error covariance of the j-th sensor node in i-th sensor group at k-th time step
ψ_j^i	energy consumption of the j-th sensor node in i-th sensor group
E_{g_i}	total energy consumption of the i-th sensor group
$g_{\min}$	sensor group which consumes minimum energy
E_T	total energy budget for T time steps

Energy Aware Sensor Group Scheduling to Minimise Estimated Error

matrix or system matrix F_{k-1} with a white Gaussian noise determined by Q_{k-1}. The F_k and Q_k are known a priori.

Let N denote the number of pre-defined sensor groups in a network and $g_i = \{s_{i_1}, s_{i_2}, ..., s_{i_{n_i}}\}$ denote the i-th sensor group. Thus the i-th sensor group has n_i sensor nodes such that $|g_i| = n_i$. We assume that the sensor measurements are linearly related to x_k and corrupted by white Gaussian noise. At time step k, the measurement from the j-th sensor node of the i-th sensor group is a column vector given by:

$$z_k^{i_j} = H_k^{i_j} x_k + v_k^{i_j}. \quad (2)$$

The noise $v_k^{i_j}$ for each sensor node is assumed to be independent of the noise from other nodes and white Gaussian, $v_k^{i_j} \sim N(0, R_k^{i_j})$. According to (2), $z_k^{i_j}$ (observed, i.e., true + noise) temperature and humidity at time step k is mapped from x_k by the observation matrix $H_k^{i_j}$ with the white Gaussian noise determined by the error covariance $R_k^{i_j}$. The assumption here is that $R_k^{i_j}$ and $H_k^{i_j}$ are known for all the sensor nodes well in advance. Furthermore, the communication between sensor node and the BS is considered as a single-hop and the power consumption ψ_j^i of a sensor node s_{i_j} during active and sleep modes is given by:

$$\psi_j^i = \begin{cases} \psi_1 + \psi_{tx}^{(i,j)} & \text{if active} \\ \psi_2 & \text{if asleep,} \end{cases} \quad (3)$$

where ψ_1 represents power consumption due to sensing and data processing, $\psi_{tx}^{(i,j)}$ represents power consumption due to transmitting data to the BS and the power required for its own timer is denoted by ψ_2. The transmission power between the sensor node and the BS is calculated as $\psi_{tx}^{(i,j)} = (\alpha_1 + \alpha_2 d_{i_j}^2)r$, where r denotes the data rate, $\alpha_1 > 0$ a constant related to the radio energy and d_{i_j} is the Euclidean distance between BS and sensor node s_{i_j} [12]. Therefore, the energy consumption due to activating the i-th sensor group is given by:

$$E_{g_i} = \sum_{j=1}^{n_i} \psi_j^i. \quad (4)$$

4.1 State Estimation

In linear Gaussian dynamics, the state x_k can be estimated recursively using the Kalman filtering technique, which produces the minimum Mean Square Error (MSE) of the estimation [13]. Let $\xi_{k|k}$ and $P_{k|k}$ denote the estimated state and its error covariance respectively. They are defined as:

$$\xi_{k|k} = \mathbb{E}\{x_k\}, \quad P_{k|k} = \mathbb{E}\{(x_k - \xi_{k|k})(x_k - \xi_{k|k})'\}.$$

If the BS selects the i-th sensor group at time step k, then the stacked measurement equation is given by:

$$Z_k = H_k X_k + v_k, \quad (5)$$

where $Z_k = \left[z_k^{i_1}, z_k^{i_2}, ..., z_k^{i_{n_i}} \right]'$, $H_k = \left[H_k^{i_1}, H_k^{i_2}, ..., H_k^{i_{n_i}} \right]'$ and $v_k \sim N(0, R_k)$. For the given stacked measurements Z_k from the i-th sensor group, the centralised Kalman filter estimates the state as:

$$\xi_{k|k} = \xi_{k|k-1} + K_k(Z_k - H_k \xi_{k|k-1}), \quad (6)$$

where K_k is the Kalman gain represented as $K_k = P_{k|k} H_k' R_k^{-1}$ and the estimated error covariance $P_{k|k}$ is given by:

$$P_{k|k}^{-1} = P_{k|k-1}^{-1} + H_k' R_k^{-1} H_k \quad (7)$$

where $P_{k|k-1} = F_{k-1} P_{k-1|k-1} F_{k-1}' + Q_{k-1}$, $\xi_{k|k-1} = F_{k-1} \xi_{k-1|k-1}$.

Since we assume that sensor nodes noises are cross-independent, the part of (7) can be represented as (8) for sensor group with n_i sensors. This is adopted from [14] and [15].

$$H' R_k^{-1} H = \sum_{j=1}^{n_i} H_k^{i_j'} R_k^{i_j^{-1}} H_k^{i_j} \quad (8)$$

We can rewrite the error covariance of the estimated state given in (7) by substituting (8) as:

$$P_{k|k}^{-1} = P_{k|k-1}^{-1} + \underbrace{\sum_{j=1}^{n_i} H_k^{i_j'} R_k^{i_j^{-1}} H_k^{i_j}}_{f(g_i)}. \quad (9)$$

Using (9), the accuracy of the estimated state can be calculated for known $H_k^{i_j}$ and $R_k^{i_j}$ of the i-th sensor group. It can also be inferred that $P_{k|k}$ is a function of $R_k^{i_j}$ and independent of $z_k^{i_j}$. Since $H_k^{i_j}$ and $R_k^{i_j}$ are known for each sensor node, $P_{k|k}$ can be calculated independent of observed measurements. Therefore the computation of error covariance can be performed off-line without the observed measurement.

In a homogeneous sensor network, the sensor nodes are identical to one another, in this case, $R_k^{i_j} = R_k^*$ and $H_k^{i_j} = H_k^*$. The $P_{k|k}$ for sensor group g_i is given by:

$$P_{k|k}^{-1} = P_{k|k-1}^{-1} + f^*(g_i) \quad (10)$$

where, $f^*(g_i) = n_i H_k^{*'} R_k^{*-1} H_k^*$.

$f^*(g_i) \to \infty$ as $n_i \to \infty$ and consequently $P_{k|k} \to 0$. This means that a sensor group with a large number of sensor nodes produces a small error covariance. However, energy dissipation will be high, due to data transmission, leading to significant reduction in network lifetime. Therefore, it is not practical to transmit data from all sensor nodes at each time step to minimise the error covariance. In the next subsection, we define our objective function with a battery energy constraint.

4.2 Error Cost Function

As mentioned earlier, the sensor network considered in this study is energy-constrained battery-operated. At each time step, the base station will identify a sensor group to transmit data. The sequence of sensor groups selected for time period T is $\mu(\text{T}) = \{u_1, u_2, ..., u_\text{T}\}$ where, u_k is the selected sensor group at time step k.

Let E_T denotes the total energy budget of a network. The energy constraint is given by:

$$\sum_{k=1}^{\text{T}} \text{E}_{u_k} \leq \text{E}_\text{T}. \tag{11}$$

This means that the sum of energy spent for the time period $\{1, 2, ..., \text{T}\}$ should be less than or equal to the total energy budget of the network. Since the base station selects a sensor group at each time step, E_T should satisfy the following condition:

$$\text{E}_\text{T} \geq \text{TE}_{g_{\min}}, \text{ where } \text{E}_{g_{\min}} = \min_{i=1,2,...,\text{N}} \{\text{E}_{g_i}\}. \tag{12}$$

The sensor group which consumes minimum energy out of N sensor groups is denoted as $\text{E}_{g_{\min}}$. If the energy budget does not satisfy the condition as defined in (12), then the BS cannot select a sensor group at each time step. Therefore, we assume that the energy budget always satisfies the condition given in (12).

We define the error cost function for the time period $\{1, 2, ..., \text{T}\}$:

$$J(\mu(\text{T})) = \sum_{k=1}^{\text{T}} \sqrt{trace(\text{P}_{u_k})}, \tag{13}$$

where P_{u_k} represents the error covariance of the estimated state associated with the selected sensor group u_k at time step k. P_{u_k} is calculated based on (9) and given by:

$$\text{P}_{u_k}^{-1} = \text{P}_{k|k-1}^{-1} + f(u_k) \tag{14}$$

Our objective is to find the optimal sequence of sensor groups which has the minimum total error cost $J(\mu(\text{T}))$ for the entire time period. The optimal sequence of sensor groups, denoted by $\mu^*(\mathrm{T}) = \{u_1^*, u_2^*, ..., u_T^*\}$, can be computed based on:

$$\mu^*(\mathrm{T}) = \arg\min_{\forall u_k} \left\{ \sum_{k=1}^{\mathrm{T}} \sqrt{trace(\mathrm{P}_{u_k})} \right\} \quad (15)$$

$$\text{subject to} \quad \sum_{k=1}^{\mathrm{T}} \mathrm{E}_{u_k} \leq \mathrm{E}_{\mathrm{T}}.$$

The BS can select a sequence of sensor groups out of maximum of N^T possible sequences. This is a combinatorial optimisation problem and it has a maximum of N^T feasible solutions. If T and N are small, then the problem can be solved by an exhaustive search method. Since the number of possible sequences of sensor groups increases exponentially as T and N increase, it is not practical to solve the problem by an exhaustive search method within an acceptable time period for larger networks. Hence, we propose methods to solve this combinatorial problem within a feasible time period.

5 Scheduling Methodologies

In this section, the proposed scheduling techniques to identify the sequence of pre-defined sensor groups to minimise the estimated error based on the energy constraint are described. The goal is to schedule the sensor groups in advance for the entire lifetime of the network. The dynamic programming technique, Particle Swarm Optimisation and the One-Step-Look-Ahead (OSLA) method are adapted to find sequences of sensor groups to communicate with the BS for a given energy budget.

The dynamic programming technique is selected to solve the proposed problem because it has the ability to provide the optimum solution for a combinatorial optimisation problem. The PSO has been successfully used to solve a combinatorial optimisation problem [16] and it was shown that it produces reasonable results with an acceptable computational cost. The OSLA method is considered for comparison and it gives a sub-optimal solution in a small duration of time. The description of the algorithms and the adaptation to solve the scheduling problem based on constraints is presented below.

5.1 Dynamic Programming

Dynamic programming (DP) is a recursive technique for finding an optimal solution to a problem that can be broken into several subproblems [17]. Each subproblem is solved optimally and the results are stored for future use. These stored results are used to construct an optimal solution for the original problem. Scheduling energy-constrained sensor nodes to minimise the tracking error was studied in [18], where the energy constraint was relaxed using Lagrangian multipliers and the problem was solved using an approximate dynamic programming technique. However, we use dynamic programming to optimally schedule the sensor groups to minimise the error of the estimation for a given energy budget.

In dynamic programming the original problem is broken into several stages and each of them are divided into many *states*. A *state* stores all information required to go from one stage to the next. Here, the stage is considered as the time step and thus the total number of stages are equal to the total number of time steps. The *state* in DP consists of two parts: the error covariance of the estimated state denoted as D_k, and the available energy budget denoted as L_k. The decision at each *state* is the selected sensor group and denoted as $\mu_k(D_k, L_k)$. Since the base station must select a sensor group at each time step, it should be aware of the available energy budget after selecting a sensor group. Therefore, the base station cannot choose a sensor group randomly such that $\mu_k(D_k, L_k) \in N_{L_k}$. However, $N_{L_k} \subseteq \{g_1, g_2, ..., g_N\}$ can be calculated based on E_k^{max} and E_k^{min}, which denotes the maximum and minimum available energy budget at the k-th stage, as defined below:

$$E_k^{max} = \begin{cases} E_T & \text{if } k = 1 \\ E_T - (k-1)E_{g_{min}} & \text{if } T \geq k > 1, \end{cases} \quad (16)$$

$$E_k^{min} = \begin{cases} E_T & \text{if } k = 1 \\ (T - k + 1)E_{g_{min}} & \text{if } T \geq k > 1, \end{cases} \quad (17)$$

let the i-th sensor group be in N_{L_k}, if the remaining energy budget $L_k - E_{g_i}$ at the k-th stage is larger than the minimum available energy budget E_{k+1}^{min} at $k+1$ then:

$$g_i \in N_{L_k} \quad \text{if} \quad (L_k - E_{g_i}) \geq E_{k+1}^{min}. \quad (18)$$

The available energy budget L_k at each stage can take values between E_k^{min} and E_k^{max} whereas D_k takes the values of all possible $P_{k-1|k-1}$. If the decision at the k-th stage is made to activate $\mu_k(D_k, L_k) = u_k$ (i.e, to activate $u_k \in N_{L_k}$ sensor group at time step k) for a given D_k and E_k, then the cost per stage is defined by:

$$G(D_k, L_k, u_k) = \sqrt{trace(P_{u_k}(D_k))}, \quad (19)$$
$$\text{where } P_{u_k}^{-1}(D_k) = P_{k|k-1}^{-1}(D_k) + f(u_k),$$
$$P_{k|k-1}(D_k) = F_{k-1}D_k F'_{k-1} + Q_{k-1}.$$

Backward DP is used to solve the sensor group scheduling problem. In this method, recursion proceeds backward from stage T to 1. The cost function of the DP approach at the last stage T is defined as:

$$J_T(D_T, L_T) = \min_{u_T \in N_{L_T}} G(D_T, L_T, u_T), \quad (20)$$

and the cost function of the DP approach at the k-th stage is defined as:

$$J_k(D_k, L_k) = \min_{u_k \in N_{L_k}} \{G(D_k, L_k, u_k) + J_{k+1}(P_{u_k}(D_k), L_k - E_{u_k})\}. \quad (21)$$

Backward DP solves the sub-problems from $k = T$ to 1 and stores the optimal value for $J_k(D_k, L_k)$ and $\mu_k^*(D_k, L_k)$. For a given initial error covariance $P_{0|0}$ of the state and a total energy budget E_T, the optimal value of the objective function in (13) is equivalent to that of the initial stage of DP:

$$J_1(D_1, L_1) = \min_{u_k \in N_{L_k}} \sum_{k=1}^{T} \sqrt{trace(P_{u_k})}, \quad (22)$$

where $D_1 = P_{0|0}$, $L_1 = E_T$.

The exact value of the *state* of DP for the first stage is known ($D_1 = P_{0|0}$ and $L_1 = E_T$) but not for the rest of the stages. Therefore, DP considers all possible values of the *state* at each stage. Since the *state* space is infinite, each element in the *state* space is discretised. The interpolation is used to approximate the value of $J_{k+1}(P_{u_k}(D_k), L_k - E_{u_k})$, if $P_{u_k}(D_k)$ or $L_k - E_{u_k}$ is not available at the $k+1$-th stage.

The finer discretisation of the *state* space reduces the approximation provided by the interpolation technique resulting in a solution closer to optimal. However, the finer discretisation leads to a higher number of *states* which in turn increases the computational cost. In DP, the computational cost depends on the number of *states* and the number of decision variables $|N_{L_k}|$ available at each stage. There will be $|N_{L_k}|$ comparisons at each *state*. For example, if the number of *states* are a_k at the k-th stage, then DP requires a total of $|N_{L_1}| + \sum_{k=2}^{T} a_k |N_{L_k}|$ comparisons. DP produces a near optimal sequence of sensor groups $\{u_1^*, u_2^*, ..., u_T^*\}$, where u_k^* is given by:

$$u_k^* = \begin{cases} \mu_1^*(P_{0|0}, E_T) & \text{if } k = 1 \\ \mu_k^*(P_{k-1|k-1}, E_T - \sum_{t=1}^{k-1} E_{u_t^*}) & \text{if } T \geq k > 1 \end{cases} \quad (23)$$

5.2 Particle Swarm Optimisation

Particle Swarm Optimisation (PSO) is a swarm-based heuristic algorithm influenced by the social behaviour of animals such as a flock of birds or a school of fish [19]. A particle in PSO is analogous to a bird or fish flying through a search (problem) space. The movement of each particle is co-ordinated by a velocity and at any instant of time is influenced by its best position and the position of the best particle in the problem space. The performance of a particle is measured by a fitness value, which is problem specific.

The PSO algorithm is similar to any other evolutionary algorithm where the population is the number of particles in the problem space. Particles represent a feasible solution to a problem and are initialised randomly. Each particle will have a fitness value, which will be evaluated by a fitness function to be optimised in each generation. Each particle knows its best position *pbest* and the best position so far among the entire group of particles *gbest*. The *pbest* of a particle is the best result (fitness value) so far reached by the particle, whereas *gbest* is the best particle in terms of fitness in an entire population. Thus, PSO relies on information sharing between the particles to generate a new population. The velocity directs the flying of the particle. In each generation the velocity and the position of particles will be updated as in (24) and (25) respectively.

$$v_i^{k+1} = \omega v_i^k + c_1 rand_1 \times (pbest_i - x_i^k) + c_2 rand_2 \times (gbest - x_i^k), \quad (24)$$

$$x_i^{k+1} = x_i^k + v_i^{k+1}, \quad (25)$$

where:

v_i^k is the velocity of particle i at iteration k
v_i^{k+1} is the velocity of particle i at iteration $k+1$
ω is the inertia weight
c_j is the acceleration coefficients; $j = 1, 2$
$rand_i$ is the random number between 0 and 1; $i = 1, 2$
x_i^k is the current position of particle i at iteration k
$pbest_i$ is the best position of particle i
$gbest$ is the position of best particle in a population
x_i^{k+1} is the position of the particle i at iteration $k+1$.

The evolutionary algorithm is generally used to solve optimisation problems without constraints. However, the problem we are addressing has an inequality constraint and need to be converted into an unconstrained optimisation in order to apply evolutionary computing techniques. There are several methods used in the evolutionary computing literature to convert constrainted to unconstrained optimisation problems [20, 21, 22].

PSO is a relatively new evolutionary computing paradigm, and researchers have adopted several techniques available in the evolutionary algorithm literature to handle constraints. Hu and Eberhart used the preservation of feasible solution method in PSO by generating particles that satisfy feasible solutions in order to make sure that the solution is always in the feasible space [23]. The main disadvantage of this method is the higher computation time required for the algorithm if the feasible space is small. In some problems, the initialisation of particles and the regeneration of them may also take longer. Parsopoulos and Vrahatis used the multi-stage penalty function method to solve the constraint optimisation problem [24]. A dynamic penalty is applied depending on the amount of violation in the fitness function. Coath and Halgamuge compare these two methods for non-linear constraint optimisation problems [25]. The results are not very conclusive; out of five test problems both methods were better than the other in two cases. We use the penalty function method to solve the inequality constraint optimisation which is depicted in (15). The reason for using the penalty function method is that it is easy to implement and external to the fitness function. However, it needs fine tuning of the penalty parameter to achieve good results. We have used the additive penalty function method given in (26):

$$Fitness(\bar{s}) = \begin{cases} f(\bar{s}), & \text{if } \bar{s} \in \text{feasible space} \\ f(\bar{s}) + P(\bar{s}), & \text{Otherwise} \end{cases} \quad (26)$$

where $f(\bar{s})$ is the objective function and $\bar{s}$ is a solution which can be either feasible or infeasible. The penalty function $P(\bar{s})$ is added to the objective function if $\bar{s}$ is infeasible, (i.e., the constraints are violated), otherwise it is zero. The original error cost function with constraint (15) is converted into an unconstrained problem by using the penalty function method. The new objective function is given by:

$$J(\mu(\text{T}), pen) = \sum_{k=1}^{\text{T}} \sqrt{trace(\text{P}_{u_k})} + pen \times \text{G}(g(\mu(\text{T}))), \quad (27)$$

where $\text{G}(g(\mu(\text{T})))) = \max[0, g(\mu(\text{T}))]^2$

$$g(\mu(\text{T})) = \sum_{k=1}^{\text{T}} \text{E}_{u_k} - \text{E}_{\text{T}} \quad : \text{Energy constraint.}$$

The *pen* in (27) is a positive scalar, it can be inferred that if a sensor group sequence $\mu(\text{T})$ does not violate any energy constraint: $g(\mu(\text{T})) \leq 0$ then (27) is equivalent to the original objective function in (15). On the contrary, a positive term will be added to the original objective function if the sensor sequence violates any energy constraint. Thus, the penalty function method discourages searching in the infeasible solution space. A good *pen* value can be found by experimentation. The multi-stage penalty function proposed in [24] was not used in this chapter because the problem addressed consists of only one constraint and a static penalty can produce reasonable results.

PSO Algorithm for Sensor Group Scheduling. The position of the particle represents a sensor group sequence, therefore the dimension of the particle is the total number of timesteps the sensor groups are scheduled in the lifetime of a network. Each dimension of a particle represents a sensor group and the evaluation of each particle is performed using the fitness function given in (27). The velocity of the particle is calculated using (24) and the particle trajectory is updated using (25). The evaluation is carried out until the stopping criteria is satisfied. The pseudo-code for the PSO algorithm in order to get a sequence of sensor groups to minimise the estimated error is given in Algorithm 1.

Energy Aware Sensor Group Scheduling to Minimise Estimated Error   295

Algorithm 1. PSO algorithm for sensor group scheduling

1: Set particle dimension as equal to the size of the time step in the lifetime of the network
2: Initialise particles' positions and velocities randomly in the search space. g_1 and g_N are the boundaries of the search space for each dimension.
3: For each particle, calculate its fitness value using (27).
4: If the fitness value is better than the previous best *pbest*, set the current fitness value as the new *pbest*.
5: After Steps 3 and 4 for all particles, select the best particle as *gbest*.
6: For all particles, calculate velocity using (24) and update their positions using (25).
7: If the stopping criteria or maximum iteration is not satisfied, repeat from Step 3.

5.3 One-Step-Look-Ahead Method

We also present a sub-optimal method based on OSLA for the sensor scheduling problem. The main reason for presenting a sub-optimal method is that DP may not produce a solution within a short time period for a large problem. Thus we use OSLA here to produce feasible solutions with lower computational cost than that of DP.

OSLA optimises only the current time step and therefore results are sub-optimal. At each time step, the OSLA finds a sensor group which minimises the current error covariance of the estimated state. OSLA produces a sub-optimal sequence of sensor groups as $\{\tilde{u}_1^*, \tilde{u}_2^*, ..., \tilde{u}_T^*\}$, where the $\tilde{u}_k^*$ is given by:

$$\tilde{u}_k^* = \arg\min_{u_k \in N_{l_k}} \sqrt{trace(P_{u_k})}, \qquad (28)$$

where $l_k = E_T - \sum_{t=1}^{k-1} E_{\tilde{u}_t^*}$ is the available energy budget at time step k. At each time step, OSLA calculates l_k and updates the set N_{l_k}. The computational cost of OSLA depends on the number of sensor groups in the set N_{l_k}. OSLA needs a total of $\sum_{k=1}^{T} |N_{l_k}|$ comparisons since it compares $|N_{l_k}|$ times at each time step.

6 Experiment and Simulation Results

The evaluation of the proposed methods are conducted via simulation. For simulation purposes, a sensor network of 40 nodes are considered to be deployed in a field of 1000 m × 1000m. The BS is located at the corner ([0,0]) of the network. The sensor nodes are divided into 6 pre-defined groups as shown in Fig. 1 and properties of the sensor groups are given in Table 2. For the purpose of simulation, error covariances of all sensor nodes in a sensor group

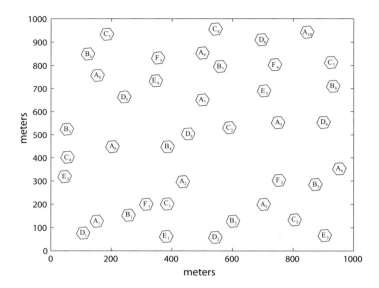

Fig. 1. Sensor groups and positions of sensor nodes in the network

Table 2. Properties of the sensor groups

Sensor Group (g_i)	Number of sensor nodes (n_i)	Energy Consumption (E_{g_i}) J	Error Covariance (R^i)
g_A	10	63	R^A
g_B	8	47	R^B
g_C	7	48	R^C
g_D	6	38	R^D
g_E	5	26	R^E
g_F	4	25	R^F

are assumed to be the same and time-invariant (i.e., $R_k^{i_j} = R^i$). However, models and the proposed methods can handle different error covariances for each sensor in a network. The error covariances of all sensor groups are given in Table 3. We assume for simulation purposes that the humidity measurement has higher noise than that of temperature.

The sensor model used in the simulation is the same as in (2) and the observation matrix H_k is assumed to be a 2×2 unit matrix. The parameters α_1 and α_2 in the energy model of (3) are set to 100 nJ/b and 1 pJ/(bm^2) respectively. Since ψ_1 & $\psi_1 >> \psi_{tx}$, only ψ_{tx} is considered in the simulation. It is assumed that the measured data size is 1 MB and the data rate is 8 Mbps.

Energy Aware Sensor Group Scheduling to Minimise Estimated Error

Table 3. The error covariances of six sensor groups considered for simulation

$$R^A = \begin{bmatrix} 1.2 & 0.0 \\ 0.0 & 3.5 \end{bmatrix}, R^B = \begin{bmatrix} 1.8 & 0.0 \\ 0.0 & 4.0 \end{bmatrix}, R^C = \begin{bmatrix} 1.4 & 0.0 \\ 0.0 & 3.8 \end{bmatrix},$$

$$R^D = \begin{bmatrix} 1.8 & 0.0 \\ 0.0 & 4.0 \end{bmatrix}, R^E = \begin{bmatrix} 2.0 & 0.0 \\ 0.0 & 4.3 \end{bmatrix}, R^F = \begin{bmatrix} 4.2 & 0.0 \\ 0.0 & 6.5 \end{bmatrix}.$$

The sensor groups transmit measurements to the BS every 0.5 hour for a total time period of 24 hours. Therefore, the BS has to schedule the sensor groups for $k = 1, 2, ..., 48$ time steps, such that $T = 48$. Energy consumption given in Table 2 is for transmitting 1 MB of data to the BS for the respective sensor groups.

The temperature and humidity of the environment vary as in (1) and the system matrix F_k is assumed to be a 2×2 unit matrix. For this simulation, the values of the process noise covariance Q_k and the initial error covariance $P_{0|0}$ are given in (29) and (30) respectively. Also for the purpose of simulation, noise in temperature and humidity are considered to be independent of each other. This is reflected in (29) and (30).

$$Q_k = \begin{cases} \begin{bmatrix} 1.0 & 0 \\ 0 & 5.0 \end{bmatrix} & \text{if} \quad 30 \geq k \geq 18 \\ \begin{bmatrix} 4.0 & 0 \\ 0 & 12.0 \end{bmatrix} & \text{if} \quad \text{otherwise} \end{cases} \quad (29)$$

$$P_{0|0} = \begin{bmatrix} 0.4 & 0 \\ 0 & 0.45 \end{bmatrix} \quad (30)$$

The results of the simulation are presented for different energy budgets in Table 4. We also consider a random selection method to schedule the sensor groups. In this method, sensor groups are selected randomly from the set N_{l_k} at each time step.

Results in Table 4 represent the mean value and standard deviation (SD) of the error cost for 30 independent runs obtained for the proposed methods. Since DP and OSLA are deterministic optimisation methods, the standard deviations are zero for all cases. Fig. 2 illustrates the variation of the cumulative error cost of the estimated state for energy budget $E_{48} = 2200$J for a single simulation. Fig. 3 shows the cumulative energy consumption of all the methods at each time step. Fig. 4 and 5 illustrates the variation of the Root Mean Square Error (RMSE) of the estimated temperature and humidity for the sensor group sequence obtained for the case $E_{48} = 2200$ J.

Table 4. Cumulative error cost $J(\mu(48))$ of the estimated state obtained by the DP, OSLA and random methods

Energy Budget (E_{48}) J	PSO (mean±SD)	DP (mean±SD)	OSLA (mean±SD)	random (mean±SD)
1300	50.1918± 0.001	50.187 ± 0.000	66.844 ± 0.000	63.770 ± 1.894
1400	49.0296± 0.000	49.060 ± 0.000	62.320 ± 0.000	61.325 ± 1.935
1500	47.8485± 0.003	47.897 ± 0.000	59.553 ± 0.000	57.644 ± 1.912
1600	46.6953± 0.002	46.778 ± 0.000	59.292 ± 0.000	54.616 ± 1.621
1700	45.6574± 0.001	45.624 ± 0.000	57.940 ± 0.000	51.592 ± 1.741
1800	44.4797± 0.004	44.548 ± 0.000	55.173 ± 0.000	48.543 ± 1.732
1900	43.4907± 0.003	43.728 ± 0.000	54.912 ± 0.000	46.583 ± 0.750
2000	42.3674± 0.001	42.728 ± 0.000	53.560 ± 0.000	45.506 ± 1.223
2100	41.1615± 0.002	41.618 ± 0.000	50.671 ± 0.000	44.863 ± 1.014
2200	40.0944± 0.000	40.793 ± 0.000	45.420 ± 0.000	44.842 ± 1.345

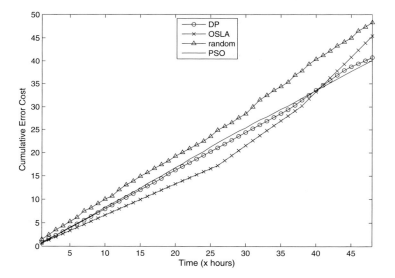

Fig. 2. Variation of cumulative error of the estimated state of the sensor network for an energy budget E_{48} = 2200 J

The results from Table 4 show that PSO performs better than all the other methods for the majority of times. DP performed better only in two instances when the energy budgets were 1300 and 1700 Joules. This was surprising because the PSO algorithm and the penalty method used in the simulation is not state-of-the-art and still it was able to outperform DP. This is due to the stochasticity of the PSO algorithm and its ability to find the global optimum.

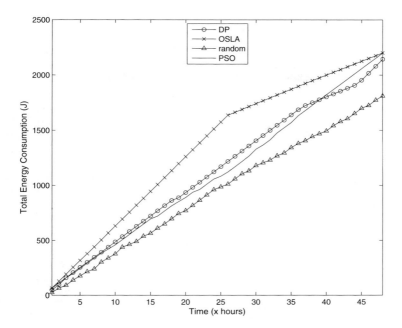

Fig. 3. Variation of the energy consumption of the sensor network for an energy budget, $E_{48} = 2200$ J

During the simulation, DP produces slightly better results when there is finer discretisation of the *state* spaces. However, the computational cost is high. In the simulation results displayed in Table 4 for DP, 340 *states* at each stage were used and the computation time is around 15 minutes for each run whereas less than a minute is needed for the OSLA and random methods. PSO took approximately 1 minute 14 seconds to simulate the problem for 500 iterations. All the simulations were run on the same computer. We can conclude that the PSO technique is superior in terms of minimising the cost estimation error and the computational cost in terms of execution time.

As explained in Section 5.2, PSO is a self-adaptive global search technique. The algorithm exhibits more stochasticity than other evolutionary algorithms due to the social behaviour of the particles in the population. There are several variants of the PSO algorithm, in the simulation time varying parameters version of PSO was used. In this PSO version, the parameters ω, c_1 and c_2 in (24) are varied over time (iterations) in the simulation. The ω was varied from 0.9 to 0.5, c_1 was varied from 2.5 to 0.5 and c_2 was varied from 0.5 to 2.5. The variation of c_1 helps the particles wander in a search space and minimise the possibility of converging to a local optima but, the variation of c_2 helps particles to converge to a global minima at the later stages of the algorithm. Due to the stochastic nature of the PSO, the simulation was

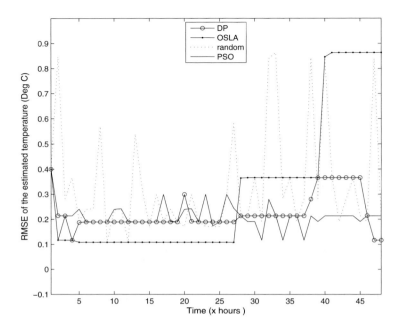

Fig. 4. Variation of the RMSE of the estimated temperature obtained for an energy budget, $E_{48} = 2200$ J

repeated a number of times and the results are the average of number of runs. Like any other method, the results in Table 4 are the average of 30 runs of the PSO algorithm. The PSO algorithm converged well with a simple penalty function method; this is partly due to the simplicity of the constraint because the objective function is minimised with only one constraint. The PSO-based sensor group scheduling algorithm was simulated by changing *pen* to 10, 50, 100, 200. But there was no significant difference in the results and the rate of convergence. The convergence of the PSO algorithm for different *pen* for an energy budget of $E_{48} = 1500$J is given in Fig. 6.

The OSLA is a myopic scheduling technique and therefore, at the initial stages of the simulation it selects a sensor group which provide less error as shown in Fig. 2. However, at the later stages it was unable to use the best sensor groups due to insufficient energy available in the network. It can also be seen from Fig. 4 and 5 that OSLA produces a lower RMSE of the estimated state than DP, PSO and the random method until $k = 27$ but not for the remaining time steps. Since DP optimally solves and stores sensor groups for all possible energy budgets at each time step, it wisely spreads energy usage over time and produces better results for the entire time period. In PSO, the problem is considered as a global optimisation. It optimises the error iteratively across the time step considering all the available sensor groups.

Energy Aware Sensor Group Scheduling to Minimise Estimated Error

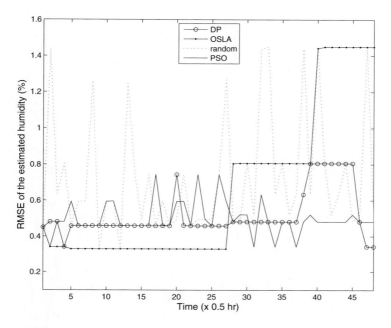

Fig. 5. Variation of the RMSE of the estimated humidity obtained for an energy budget, $E_{48} = 2200$ J

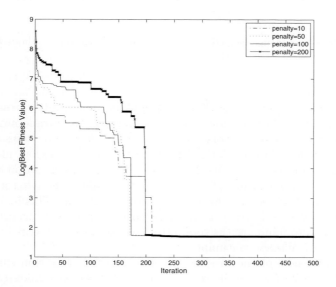

Fig. 6. The convergence of the PSO algorithm for different penalty values *pen* for an energy budget, $E_{48} = 1500$J

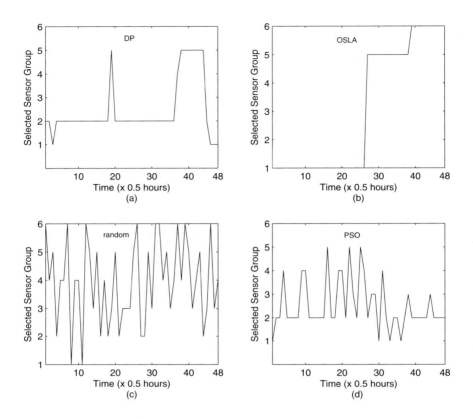

Fig. 7. The sensor group selected at each time step for all four methods for energy budget, $E_{48} = 2200J$

Most often the random selection method produces better results than OSLA. But, since the selection of sensor groups is purely random, the random method may produce worse results than OSLA in some cases as shown in Fig. 2. The standard deviation of the results for the random method indicates the instability of the method. It can be seen in Fig. 3 that the PSO, DP and OSLA used nearly the entire energy budget whereas the random technique used lesser energy for the same total time step. This is due to the lack of objectivity in the algorithm. In the other algorithms the objective of the problem is to minimise the total error cost for a given energy budget not to minimise the energy consumption.

The randomness of the random selection method algorithm can also be noticed in Fig. 7(c) where the sensor groups are selected randomly. The selected sensor group $\{1,..,6\}$ represents $\{g_A, ..., g_F\}$ in Table 2. The Fig. 7(b) reflects the working of OSLA, the best sensor group to minimise the error was chosen until all the nodes ran out of battery power. The sensor group selection of DP and PSO (Fig.7(a) and (d)) are quite similar, both the algorithms selected g_A four times and g_B was selected at the majority of the time steps. The PSO fared slightly better than DP due to its stochastic nature which gave particles an opportunity to explore the problem space more before converging to a global minima. The finer discretisation of the *state* spaces in DP can provide an optimum result, but the computational cost in terms of time to execute the algorithm will be high.

7 Conclusion

In this chapter, the problem of scheduling sensor groups to estimate environmental phenomena with minimal error for an energy-constrained sensor network was addressed. It is widely known that none of the sensor measurements are error free however, the estimation of the phenomenon is carried out based on the noisy measurements.

Several methods were proposed to minimise the error in the estimated state. The state of the phenomena consists of temperature and humidity which linearly vary with Gaussian noise. The sensor measurements are assumed to be linearly related to the state of the phenomena and corrupted by white Gaussian noise. The objective of the base station is to select a sequence of sensor groups to minimise the total error in the estimation for a given energy budget. The proposed solutions to the problem are based on the dynamic programming technique, particle swarm optimisation and the one-step-look-ahead method.

The PSO technique provides better results for most energy budgets compared to DP and OSLA, but DP has an ability to provide an optimum solution where other methods cannot. However, the computational cost is high for DP to obtain this optimum result. The PSO technique is useful because it can provide good results in a very reasonable amount of time and the algorithm is simple to understand and implement when compared to DP.

Acknowledgements. Authors would like to thank Greg Timms for his comments on the manuscript. Part of this work was conducted when Suhinthan was an Industrial Trainee at the CSIRO Tasmanian ICT Centre.

This project is jointly funded by CSIRO Water for a Healthy Country flagship and the Tasmanian Government. The CSIRO Tasmanian ICT Centre is jointly funded by the Australian Government through the Intelligent Island Program and Australia's Commonwealth Scientific and Industrial Research Organisation (CSIRO). The Intelligent Island Program is administered by the Tasmanian Department of Economic Development, Tourism and the Arts.

References

[1] Nakamura, E.F., Loureiro, A.A.F., Frery, A.C.: Information fusion for wireless sensor networks: Methods, models, and classifications. ACM Comput. Surv. 39(3), 9 (2007)

2. Lin, T.Y., Sehgal, V., Hamid, H.S.: Sensoclean: Handling noisy and incomplete data in sensor networks using modeling. Technical report, University of Maryland (2005)
3. Takruri, M., Rajasegarar, S., Challa, S., Leckie, C., Palaniswami, M.: Online drift correction in wireless sensor networks using spatio-temporal modeling. In: 2008 11th International Conference on Information Fusion, pp. 1–8 (2008)
4. Elnahrawy, E., Nath, B.: Cleaning and querying noisy sensors. In: WSNA 2003: Proceedings of the 2nd ACM international conference on Wireless sensor networks and applications, pp. 78–87. ACM, New York (2003)
5. Bychkovskiy, V., Megerian, S., Estrin, D., Potkonjak, M.: A collaborative approach to in-place sensor calibration. In: Proceedings of the Second International Workshop on Information Processing in Sensor Networks IPSN, pp. 301–316 (2003)
6. McCulloch, J., Guru, S.M., McCarthy, P., Hugo, D., Peng, W., Terhorst, A.: Wireless sensor network deployment for water use efficiency in irrigation. In: REALWSN 2008: Proceedings of the workshop on Real-world wireless sensor networks, pp. 46–50. ACM, New York (2008)
7. Benth, J.S., Benth, F.E., Jalinskas, P.: A spatial-temporal model for temperature with seasonal variance. Journal of Applied Statistics 34(7), 823–841 (2007)
8. Intel Data. Intel Lab Dataset (2009), http://db.csail.mit.edu/labdata/labdata.html/
9. Maheswararajah, M., Halgamuge, S., Premaratne, M.: Sensor scheduling for target tracking by sub-optimal algorithms. IEEE Transactions on Vehicular Technology 58(3), 1467–1479 (2009a)
10. Maheswararajah, S., Halgamuge, S.K., Premaratne, M.: Energy efficient sensor scheduling with a mobile sink node for the target tracking app. Sensors 9, 696–716 (2009b)
11. Evans, J., Krishnamurthy, V.: Optimal sensor scheduling for hidden Markov models. In: Proceedings of the 1998 IEEE International Conference on Acoustics, Speech, and Signal Processing, 1998. ICASSP 1998, vol. 4, pp. 2161–2164 (1998)
12. Rappaport, T.: Wireless Communications: Principles & practice. Prentice-Hall, Inc., New Jersey (1996)
13. Kalman, R.E.: A new approach to linear filtering and prediction problem. Transaction of ASME, Journal of Basic Engineering on Automatic Control 82, 35–45 (1960)
14. Song, E., Zhu, Y., Zhou, J.: The optimality of Kalman filtering fusion with cross-correlated sensor noises. In: 43rd IEEE Conference on Decision and Control, 2004. CDC, vol. 5 (2004) ISSN 0191-2216
15. Song, E., Zhu, Y., Zhou, J., You, Z.: The optimality of Kalman filtering fusion with cross-correlated sensor noises. Automatica 43, 1450–1456 (2007)
16. Tchomte, S.K., Gourgand, M.: Particle swarm optimisation: A study of particle displacement for solving continous and combinatorial optimisation problems. International Journal Production Economics 121(1), 57–67 (2009)
17. Bertsekas, D.P.: Dynamic Programming and Optimal Control. 3rd edn., vol. 1, Athena Scientific (2005)
18. Williams, J.L., Fisher, J.W., Willsky, A.S.: Approximate dynamic programming for communication-constrained sensornetwork management. IEEE Transactions on signal Processing 55(8), 4300–4311 (2007)

[19] Kennedy, J., Eberhart, R.: Particle swarm optimization. In: Proceedings of IEEE International Conference on Neural Networks, vol. 4, pp. 1942–1948 (1995)
[20] Homaifar, A., Qi, C.X., Lai, S.H.: Constrained optimization via genetic algorithms. Simulation 62, 242–254 (1994)
[21] Yeniay, O.: Penalty function method for constraint optimisation in genetic algorithms. Mathematical and Computational Applications 10(1), 45–56 (2005)
[22] Carlos, C.C.: Theoretical and numerical constraint-handling techniques used with evolutionary algorithms: a survey of the state of the art. Computer Methods in Applied Mechanics and Engineering 191(43), 1245–1287 (2002)
[23] Hu, X., Eberhart, R.: Solving constrained nonlinear optimization problems with particle swarm optimization. In: 6th Multiconference on Systems, Cybernetics and Informatics (SCI 2002), vol. 5 (2002)
[24] Parsopoulos, K.E., Vrahatis, M.N.: Particle swarm optimisation method for constrained optimization problems. In: Intelligent technologies - theory and applications: new trends in intelligent technologies, vol. 76, pp. 214–220 (2002)
[25] Coath, G., Halgamuge, S.K.: A comparison of constraint-handling methods for the application of particle swarm optimization to constrained nonlinear optimization problems. In: The 2003 Congress on Evolutionary Computation. CEC 2003, vol. 4, pp. 2419–2425 (2003)

Smart Home for Elderly Using Optimized Number of Wireless Sensors

A. Gaddam, S.C. Mukhopadhyay, and G. Sen Gupta

School of Engineering and Advance Technology
Massey University,
Palmerston North, New Zealand
S.C.Mukhopadhyay@massey.ac.nz

Abstract. In this chapter, we are reporting a novel in-home monitoring system designed to for elder-care application. The statistics shows that there is increasing number of elderly people around the world and this isn't going to change. We developed a smart system consists of optimum number of wireless sensors that includes current, bed, and water flow sensors. The sensors provide information that can be used for monitoring elderly by detecting abnormality pattern in their active daily life. The system will generate early warning message to care giver, when an unforeseen abnormal condition occurs. It will also, analyze the gathered data to determine resident's behavior. Instead of using many number of sensors, the importance of positioning the optimal number of intelligent sensors close to the source of a potential problem phenomenon, where the acquired data provide the greatest benefit or impact has been discussed.

1 Introduction

The elderly population of the world is growing rapidly, creating the need to increase elderly care. In New Zealand, the population of elderly people is increasing at a much higher rate compared to population under 65 years as is shown in figure 1[1]. The situation is veri similar in other parts of the world.

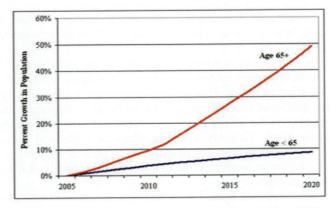

Fig. 1. Population growth in New Zealand, from 2000 to 2020

On other hand, the average human age i.e. the life expectancy of humans has increased over the past decade. This means that there are more elderly people living in the world today then there were 10 – 20 years ago [2]. The world population growth can be seen in the figure 2 below.

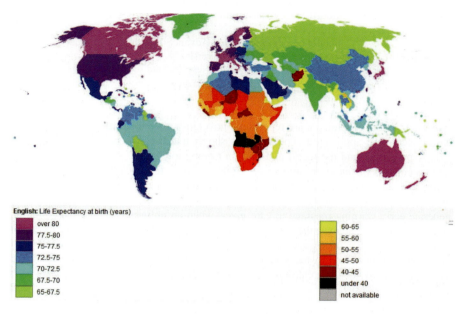

Fig. 2. Life Expectancy of the World [2]

Though this is really good for human civilization but it puts a lot of strain on the resources of a nation. The cost of elder care is only escalating. More and more elderly people are choosing to stay alone, independently, rather than in a retirement village or old people's home. Such people, often frail and infirm, do however require constant monitoring so that medical help can be provided immediately in times of dire needs. Considerable research efforts have been focused towards in-home monitoring of people [3], often using wireless personal area networks [4].

Due to the advancement in wireless communications and electronics over the past years, the development of low cost, low power multi functional sensors is receiving increasing attention. This sensing technology is playing a vital role in enhancing the quality of life for elderly. One of the most important features that human beings have is the ability to sense. We use our senses in our daily life to perform various tasks. Similarly a sensor acquires a physical parameter and converts it into a signal suitable for processing. As sensing technology continues to evolve, it is playing an important role in improving the quality of life for elderly people and their families.

A wireless sensors based home monitoring system, especially for elder people, has been undertaken and a prototype system has been developed, which detects the usage of household appliances. By integrating sensing in the house hold devices would allow people to be monitored constantly. The key benefit by this approach is to help increase health monitoring of elderly people, at risk patients, thus providing an ambient awareness. Moreover, this will help to cut down raising health care costs by increasing the health monitoring and improving patient to doctor efficiency.

2 Motivation – Need for Early Detection of Aging Changes

We aimed to develop a framework, dealing with the design intricacies and implementation issues of novel sensors, targeted to achieve a Smart Home specifically for the elderly. It aims to address the research issues and bridge the gaps by introducing prototypes, in order to achieve a smart home to monitoring elderly people. The smart homes equipped with the wireless sensor networks will benefit elderly people living alone. By monitoring continuously and reporting any abnormal situation, the system frees human labor and thus reducing labor costs and increasing efficiency by notifying doctors or health care provider quickly. The data collected from various sensors over the network in a smart home can be stored and integrated into comprehensive health record. Other issues like quality of life for elder people, such as privacy, independence, dignity and convenience are supported and enhanced by the ability to provide services in the patients own home. So the research hypothesis has been built on understanding the impediments in designing and implementing a wireless sensor network for elderly monitoring. The proposed system will be an intelligent, pervasive information handler using an efficient communications technology. Thus leads to emerge in to the users' environment so that the user can experience a rich variety of heterogeneous services and systems. By using this wireless sensing technology can actually lead to a safe, sound, secured environment for the elderly.

3 Literature Survey

The published works on the home monitoring discussed about the approach or building prototype, which will work in real world with limited practicality. Nonetheless, some of these studies had demonstrated the feasibility of remotely monitoring activities in the home [5]. Camera based system for home monitoring has not been popular to old people [6]. Many documented works has been reported to develop a system and method for monitoring an individual, especially for an old age person living independently, by distinguishing his/her normal activity from abnormal activity. This technological assistance or monitoring of a person in the home is achieved using various kinds of sensors, which are centralized in structure and distributed around the house. These sensors capture the activity of the person and the collected data is communicated to a remote monitoring centre using conventional data transfer techniques or by wireless data transfer techniques. In this system the types of sensors used include motion sensors, inside door sensors, cabinet sensors, kitchen appliances sensors and any other sensor suitable for collecting and communicating data regarding activities on-going in the home [3, 6]. But the system has at least one sensor at any point, which captures the activity of the resident. The sensor data is sent to the monitoring centre, where in it adds the data into a database. The collected data points are collated at least three slots per every twenty-four hour cycle for determining activity of the resident. Since one of the aspects of this invention is to distinguish between normal and unusual activity in a home, by evaluating the new data points, against the collated historical activity data present in the database, this determines whether the new data points indicate normal or unusual activity in the home. Even though many intelligent wireless sensor networks developed for specific applications have been reported [7], they have very limited practical applicability. In terms of installation and complexity, reports use of a huge number of sensors, require several experts many hours to install [8]. This makes the system commercially not viable and practically not feasible. Moreover, the system only monitors the daily activities of a person but does not generate pre-warnings or predict impending disaster/mishap [9]. Some abnormal events which can occur to an old person in a home often leads to more serious illnesses or even death. But in a case of reduced mobility or other factors which leads to this kind of situation should also be considered for effective monitoring, which will make a significant impact on the health of the elderly people. Similarly, the detection of poor medication compliance, changes in sleep pattern, changes in physiological parameters, and even changes in the cognitive abilities, are essential to managing the changing health status of the elderly person. In US06796799 [9], the system uses a detective approach to determining uncharacteristic behaviour of a monitored subject This system would provide an excellent method of determining whether the monitored subject is presenting normal behaviour or not. However, some potential problems in the system include, cost of the system, use of sensors such as cameras and toilet flushing sensors exist in it. Also, costs to install the system as each sensor utilises a different physical property and requires individualised installation is also high. Finally, the system provides such an insight into the subject's life that they could easily feel over monitored and uncomfortable.

In US06002994 [10], the system monitors activities throughout the home with individualised sensors such as fridge door open sensors and pressure sensitive mats. The variation for other patents is the number of sensors, as with patent US06796799 [11] would inevitably increase the complexity and the cost of the system.

Furthermore to implement wireless sensor networks for in home elderly monitoring, there is need to consider the following potentials and challenges:

(a) Interoperability and Interference, since sensors nodes are present in an in home wireless sensor network, there is a need to limit or avoid interference between sensors and various RF devices. The home care network must provide middle ware interoperability between disparate devices, and support unique relationship among various devices such as implants' and their outside controllers. In home situation, the system will have more interference due to wall and other obstructions.

(b) Real time data acquisition and processing: There is a critical need for efficient communication and processing the data over the sensor network. Some techniques which come handy for this kind of situation are event ordering, time stamping, synchronisation, and quick response in emergency situations.

(c) Reliability and robustness: Since the sensor network is not meant for frequent maintenance and not situated in a controlled environment, these devices and other sensors must be operated with good reliability, so as to provide reliable data for picking up and diagnosis any abnormalities in an elderly house hold.

(d) Data privacy and security: There is a need to protect the privacy of the person who is being monitored using wireless sensors network, the collected data may be sensitive. Data is only accessible during an emergency situation, but the accesses should have encryption, so abuses can be detected.

(e) Comfort and unobtrusive operation: The sensors which are designed for this task should be almost invisible and should be stealthy.

4 Wireless Sensors Based In-home Monitoring Using Optimized Number of Sensors

The advances in the sensor technology made us to have sensors which are very small and easy to deploy, they still present a momentous challenges in a smart home setting. The reason is the variation of the occupant's needs and the appliances which are used by them. A rather simple solution to overcome these challenges is to install huge number of sensors to monitor in a smart home environment [12, 13]. Even though the system will be very efficient in identify the lifestyle of the resident, it causes a new problem. As more sensors are added to meet these needs, the whole system becomes complicated to handle and the task of maintaining them can become more of a challenge. The cost of implementation will drastically increase with the rise of number of sensors [14]. Compared to the use of a large number of wireless sensors [13], the strategy of deploying a few but highly accurate numbers of inexpensive smart and intelligent sensors has significant advantages in case of home monitoring. The advantages include uniform coverage, small obtrusiveness, ease of deployment, reduced energy consumption, low on maintenance and consequently more acceptability by the elderly community [16].

The main point is to position sensors close to the source of a potential problem phenomenon, where the acquired data is likely to have the greatest benefit or impact. All sensors in the network must justify their existence and purpose in the system. To achieve this, a sensor must satisfy a quantifiable and verifiable set of requirements that will demonstrate its benefit and purpose to the system. In addition, any risk to the system's performance must also be identified. Therefore, substantial benefit to the design and development process can be realized by utilizing a sensor selection methodology, specifically directed toward system performance assessment. The sensor selection process can be divided into two parts; one is evaluation module and the other is optimization module. In Evaluation module, in - depth performance parameters of a particular type of sensor node or network under review are studied and the optimization module searches through the space of all possible sensor nodes for further evaluation. The process is continued until the "best" or the optimum sensor number is obtained. The aim of the research is to select limited number of sensors that fulfill specified performance requirements within a set of system constraints. These performance requirements are defined as the Figures of Merit (FOMs) of the system [19]. The basic components to build an algorithm based on FOM are Observability, Sensor Reliability/Sensor Fault Robustness, Fault Delectability/Fault Discriminability and Cost. Therefore, it is intended to develop an intelligent technique/algorithm or a mathematical model based on the above mention factors, to make an efficient and simple smart home monitoring system.

Many techniques have been developed for general optimized solution searches for many other applications, but none for in‐home health monitoring. The proposed monitoring system presented in this chapter, is based on the integration of different sensors, which has the capability of transmitting the data via wireless communication. The data are collated by a central processor, which saves all data for processing as well as future use. The lifestyle of the person under care is understood by the system by collecting data and comparing it with the stored pattern and depending on the situation, the actions are defined as unusual or normal. In addition, by considering issues related with quantitative approaches over the qualitative approaches, a maximum level of fidelity available in establishing the justification for sensor selection can be achieved. The methodology will be flexible enough to incorporate the best system design for home monitoring. The methodology incorporates other metrics, such as sensor reliability, communication techniques and sensor robustness, as needed.

5 System Description

While several sensors are readily available off the shelf, making them "intelligent" in the context of a specific application (such as monitoring of the elderly) is always a challenging task. For example, an intelligent wireless sensor system will not only detect the usage pattern of the daily appliances, it will have capabilities to collate the data and flag out anomalies.

Since it is a known fact that the use and the way of implementation of wireless sensor networks drastically change from one application to other application, the need for in-depth study on performance parameters is vital [20]. To achieve this we developed and implemented a SMART component-based system – S.A.M, Selective Activity Monitoring by integrating various sensors and communicate via standard RF protocols. The system depends on a set of selected number of wireless intelligent sensors and controller which relies on inputs from sensors. This system will consist of a proof-of-concept that we developed is feasible, reliable, practical and scalable.

Figure 3 shows the functional block diagram of the Selective Activity Monitoring (SAM) system [21]. The hardware components of the system are the Sensor Units, the Central Controller unit, a PC and a cellular modem.

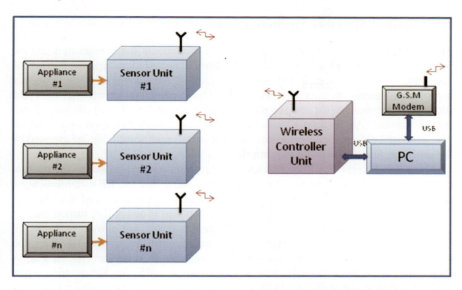

Fig. 3. Functional diagram of wireless sensor based home monitoring system

This wireless sensor system consists of three sensors for detecting use of electrical appliances, one sensor for bed use, one sensor for water use and a panic button for emergency need. The selected sensors were installed to monitor television, reading lamp, toaster, microwave oven, bed, among other locations as shown in figure 4.

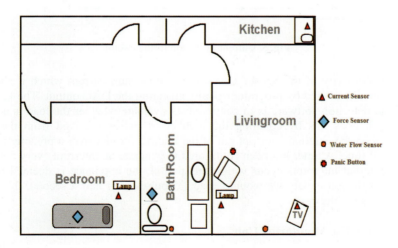

Fig. 4. Sensor's placement in a typical house-hold

5.1 Electrical Appliance Monitoring Unit

The electrical appliance monitoring unit consists of a current transformer to detect the flow of current in the phase line of the AC mains connected to an appliance. When an appliance is turned on, the current drawn generates an AC voltage of 50Hz frequency across the current transformer windings. This voltage is compared with the programmable voltage of the Digital-to-Analog converter (DAC) output of the microcontroller. An op-amp based comparator generates a transition at its output. The generated pulses, used as external interrupts to the microcontroller, are counted by the program running on the microcontroller.

There are 3 Light Emitting Diodes – one to indicate that power to the current detecting sensor unit is powered on, another to show that the RF communication with the current detecting sensor unit is currently in progress and the third to indicate that the appliance connected to the sensor unit is active, i.e. turned on.

The current transformer and the associated circuitry are shown in figure 5. Pulses, used as external interrupts to the microcontroller, are counted by the program running on the microcontroller. It is possible to know the type of load as well as duration of use with the help of medication of software. Fabricated current detecting sensor unit prototype is shown in figure 6.

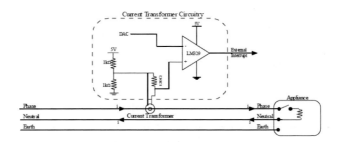

Fig. 5. Current flow monitoring circuit

The sensitivity of the Sensor Unit, i.e. the minimum current which it should detect, can be altered by programmatically changing the DAC output. The task of the load detection software is to continuously evaluate whether the load is active or not. It does this by counting the number of pulses received from the CT circuitry on the external interrupt pin of the microcontroller over a predetermined period of time. A timer has been configured to generate an interrupt every 500ms. If 10 external interrupts occur within this period, the appliance is deemed to be active. Although ideally we would expect 25 pulses over half a second, this was sometimes not the case.

5.2 Water-Use Monitoring Unit

After electricity the next most common consumable that a household uses is water. Monitoring the water use in the house will give a general overview of when

Fig. 6. Fabricated electrical appliance monitoring unit prototype

and how much water is being used in the home. The use of shower can be monitored with the help of water-use monitoring unit too. A flow sensor is the best way to monitor water use. A shower or bath is also of concern for a user as a slip in the shower can end very badly. And a bath is worse which can result in, if the person becomes unconscious or incapable of keeping their head above the water in a bath tub.

The shower can have the flow sensor installed after the shower mixer. If the shower is on for a longer than average period then the system will flag something could be wrong. As the system will still be running it is vital that the situation is dealt with more priority than an electrical device as it is less likely that a person will forget to turn the shower off. If the occupant has had a fall in the shower this could result in slip/fall and at minimum the occupant not being in a comfortable position as shower falls rarely end well.

For the prototype system an inline flow transducer [20] as shown in figure 7.

Fig. 7. Flow Transducer

This Flow sensor worked very satisfactorily for the prototype outputting a square wave that increased or decreased in frequency as the flow rate increased and decreased. The sensor can be positioned in many positions in the home, for example, it could go after the incoming water pipe or the hot water cylinder. The hot water cylinder gives an idea of when appliances like dishwashers and washing machines are being used (if they exist and, in the case of the washing machine, use a hot/warm wash). The sensor output is then connected to the microcontroller, as shown in Figure 8.

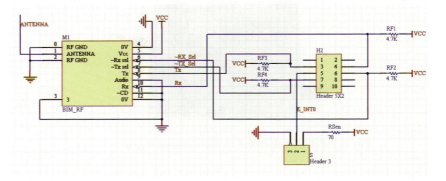

Fig. 8. Schematic diagram of water monitoring circuit

The flow sensor is connected to the same port, on the microcontroller, as the wireless module shown in Figure 8, as it uses the external interrupt which runs on the same port as the UART which is used for communication between the microcontroller and the wireless module.

The microcontroller has an external interrupt that was setup to be triggered when it saw an edge pulse on the port that it monitored. It provides the microcontroller with information that it can use to determine if the flow sensor has water flowing through it or not. The water monitoring wireless device communicates wirelessly with the controller unit. The fabricated prototype of the water monitoring unit is shown in figure 9.

5.3 Bed Monitring Unit

One important sensor used in the smart home is the intelligent bed monitoring unit. This sensor is used to monitor the use of bed and sleeping behavior of the user. This sensor is in particular useful to monitor elderly people living alone at home or people living at homes with disabilities. The monitoring unit use weight measuring technique to indicate if a person is on the bed or not. The bed monitoring unit can prove to be a life saving device for many users. There have been many different types of bed sensors made using wired and some use wireless technology [20], such as bed sensor mats and pads available in the market. A bed is a place where a person spends most of their time in the house. We consider three scenarios in which the bed sensor will be useful.

Fig. 9. Fabricated water monitoring unit

- A person who has a habit of waking up at a particular time in the morning sleeps beyond his usual wakeup time.
- A person who has a particular sleeping time and he/she is not in bed past their usual bed time.
- If a person is restless, while on bed.

The first two situations mentioned above, indicate that the person might have had a health issue or some abnormality happened. The third scenario may indicate that the person has become unwell or medically distressed.

This intelligent bed monitoring sensor system cannot only be used for elderly and patients but it can also be used to monitor the sleeping patterns of anyone. We can monitor people who have restless sleep at night and gather the results for that. It can also be helpful in monitoring someone who has a habit of sleep-walking. A prototype of a bed monitoring device using wireless sensor has been developed, which has demonstrated its feasibility for monitoring the activity of a user in bed. This system is different from the sensor pad-alarm systems. These sensors offer the monitoring of bed at low cost, less complexity and good accuracy. The system uses a Flexi-Force sensor [21] to measure the force and give the results according to that. The sensor is connected to a driving circuit which in turn is connected to a microcontroller to convert the analog readings from the circuit into digital. The microcontroller can send this information using RF (Radio Frequency) signals wirelessly to the controller unit. The sensor is placed under leg of a bed so they provide monitoring without any disturbance or discomfort to the user. The bed sensor system uses the Tekscan's Flexi-Force Sensors as shown in figure 10. The Flexi-Force sensors were chosen because of their reliability, cost-effectiveness and ease of integration.

Fig. 10. Tekscan's Flexi-Force sensor

The Flexi-Force sensors were strategically placed under the legs of the bed to determine if a force is being exerted on the bed i.e. someone is lying on the bed. The schematic representation of the four sensors placed under the bed is shown figure 11. S1, S2, S3 and S4 represent the four sensors Sensor 1, Sensor 2, Sensor 3 and Sensor 4 respectively.

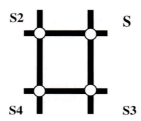

Fig. 11. The position of four sensors under the four legs of a bed

The block diagram representation of interfacing the sensors to the microcontroller is shown in as shown in figure 12.

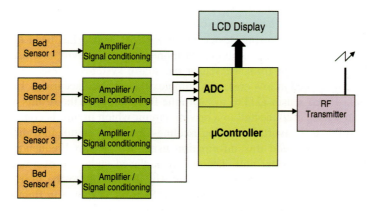

Fig. 12. Block Diagram representation of interfacing of force sensors to micro-controller

The necessary electronic circuit for signal processing of the force sensor is shown in figure 13. This circuit is driven by a 5V DC excitation voltage. The circuit uses an inverting amplifier to produce analogue output based on the sensor resistance and a fixed reference resistance (R_F). The dynamic force range (sensitivity) of the sensor can be changed by changing the value of R_F or by changing the drive voltage V_T. A lower reference resistance and/or a drive voltage will make the sensor less sensitive and increase its active force range. A -5V is also connected to drive the sensors.

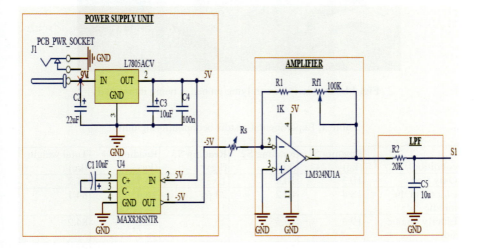

Fig. 13. The Analog circuit for the force sensor interfacing

The output signals from the microcontroller are used to determine all the information related to person's posture, quality of sleep and so on.

The experiments to test the bed sensor system were conducted with three inhabitants, one elderly person of 74 years of age as shown in figure 14 and around 42 kg of weight, one 5 year old child with 20 kg of weight and one adult with 68.5 kg of weight.

The results are obtained from the experiment are shown in table 1 below. We can see from these results that under normal situation all the four sensors read very similar weights. The sensors only measure the weight of the bed in this situation, so each sensor shows roughly one-fourth of the weight of the bed. When a person lies on the bed the sensors read different readings depending on the amount of weight shared by that particular leg of the bed. The total weight measured by the four sensors is roughly equal to the weight of the bed and the weight of the person. From table 1 it can be easily said whether the bed is occupied by an elderly or a child from weight consideration. The position of the head whether the head is at (S1, S2) side or (S3, S4) side can easily be said by checking the difference of the sensors signals. The pair of sensors with the head side on them would show higher reading as compared to the other sensors.

Fig. 14. An elderly person lying on bed is being monitored

Table 1. Experimental results bed monitoring unit

Test condition	Reading S1 (kg)	Reading S2 (kg)	Reading S3 (kg)	Reading S4 (kg)	Total weight (kg)
Only Bed	10.3	10.4	10.3	10.5	41.2
Elderly in the middle	18.4	15.2	23.6	26.9	84.1
Elderly on one side	19.3	16.5	26.2	22.2	84.2
Elderly on another side	14.0	21.3	18.5	30.5	84.3
Child in the middle	13.5	14.3	16.9	17.0	61.7
Child on one side	14.6	10.5	21.6	15.1	61.8
Child on another side	12.2	13.4	13.7	22.4	61.7
Adult in the middle	18.7	20.8	36.7	33.8	110
Adult on one side	25.2	14.2	45.5	25.3	110.2
Adult on another side	13.7	22.5	25.8	48.1	110.1
Adult lying diagonally (S4-S1)	17.2	20.5	28.2	44.3	110.2
Adult lying diagonally (S3-S2)	21.5	17.4	44.8	26.5	110.2

The following observations are made with reference to figure 11:

S1 : Signal from Sensors#1.
S2 : Signal from Sensors#2.
S3 : Signal from Sensors#3.
S4 : Signal from Sensors#4.

Savg = (S1 + S2 + S3 + S4)/4; Savg is the average signal.

If (S1 + S2) > 2* Savg; the head is at (S1, S2) side.

If (S3 + S4) > 2* Savg; the head is at (S3, S4) side.

If (S1 + S3) = (S2 + S4); the person is sleeping in the middle of the bed.

If (S1 + S3) > (S2 + S4); the person is sleeping in the right side of the bed.

If (S1 + S3) < (S2 + S4); the person is sleeping in the left side of the bed.

If the sleep quality is good, the four sensors will provide signals which are quite steady. If the person is suffering from lack of sleep, the sensors will not provide steady signals. The signals from the sensors can be studied and conclusions can be drawn based on sensors' information. If the person comes out of the bed at night, the time duration for which the bed is not used is monitored. The time information can be used for making some decision. Since the bed sensing system has been configured around a microcontroller, the processor can do some amount of processing and the conclusions along with the sensors output is sent to the central controller. The central controller can do another level of processing, to take final decision.

5.4 Emergency Button

In case of emergency or an urgent assistance, there is an option for the users to use the emergency button or panic button. It is integrated in to SAM system. The panic button as shown in figure 15, has the highest priority than the other sensor units. Once the emergency button is set by the user, the controller reads the status of the unit and sends message (SMS) to the concern person or emergency services.

Fig. 15. Emergency Button

5.5 The Cellular Modem

Wavecom Wismo cellular modem [22] used within the S.A.M. It is a GSM cellular modem in figure 16, which runs the same protocol as Vodafone New Zealand, simply by inserting a Vodafone New Zealand SIM card we are able to connect into the New Zealand network thus enables the SAM to send SMS to any cell phone around the world. The cellular modem comes with its own power supply and only has one connection to the personal computer which is completely independent of the SAM controller's connection. The cellular modem receives commands from the Personal computer through USB interface. In the future prototype, the modem will be integrated inside the controller unit.

Fig. 16. Cellular Modem and its interface with PC

5.6 Radio Frequency Communication Protocol

All the sensor units communicate with the central controller unit on the same frequency, it is important that a protocol be established to avoid data collisions. Data collisions occur when multiple devices try to communicate simultaneously on the same channel. The basic principle behind the protocol used in the SAM system is that sensors will respond only when told to do so. This means the controller is the only device that can transmit on the radio channel, unprompted. The controller is in fact the master which decides which sensor unit should transmit. Figure 17 below explains the sequence of actions that take place when the sensor unit and controller unit communicate with each other.

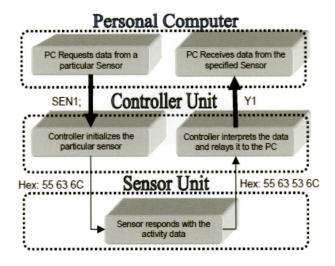

Fig. 17. Communication Protocol between controller and sensor unit

The personal computer determines which sensor it requires data from, in this case it requires data from sensor one. It therefore sends the command "SEN1;" serially over the RS232 interface to the controller. The controller receives this command and converts it to an appropriate hexadecimal / binary packet for radio transmission. The packet is received by all the sensor unit's but only the specified sensor unit will respond. This ensures that only one SU 'talks' at any time. The packet sent back by the sensor unit contains the status byte indicating whether the appliance is active or inactive. The controller receives this response and transmits it to the personal computer, via the RS232 link. The computer then confirms that the correct sensor has responded and continues with the monitoring process, querying the next sensor unit.

5.7 Interface and Control Software

The interface and control software has been written in Visual Basic and runs on the PC. It allows the user to configure the sensor units; up to 6 six sensors may be selected and associated with different appliances. The Graphical User Interface (GUI) to accomplish this is shown in figure 18. Once the sensors have been set up, the monitoring can start.

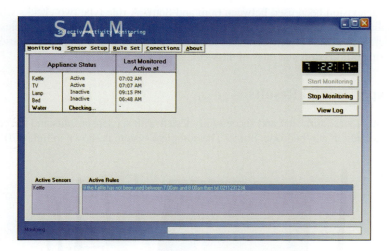

Fig 18. Graphical User Interface (GUI)

Once the sensors have been configured, the rule creation can begin. There are two methods of creating a rule. The user can use an elaborate rule creation wizard and be guided through the rule creation process with examples at each step. Or, the user can bring up an immediate window which allows for a much faster rule creation process. The rule creation wizard would be the recommended method initially, until the user has come to terms with the software. The GUI of the rule creation wizard is shown in figure 19.

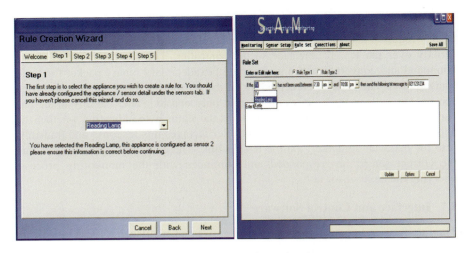

Fig. 19. Rule Creation Wizard

The user can select a rule to either be a rule type one or rule type two. A rule type one checks whether an appliance is activated within a given time period. For example, has the television been used between 6pm and 8pm? A rule type two checks whether an appliance is still active after a predetermined time. For example, has the television been left active after 11pm? The rules are based around a 24-hour cycle. The user can turn a rule off on particular days of the week.

6 Experimental Results

The system was fabricated and successfully tested with six sensor units and the central controller unit in a 3- bedroom home. The complete system is shown in figure 20.

Fig. 20. Developed home monitoring system

The sensor unit was successfully tested for various conditions, and load. It is observed from this real time testing, that the communication between the sensor units and the controller is very reliable. A Wavecom Wismo cellular modem was used for sending text messages to a mobile phone.

During this experimental stage, the reliability and efficiency of the sensors are tested for various conditions, and loads. By considering various factors like reflections due to walls, noise caused by household appliance the sensor node RF communication reliability was also tested and the results obtained are satisfactory. Figure 21, Shows the communication success rate as a function of distance in a home environment.

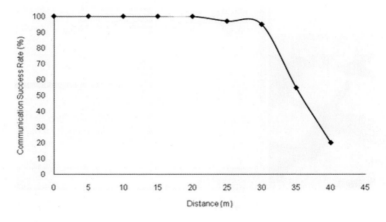

Fig. 21. Communication success rate as a function of distance in home environment

As part of the system testing, a data logger was developed and integrated into the SAM control software. It logs every activity, which can be viewed later. Figure 22 shows the activity log viewer.

In this real time scenario we have connected the sensor to the following appliances – microwave oven, hot water kettle, television, toaster, and bed. Various rules are set on the supervisory program and the system is left to collect the data for over a week. From figure 23, the daily activity pattern of the resident/s can be observed and understood.

This pattern helps us to identify the lifestyle of the person effectively and set the rules in system effectively. Since the sensors are almost invisible to the user and no cameras are used, the acceptance level to use the system in a house hold environment is good and privacy of the resident will not be hindered.

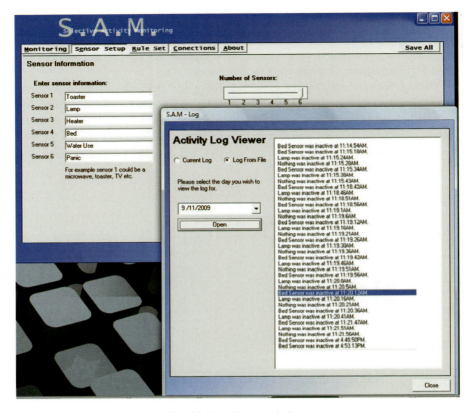

Fig. 22. Data logger window

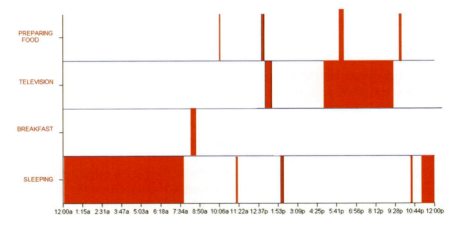

Fig. 23. Daily Activity pattern

7 Conclusions and Future Work

The work described here is on-going but demonstrates that smart, intelligent, wireless sensor devices can be used to recognize use of appliances necessary for daily living and therefore determine the life style of elderly person living alone. The system developed in this work uses a limited number of sensors and can be installed/ maintained in residential environments with ease. Moreover, the proposed sensing system presents an alternative to sensors that are perceived by most people as invasive such as cameras and microphones. The system can be easily installed in an existing home environment with no major modifications or damage. In future, a lot more functions are planned to be incorporated in the sensor itself.

References

[1] Bowman, R.: Doctors are where patient's are, http://www.dailyyonder.com/doctors-are-where-patients-arent/2009/02/12/1924 (Retrieved October 23, 2009)
[2] Koontz, D. Life expectancy, http://en.wikipedia.org/wiki/Life_expectancy (Retrieved October 24, 2009)
[3] Dittmar, A., Axisa, F., Delhomme, G., Gehin, C.: New concepts and technologies in home care and ambulatory monitoring. Studies in Health Technology and Informatics, 9–35 (2004)
[4] Jovanov, E., Raskovic, D., Price, J., Chapman, J., Moore, A., Krishnamurthy, A.: Patient Monitoring Using Personal Area Networks of Wireless Intelligent Sensors. Biomedical Sciences Instrumentation, 373–378 (2001)
[5] Ohta, S., Nakamoto, H., Shinagawa, Y., Tanikawa, T.: A health monitoring system for elderly people living alone, PMID: 12097176 [PubMed - indexed for MEDLINE]
[6] US20030058111, Computer vision based elderly care monitoring system
[7] Maki, H., Yonczawa, Y., Ogawa, H., Sato, H., Hahn, A.W., Caldwell, W.M.: A welfare facility resident care support system. Biomedical Sciences Instrumentation, 480–483 (2004)
[8] Munguia Tapia, E., Intille, S.S., Larson, K.: Activity recognition in the home setting using simple and ubiquitous sensors. In: Ferscha, A., Mattern, F. (eds.) PERVASIVE 2004. LNCS, vol. 3001, pp. 158–175. Springer, Heidelberg (2004)
[9] US Patent US06796799, Behavior determining apparatus, care system, care residence and behavior information specifying apparatus and system
[10] Callaway, E., Gorday, P., Hester, L., Gutierrez, J.A., Naeve, M., Heile, B., Bahl, V.: Home Networking with IEEE 802.15.4: A Developing Standard for Low-Rate Wireless Personal Area Networks. IEEE Communications Magazine, 69–77 (August 2002)
[11] US Patent 6002994 - Method of user monitoring of physiological and non-physiological measurements
[12] US Patent 4990893 - Method in alarm system, including recording of energy consumption

[13] Eriksson, H., Timpka, T.: The potential of smart homes for injury prevention among the elderly. Injury Control and Safety Promotion 9(2), 127–131 (2002)
[14] Dengler, S., Awad, A., Dressler, F.: Sensor/Actuator Networks in Smart Homes for Supporting Elderly and Handicapped People. In: Proceedings of the 21st International Conference on Advanced Information Networking and Applications Workshops 2007, vol. 2, pp. 863–868 (2007)
[15] Joshi, S., Boyd, S.: Department of Electrical Engineering, Stanford University Sensor Selection via Convex Optimization. IEEE Transactions on Signal Processing, 321–325 (November 2007)
[16] Giraud, C., Jouvencel, B.: Sensor selection: A geometrical approach. In: Proceedings of the IEEE/RSJ International Conference on Intelligent Robots and Systems, vol. 2, pp. 45–49 (1995)
[17] Zhang, H., Zhang, H.: Node Selection Algorithm Optimized for Wireless Sensor Network. In: Proceedings of 2008 Workshop on Knowledge Discovery and Data Mining, pp. 481–484. IEEE Computer society, Los Alamitos (2008)
[18] Figueredo, M., Dias, J.: Mobile telemedicine system for home care and patient monitoring. In: Proc. 26th Annual International Conference of the IEEE Engineering in Medicine and Biology Society, IEMBS 2004, vol. 2, pp. 3387–3390 (2004)
[19] Mukhopadhyay, S.C., Gaddam, A., Gupta, G.S.: Wireless Sensors for Home Monitoring - A Review. Recent Patents on Electrical Engineering 1, 32–39 (2008)
[20] Inline Flow Transducers, Sensor,flow,liquid,0.25-6.5L/min,pulse O/P, http://newzealand.rsonline.com/web/search/searchBrowseAction.html?method=getProduct&R=0257149
[21] Gusakov, I.: Bed patient position monitor, United States Patent 5184112 (1993)
[22] Tekscan. FlexiForce force sensors, http://www.tekscan.com/Flexi-Force/Flexi-Force.html (Retrieved October 25, 2009)

Estimation of Packet Error Rate at Wireless Link of VANET

Hao Jiang, Yang Yang, Jun Xu, and Lin Wang

School of Electronic Information, Wuhan University
jianghaow@263.net, gmame@163.com,
xujunlinda@126.com, 545073261@qq.com

Abstract. Node motion and complex radio environment make packet loss estimation in VANET difficult. However, packet loss estimation impacts routing protocol and transmission control algorithm of VANET, so it is an important issue. In this chapter, we measured packet error of VANET in real urban road, and analyzed the characteristics of packet error in VANET, described packet error by a packet-level Makov (PLM) model and used GMM (Gaussian Mixture Model) to present probability density of packet error, and then proposed two methods to estimate packet error, one is based on PLM model, and another is RPEE (real-time packet error estimation) which adopts GMM of probability density of packet error in VANET.

1 Introduction

VANET is a special kind of MANETs (Mobile Ad hoc Networks). Besides some common characteristics (e.g., moving nodes and open channels) of MANETs, VANET does have its own unique features [1], such as

1) Highly dynamic topology
2) Frequently disconnected network
3) Sufficient energy and storage
4) Geographical type of communication
5) Mobility model and predication
6) Delay constraints
7) Interaction with on-board sensors

VANET can be used for vehicle to vehicle (V2V) communications and vehicle to infrastructure (V2I) communications. It has a list of applications that can be categorized as follows [2]:

1) Safety: traffic signal violation warning, curve speed warning, emergency electronic brake light, pre-crash sensing, cooperative forward collision warning, left turn assistant, lane-change warning, and stop sign movement assistant.

2) Transport efficiency: route guidance and navigation, green light optimal speed advisory, and lane merging assistants.
3) Information/entertainment: tolling (one of the initial motivations for vehicle-to-infrastructure communications), point-of-interest notifications, fuel consumption management, traffic information broadcasting, and multi-hop wireless Internet access.

1.1 Related Works

Vehicular communications is regarded as being a major innovative feature for the in-car technology. At present, all major industrialized countries are actively promoting research focusing on this issue. Automobile manufacturers are pushing product development through in-house research and standardization efforts. Many projects on VANET, such as CarNet [3], CarTalk [4], CarTel [5], E-Road [6], C3 [7] and Vehicular Networking project of WinLab [8], have been developed. Meanwhile, a large amount of research efforts have been devoted into the MAC layer design, routing, transmission control protocol, congestion control and wireless link of VANETs.

MAC

The communication channel of VANET is quite different from the conventional wireless networks due to the high dynamics and the mobility of nodes, the frequent topology change, the high variability in nodes density and neighborhood [9]. Hence, the conventional IEEE 802.11 standard is not suitable, and IEEE 802.11p (Wireless Access in the Vehicular Environment (WAVE)) [10] is proposed to support Intelligent Transportation Systems (ITS), which include data exchanges between high-speed vehicles and roadside infrastructures. In IEEE 802.11p, IEEE 802.11 MAC functions are improved for rapidly changing communication channels. MAC sub-layer authentication and association is not required any more before data transfer in WAVE Mode Basic Service Set (WBSS). As described in [11], there are still some challenges, such as stateless channel access, caching for handoff and opportunistic frame scheduling.

Routing

Finding and maintaining routes is very challenging in VANET since the network topology and the characteristics of communication channels may vary frequently. The protocols of data dissemination thus play an important role for effective realization of VANET applications. Routing in VANETs has been studied extensively, and any solutions have been proposed, such as Global state routing (GSR) [12], Anchor-based Street and Traffic Aware Routing (A-STAR) [13], and Greedy Perimeter Coordinator Routing (GPCR) [14]. GSR [12] combines the position-based routing with the topological knowledge. It is a promising routing strategy for VANETs in urban environments. A-STAR is a position-based routing scheme designed specifically for inter-vehicular communications in urban environments. GPCR is designed to deal with the challenges of city scenarios, and the main idea of GPCR is to forward data packets using a restricted greedy forwarding procedure. Papers [15-16] propose routing protocols suitable for city environments. Paper [17] proposes a routing protocol for VANET for rural areas.

Estimation of Packet Error Rate at Wireless Link of VANET 331

Transport protocol
About TCP in VANET, paper [18] studies path characteristics of VANET in highway scenarios. Through analysis, it derives the upper bounds of the expected connectivity and disruption duration. It also briefly outlines a preliminary VTP design that incorporates these analytical results. Paper [19] proposes an approximately modified version of TCP Snoop, called on-board TCP (obTCP) for vehicular on-board IP networks. The main idea is to shield the sender from the non-congestion related losses without changing the existing TCP implementation at the end hosts. Some works have been done to analyze the effect of tuning transmission power on the performance of the transport layer protocols. Paper [20] specifically looks at the effect of tuning transmission power on the UDP throughput. Paper [21] analyzes the performance of multi-hop TCP in a multi-lane highway where vehicles are configured as clients or routers trying to reach a fixed access point. In particular, it studies the effects of tuning transmission power on the TCP throughput and latency in dense and sparse roads, respectively.

Congestion control
The congestion control methods for VANET can be classified into two categories: end-to-end and hop-by-hop. The first addresses the flow fluidity between senders and receivers, while the second addresses flows fluidity between a sender and a receiver [22]. Torrent-Moreno [23] et al. find the relationship between the transmit power and the channel load. They then proposed fair power adjustment for vehicular (FPAV), by which the channel congestion can be avoided. It should be noted that FPAV requires tight synchronization among the nodes and a global knowledge of the channel load, which are not easily to obtain. A utility-based congestion control approach is proposed in [24], which the bandwidth is dynamically assigned according to the utility value of the messages in each device.

Wireless link
The high mobility of nodes in VANET results in frequent fragmentation and short-life links. The environment of VANET is complex. The link quality is affected by many factors, such as buildings, trees, road surfaces and even the weather. Having a good knowledge of VANET link is important for the design of the upper layer protocols. The use of link quality feedback in terms of physical layer information for the routing protocol has been considered in many instances [25]. The smooth SNR value measured at links is used as a metric for routing decision and as a trigger for hands-off initiation in [26]. Various methods of predicting the link quality are proposed. Paper [27] proposes using the changing rate of the received signal strength to predict the link available time. Paper [25] proposes a cross-layer approach, where the received power metric logged at the physical layer is used to produce estimates of the links' residual lifetime. Paper [28] proposes to use a dominant pruning (DP) algorithm, instead of using the conventional blind flooding in address auto-configuration protocol (AAPs), to reduce the broadcasting redundancy in MANET AAP in the status of link change. With help of the model MM (mobility models), paper [29] identifies the MANET topology and its live time. There are a few prior efforts in the wireless domain that have tried to diagnose the wireless packet losses. Paper [42] et. al. proposes a promising technique called

Collision Inferencing Engine(COLLIE), which employs a direct approach by using explicit feedback from the receiver to immediately determine the cause of the packet loss. Paper [43] propose a novel scheme to detect collision, which utilizes transmission time information and RF energy duration on the channel. Paper [44] describes the Multi-Radio Diversity (MRD) wireless system, which uses path diversity to improve loss resilience in wireless local area networks (WLANs). Paper [45] present a system called Jigsaw that uses multiple monitors to provide a single unified view of all physical, link, network and transport-layer activity on an 802.11 network.

1.2 Our Works

In this chapter, we measured packet error of VANET in real urban road. From the measurement result, we analyzed the characteristics of packet error in VANET, described packet error by a packet-level Makov(PLM) model and used GMM (Gaussian Mixture Model) to present probability density of packet error. Based on analysis of packet error, we proposed two methods to estimate packet error, one is based on PLM model, and another is called real-time packet error estimation (RPEE) which adopts GMM of probability density of packet error in VANET.

In this chapter, we analyze distribution of packet error rates and their dependence of gap length and burst length. In this chapter,the gap length represents the length of consecutive received packets, the burst length represents the length of consecutive lost packets. Fig 1 shows the gap length and burst length.

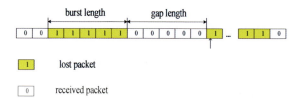

Fig. 1. Gap length and burst length

It is found that node-to-node distance has an impact on the quality of link in VANET. The link quality is not varying with time. No matter the car traffic is heavy or light, the packet error rates under the same scenario share the same probability distribution if the node-to-node distance is in the same range.

The gap length and burst length are dependent on each other. Based on the statistical dependence between burst lengths and gap lengths, we propose a packet-level Markov (PLM) model which is capable of describing the measured statistics of Packet Error Rate properly. The parameters of the model are obtained from the packet error statistical characteristics. And we demonstrate that the accuracy and the efficiency of packet error estimation based on PLM model are better than other popular markov models.

We also propose real-time packet error estimation (RPEE) method to estimate packet error rates of VANET. RPEE can estimate the packet error rates correctly

Estimation of Packet Error Rate at Wireless Link of VANET 333

and quickly with a few probe packets. It can be used to grasp the overall characteristics of the link, and support the data transmission control and routing algorithms in network protocols.

2 Measurement of Packet Error Rate in Vanet

Two vehicles are used in our experiments. One is configured as a transmitter (T Vehicle) and the other as a receiver (R Vehicle). Each vehicle contains embedded-PCs (EPC) which are built with a mini-itx board, an automotive power supply, a celeron processor, storage devices and a high-powered 802.11b miniPCI Wi-Fi card (Atheros AG5005G) with an external omni-directional antenna (8dBi). The GPS devices record the latitude and longitude of the vehicles every second.

Table 1 shows the measurement configurations

Table 1. Measurement configurations

Parameters	Values
Antenna Type	Omni-directional
Antenna Height	3.2 meters
Antenna Gain	8 dBi
Wireless adapter	Intel® 2200BG
Wireless adapter power	100 mW
Networking mode	Ad-hoc
MAC protocol	802.11b/g
Operating System	Linux

The transmitter broadcasts 100 UDP packets per second at 1 Mbps (the lowest rate). The UDP payload consists of a 32-bit sequence number that is incremented by the transmitter for every successive packet and local GPS record,the UDP packet format can be seen in Table 2. The receiver sniffs the packets from the wireless interface and keeps a trace of the lost ones by checking the UDP packet sequence numbers. Fig 2 shows the measurement method.

The following measurements are taken in urban of Wuhan city, China. The two vehicles were both in motion, their velocities were limited to less than about 60km/h, and inter-vehicle distance were from approximate a few meters to 110

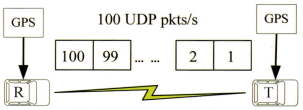

Fig. 2. Measurement method

Table 2. Format of data

Experiment times	Message	Sequence number	Payload

meters. The environment between the two vehicles is disturbing and unstable due to blockage of the line of sight (LOS) signal by large vehicles such as trucks and buses.

The experiments were conducted in Luoyu Road and Zhongnan Road, Wuhan, China. There are many tall buildings and vehicles, and they are typical urban environment. In the urban environment, the typical communication range between vehicles is within 110 meters, Fig 3 shows the road of experiments. The experiments were conducted at differrent times in a day, every three days from January 2008 to April 2008, and each experiment lasted 2 hours. About 60 sets of samples were obtained.

Fig. 3. Environments of experiments

3 Analysis of Packet Error in VANET

3.1 Burst Length and Gas Length

FINITE-STATE Markov modeling of communication channel is a simple and effective approach for communication channel description [30]. The first model of burst channel was presented by Gilbert [31], which is a two-state Markov model. Elliott [32] then enhanced it to Gilbert-Elliott's model. The parameters for these forenamed two models can be calculated directly from the measured error trace by using the mean burst length (the error run) and the mean gap length (the error-free run). Furthermore, both burst length and gap length are geometrically distributed. Fritchman [33] expanded the Gilbert-Elliott's model to a partitioned Markov model with several error-free and error states. Being a renewal model, the gap length distribution uniquely specifies the model. G.T. Nguyen [34] presented a two-state model with segmented exponential distributed burst lengths and gap lengths.

The packet error sequence may be represented as a binary sequence $\{X_k\}$ where $X_k=1$ if the kth packet is in error, and $X_k=0$ otherwise. For clarity, the definitions of the used terms [35] in relevance to the packet error sequences are introduced in this subsection. A burst is defined as a string of consecutive ones between two zeros and the burstlength is defined to be the number of ones. Likewise, a gap is defined as a string of consecutive zeros between two ones and the gaplength is defined to be the number of zeros. $p_f(m)$ represents the probability mass function (PMF) of the burstlengths for all positive integers m, $p_f(m) = P(1^m 0 | 0)$. The unconditional cumulative distribution function (CDF) is then defined as $F_B(m) = P(m_b \leq m) = \sum_{m_b=1}^{m} P(1^{m_b} 0 | 0)$, where m_b represents the burst length. In the same way, the unconditional CDF of gap lengths, $F_G(m)$, is $F_G(m) = P(m_g \leq m) = \sum_{m_g=1}^{m} P(0^{m_g} 1 | 1)$, where m_g is the gap length.

It is noted that the burst and gap are alternant in a packet error sequence. Hence the CDFs of burstlengths conditioned on a certain previous gaplengths, $F_{B|G}(m|n)$, is very helpful for investigating the correlations between burst and gap, which is defined as:

$$F_{B|G}(m|n) = P(m_b \leq m | \hat{m}_g = n) = \sum_{m_b=1}^{m} P(1^{m_b} 0 | 0^{\hat{m}_g}) \qquad (1)$$

where m_b is the length of current gap and $\hat{m}_g$ represents the length of previous burst. Similarly, the CDF of gap lengths conditioned on previous burst lengths is

$$F_{G|B}(m|n) = P(m_g \leq m | \hat{m}_b = n) = \sum_{m_g=1}^{m} P(1^{m_g} 0 | 0^{\hat{m}_b}) \qquad (2)$$

where m_g is the length of current gap and $\hat{m}_b$ represents the length of previous burst.

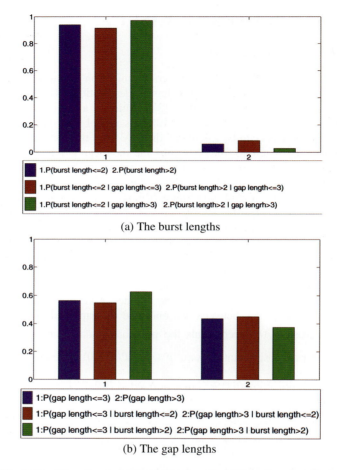

Fig. 4. Conditional probabilities of the burst lengths and gap lengths

To analyze the statistical dependency between burst and gap, the simple and conditional probabilities of burst (conditioned on the length of the previous gap) of having a short gaps (gap length ≤ 3) and long gaps (gap length > 3), are shown in Fig 4(a). As the simple and conditional probabilities are different, there is no statistical independence between gap lengths and burst lengths. Similar results about gap lengths conditioned on the previous burst lengths can be shown in Fig 4(b). Therefore, some conclusions that the gap and burst are likely statistical dependent can be drawn.

In addition, our intention is to keep the complexity of the model as small as possible. Therefore, when calculating the conditional CDF of burstlength, the conditioned variables are grouped into two sets (short bursts and long bursts). The new set of conditional CDF is shown in Fig 5(a). Similarly, while calculating the conditional CDF of gap length, the conditioned variables are grouped into short

Estimation of Packet Error Rate at Wireless Link of VANET 337

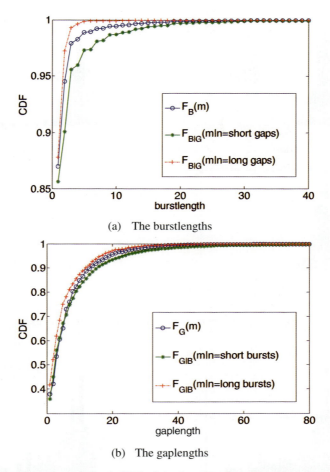

Fig. 5. The new set of conditional CDF of the burstlengths and gaplengths in dynamic. scenarios.

bursts (burstlength ≤ 2) and long burst (burst length > 2). The new set of conditional CDF is shown in Fig 5 (b). In order to capture these statistical dependencies appropriately we propose the PLM. The detail of PLM is shown in subsection 4.1.

3.2 Statistical Properties of Packet Error Rate

A large amount of measurement data were obtained from real road testing. Fig 6, 7 and 8 plot the probability density functions (PDF) of the packet error rate at different distance ranges. Fig 6 shows the PDF of PER when the node-to-node distance is within 10-20 meters (m). Fig 7 and 8 show the PER's PDFs at the distance ranges of 20-30 m and 30-40 m, respectively. It can be seen that.

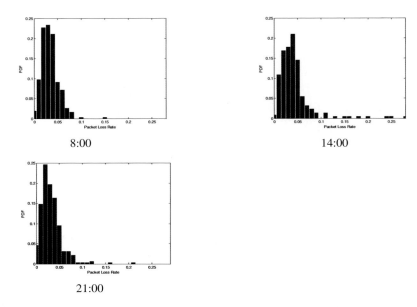

Fig. 6. The distribution of the packet error rate at 10-20 m

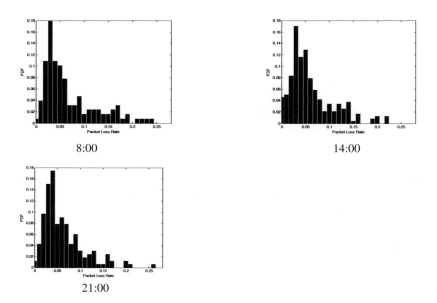

Fig. 7. The distribution of the packet error rate at 20-30 m

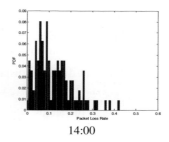
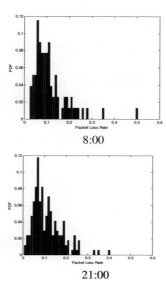

8:00   14:00

21:00

Fig. 8. The distribution of the packet error rate at 30-40 m

1) The node–to-node distance impacts the link quality of VANET. The packet error rates will increase with distance between nodes, and the range of PER also increase with distance. If the distance is less than 20 meters, the packet loss rate will be under 0.05, and it will become larger obviously if the distance is more than 30.
2) The link qualities at different times are more or less similar.. The packet error rates with same distance almost follow the same distribution, no matter the traffic is heavy (8:00-14:00), or light (21:00).

Based on measurements and estimations, the PDF of packet error rates at different distances in the urban area were obtained. Due to the different distribution rules of PDF, we use GMM (Gaussian Mixture Model) [40] to model the packet error rates at different distances. GMM is a linear combination of multiple Gaussian distributions, when the Gaussian mixture degree is large enough, it can be used to approximate any distribution.

The reasons of choosing GMM to model the PDF of packet error rates is described below:

1) Accuracy. Many factors affect the link quality of VANET, when the distance between two vehicles are father than 20 meters; the PDF of packet error rate is very different from Gaussian distribution, while GMM can predict the distribution of packet error rate.

2) Manageable. As to a known PDF of packet error rates, there is always a character to converge the GMM.
3) Scalability. One can use GMM to describe the PDF of packet error rate at different distances.

In this passage, at different distances, we build the GMM of PDF of packet error rates, like $G_d(x) = \sum_{i=1}^{k} \omega_i N(x | \mu_i, \xi_i)$, and d is the distances between two vehicles, k is the number of the single Gaussian, ω_i is the weight of the i-th Gaussian, μ_i is the mean of each parameter, ξ_i is the covariance matrix. All the ω_i are positive and the sum of ω_i is 1. ω_i, μ_i, ξ_i are parameters to be calculated.

The Chi-square[36] test is used to prove that the packet error rates at different times follow the same probability distribution if the distance between nodes within a certain range. If the theoretical distribution of packet error rates is $F_0(x)$, and the observed distribution is $F(x)$, to the objective is to prove $F(x) = F_0(x)$.

The N observed PDF samples are divided into k cells, Let the theoretic frequencies of various cells be p_1, p_2,......, p_k, with observed frequencies of each cell v_1, v_2,......, v_i. The quantity

$$\chi^2 = \sum_{i=0}^{k} \frac{v_i^2 - N*p_i}{N*p_i}$$

Where

χ^2 = the test statistic that asymptotically approaches a χ^2 distribution
v_i^2 = an observed frequency
$N*p_i$ = an expected (theoretical) frequency
N = the number of possible outcomes of each event.

The chi-square statistic can then be used to calculate a r-value by comparing the value of the statistic to a chi-square distribution. The number of degrees of freedom is equal to the number of cells k, minus the reduction in degrees of freedom r.

Then, we choose the level of significance α to be 0.05, if α, the assumption is correct, however the assumption is incorrect.

According to the chi-square test described above, we find that the packet error rates with the same distance at different times follow the same distribution. With the distance of 10m-20m, the PDF of packet error rates is the single-Gaussian distribution within the allowable error range, the distribution at different time is shown in Fig 9, With the distance of 20m-30m, the PDF of packet error rates can be modeled by Gaussian mixture model (GMM) with k equals to 2. the distribution of the PDF of packet error rates of distance 20m-30m is shown in Fig 10. With the distance of 30m-40m, the PDF of packet error rates can be described by GMM (Gaussian Mixture Model) with k equals 4, the distribution of the PDF of packet error rates of distance 30m-40m is shown in Fig 11.

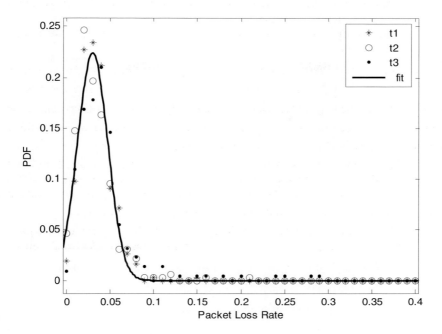

Fig. 9. The PDF of the distance range being 10-20 m

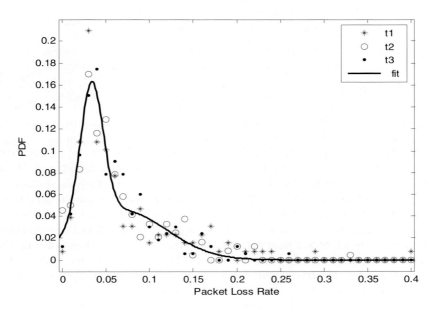

Fig. 10. The PDF of the distance range being 20-30 m

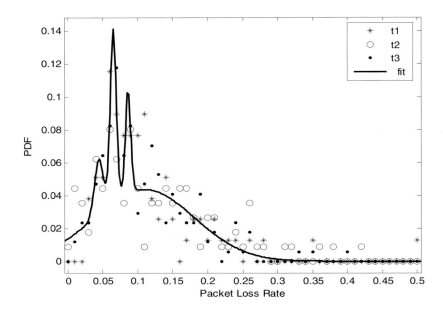

Fig. 11. The PDF of the distance range being 30-40 m

4 Estimation of Packet Error Rate in VANET

Network measurement,which is according to certain methods and techniques, using software packages or hardware tools to test network performance.

The main purpose of network measurement is to measure the network transfer delay, packet error rate and other parameters through testing information packets while the network in normal operation. Network measurement, depending on whether to send measurement probe packets can be divided into active measurement and passive measurement.

Active measurement produces the probe packets and inject into the network at the selected measurement point, and by measuring the transmission of probe packets to analyze network performance.

Passive Measurement refers to sniff the packets on the link or device, such as routers, switches and other equipments. Because it analyzes the network performance by gathering the existing network packets, it does not impose any extra traffic into the network.

Active measurement and passive measurement has its own advantages and disadvantages, but for the different performance parameters, the active measurement and passive measurement also has its own use. Therefore, the combination of active measurement and passive are the trends of development of network performance measure.

We propose two methods to estimate the packet error here. One method is a kind of passive measurement method which is based on PLM; the other is RPEE that is kind of active measurement method. These two methods can be used for different objectives, packet error estimation based on PLM can estimate wireless link quality when data traffic is large, and it does not inject load into network. RPEE can be used to estimate packet error rate while some important message need to be sent, such accidence and safety message.

4.1 Estimate Per Using Plm

4.1.1 PLM Model

In the Gilbert-Elliott model [37], the produced burst length and gap length follow a geometric distribution. However, these statistics of burst lengths and gap lengths of the measured data cannot be met via a geometric distribution. A two-parameter Weibull distribution can perfectly fit the measured distribution of the burst lengths and gap lengths. The reason for using a Weibull distribution in modeling gap and burst statistics is the high flexibility of the two-parameter Weibull distribution. By proper adjustment of the shape parameter, exponential, Rayleigh, normal- and even other distributions can be met or approximated [38].

The two-parameter Weibull CDF is given by $F(x) = 1 - e^{-(\frac{x}{a})^b}$, where a and b are scale and shape parameters, respectively. $F_{sg}(m)$ is the Weibull CDF modeling the CDF of burst lengths conditioned on the previous short gaps. The Weibull CDF modeling the CDF of burstlengths conditioned on the previous long gaps is defined as $F_{lg}(m)$. Similarly, The Weibull CDF modeling the CDF of gap lengths conditioned on the previous short bursts and long bursts are defined as $F_{sb}(m)$ and $F_{lb}(m)$, respectively. The reason for using a Weibull distribution in modeling burst and gap statistics is the high flexibility of the two-parameter Weibull distribution [39].

Fig 12 shows that these statistics of burst lengths and gap lengths can all be perfectly matched by Weibull distributions. These parameters of the Weibull distributions are shown in Table 3.

Table 3. Parameters of the Weibull distributions

	a	b
$F_{sg}(m)$	0.1683	0.3655
$F_{lg}(m)$	0.3793	0.7678
$F_{sb}(m)$	4.3172	0.6274
$F_{lb}(m)$	3.0736	0.6415

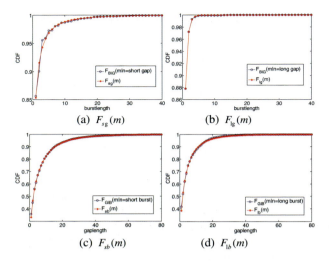

Fig. 12. Measurement vs. simulation of the burst lengths and gap lengths in dynamic scenarios

Following the explained method of describing adequate statistics for burst lengths and gap lengths we arrive at PLM model (four-state Markov model) as shown in Fig 13, where in 'short-gap' state, short gaps are generated; in 'long-gap' state, long gaps are generated; in 'short-burst' state, short bursts are generated and in 'long-burst' state ,long bursts are generated. The corresponding transition probabilities are the conditional probabilities out of Table 4. After entering a certain state, the burst length or gap length is calculated via a Weibull distributed random number according to the previous state and the current state. For example, we assume that the previous state is 'short-gap' state, if the current state is 'short-burst' state, the burstlength is calculated via a Weibull $F_{sg}(m)$ distributed random number with a probability of $p_{s_g s_b}$ to produce short bursts.

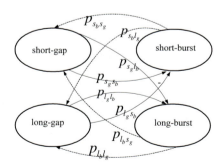

Fig. 13. The PLM model

Table 4. The transition probabilities of the PLM model

$$p_{s_g s_b} = F_{sg}(3) \qquad p_{s_g l_b} = 1 - F_{sg}(3)$$

$$p_{l_g s_b} = F_{lg}(3) \qquad p_{l_g l_b} = 1 - F_{lg}(3)$$

$$p_{s_b s_g} = F_{sb}(2) \qquad p_{s_b l_g} = 1 - F_{sb}(2)$$

$$p_{l_b s_g} = F_{lb}(2) \qquad p_{l_b l_g} = 1 - F_{lb}(2)$$

Throughout this document the following notation is used to describe the error process. An erroneously received packet is indicated by '1' while '0' means error-free transmission. A positive integer in the exponent determines the number of consecutive erroneous or error-free packets. We can see in Fig 14 that the sequence can be written as '15051'.

The conditional link error probability $\Pr(1|1^n 0^m)$ can be expressed by

$$\Pr(1|1^n 0^m) = \frac{\Pr(1^n 0^m 1)}{\Pr(1^n 0^m)} = \frac{\Pr(0^m 1|1^n)\Pr(1^n)}{\Pr(0^m | 1^n)\Pr(1^n)} = \frac{\Pr(0^m 1|1^n)}{\Pr(0^m | 1^n)}$$

If $n \leq 3$, $\Pr(0^m | 1^n) = \sum_{k=m}^{\infty} \Pr(0^k 1|1^n) = 1 - F_{sb}(m) + \Pr(0^m 1|1^n)$

If $n > 3$, $\Pr(0^m | 1^n) = \sum_{k=m}^{\infty} \Pr(0^k 1|1^n) = 1 - F_{lb}(m) + \Pr(0^m 1|1^n)$

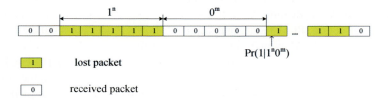

Fig. 14. Illustration of the prediction of burst length

The performance of such an error estimator can be seen in Fig 15 We estimate the measured conditional link error probability $\Pr(1|1^n 0^m)$ by according to table3 and with Weibull scale parameter a and shape parameter b. It can be observed that the estimator meets the measured conditional link error probability properly.

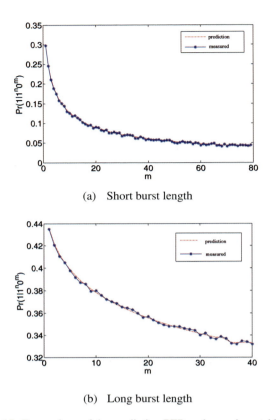

(a) Short burst length

(b) Long burst length

Fig. 15. Comparison of the prediction PER and experimental PER

4.1.2 Simulation Results

In this subsection, PLM model and Gilbert-Elliott's model are simulated. The parameters of the PLM model are taken from Table 3 and Table 4. The Gilbert-Elliott's model parameters can be calculated by using the mean burst length (1.31) and the mean gap length (6.72). Fig 16 shows a comparison between the simulated data and the measured data, in terms of the cumulative distribution of the bursts and gaps. We can see that the PLM model is able to describe the measured statistics quite well.

Table 5 shows a comparison between PLM model and Gilbert-Elliott's Model, which is widely used for digital wireless channel modeling. The results in the table imply that PLM model reaches a more precise characterization of the burst and gap properties in VANET than the Gilbert-Elliott Model. Therefore, we conclude that the presented PLM model with a proper setting of parameters may serve to well characterize the bursts and gaps nature of VANET wireless channels.

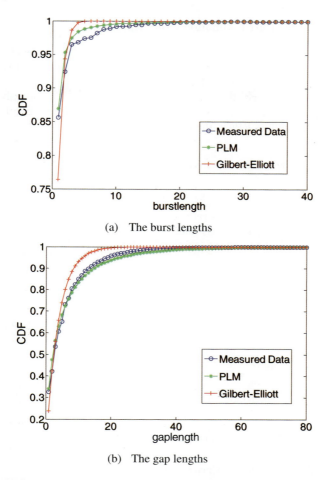

(a) The burst lengths

(b) The gap lengths

Fig. 16. Comparison of the burst lengths and gap lengths measured, PLM model and Gilbert-Elliott Model, dynamic scenario

Table 5. comparison between PLM Model and Gilbert-Elliott's Model

	Measured Data	PLM Model	Gilbert-Elliott's model
Packet Error Rate	0.1630	0.1676	0.2359
Average burst length	1.31	1.31	1.31
Std. of burst length	0.017	0.016	0.006
Average gap length	6.72	6.52	4.24
Std. of gap length	0.132	0.098	0.033

4.2 Packet Error Rate Estimation Using Rpee

In Section 3, it has been proved that the PDFs of packet error rate at different times is the same if the node-to-node distance ranges are the same. In addition, the PDF can be described by a GMM model.

Based on lots of measurements, we use GMM to present probability density model which describe packet error rates at different distances. Fig 17 shows the principle of the method RPEE, the two main steps are as below.

Step1
Building the model of packet loss rate basing on the samples obtained through experiments.

Step2
Estimating the link packet loss rate according to the model built in step1 and the real-time detection of packet loss rate.

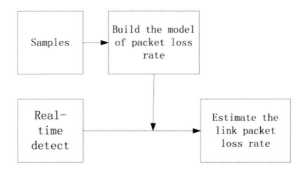

Fig. 17. Principle of estimating the packet error rates

4.2.1 Modeling the Packet Error Rates

EM (Expectation Maximization) algorithm [41] was used to calculate the parameters of GMM. EM is an iterative method which alternates between performing an expectation (E) step and a maximization (M) step,as shown in Fig 18. E step computes an expectation of the log likelihood with respect to the current estimate of the distribution for the latent variables. M step computes the parameters which maximize the expected log likelihood find on the E step. These parameters are then used to determine the distribution of the latent variables in the next E step.

Given ω_i, μ_i and ξ_i, calculating the statistical probability of x in k-Gaussian, then use the statistical probability to calculate the expectations for each Gaussian, then the expectations maximize the parameters in GMM, obtain the value of ω_i, μ_i, ξ_i.

E-step:

$$h_j^{(k)}(x_n) = \frac{\omega_{ij}^{(k)} N_{ij}^{(k)}(x_n | \mu_i, \xi_i)}{\sum_{j=1}^{k} \omega_{ij}^{(k)} N_{ij}^{(k)}(x_n | \mu_i, \xi_i)}$$

Estimation of Packet Error Rate at Wireless Link of VANET

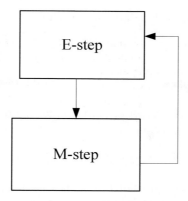

Fig. 18. EM algorithm

M-step:

$$\omega_{ij}^{(k)} = \frac{1}{N}\sum_{n=1}^{N} h_j^{(k)}(x_n)$$

$$\mu_{ij}^{(k+1)} = \frac{\sum_{n=1}^{N} h_{ij}^{(k)}(x_n)x_n}{\sum_{n=1}^{N} h_{ij}^{(k)}(x_n)}$$

$$\xi_{ij}^{(k+1)} = \frac{\sum_{n=1}^{N} h_{ij}^{(k)}(x_n)[x_n - \mu_{ij}^{(k+1)}][x_n - \mu_{ij}^{(k+1)}]^T}{\sum_{n=1}^{N} h_j^{(k)}(x_n)x_n}$$

Repeating the above steps until the function convergence, one can obtain parameters of Gaussian function.

In order to estimate the real-time packet error rates, by sending a few probe packets, we get the current distance and the packet error rate at this distance. Based on the GMM at the certain distance, using RPEE, real-time packet error rate can be obtained.

The process of real-time estimation is described below.

Suppose two vehicles A and B are moving in the same direction, A sends N packets during time t (t<1s), B receives N_{recv} broadcasting packets, then the packet error rate of link A → B is $x(t)$, $x(t) = \frac{N_{recv}}{N} \times 100\%$.

$x(t)$ is measurement value of the packet error rate during time t (t<1s), in order to get the packet error rate $\max\{N_{MAP}(x_t | \mu_i, \xi_i)\}$ in the current second, we should get the distribution $N_{MAP}(x_t | \mu_i, \xi_i)$ which $x(t)$ follows, and MAP was used to achieve this.

A MAP estimate is a mode of the posterior distribution. The MAP can be used to obtain a point estimate of an unobserved quantity on the basis of empirical data.

Based on MAP, $N_{MAP}(x_t | \mu_i, \xi_i) = \arg\max_N \frac{f(x_t | N)g(N)}{\int f(x_t | N')g(N')dN'}$, N is a Gaussian distribution function.

$$N_{MAP}(x_t \mid \mu_i, \xi_i) = \arg\max_N \frac{f(x_t \mid N) g(N)}{\int f(x_t \mid N') g(N') dN'}$$

$$\Rightarrow \mu_{MAP} = \arg\max_\mu \frac{f(x_t \mid \mu_j) g(\mu_j)}{\sum_{i=1}^{k} f(x_t \mid \mu_i) g(\mu_i)}$$

where $g(\mu_i)$ represents the weight ω_i of i-th Gaussian distribution in GMM, and $f(x_t \mid \mu_i)$ represents the probability of $x(t)$ belong to $N(x_t \mid \mu_i, \xi_i)$.

$$\hat{x}(t) = \mu_i - \xi_i \ , (\ x_t \subseteq [0, \infty]\ ).$$

From Fig 6, 7 and 8, the distribution of PDF has a long tail, it is generated mainly by the Gaussian distribution with large variance. In this subsection, we estimate the PDF of packet error rates like that : If the variance of the Gaussian distribution is small, we consider the parameter μ_i in $\max\{N_{MAP}(x_t \mid \mu_i, \xi_i)\}$ as estimating packet error rate $\hat{x}(t)$ in current second. If the variance of the Gaussian distribution is large, and $x(t) \leq \mu_i - \xi_i$ or $x(t) \geq \mu_i + \xi_i$, then we consider $\hat{x}(t) = \mu_i - \xi_i$ or $\hat{x}(t) = \mu_i - \xi_i$. Fig 19 shows the process.

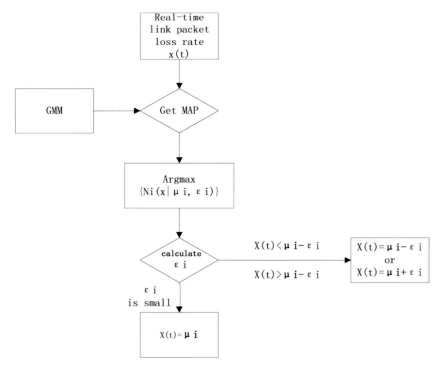

Fig. 19. Estimation of packet error rate based on MAP

4.2.2 Experiment Results

Lots of experiments were carried under the same scenario, datasets of packet error rates were obtained. We classified these datasets according to node-to-node distance range, and got PLR sub dataset of 10-20 meters, 20-30 meters, 30-40 meters, 40-50 meters. Using half of sub datasets, we obtain the PDF of packet error rates at specific distance range, then get the GMM $G_d(x) = \sum_{i=1}^{k} \omega_{id} N(x | \mu_{id}, \xi_{id})$, then use rest of data to test RPEE.

We get the PDF and GMM of packet error rates at distances range 10-20 meters, 20-30 meters, 30-40 meters, 40-50 meters after 10 times experiments. As shown in Fig 20. In this subsection, 20 meters represents 10-20 meters, 30 meters represents 20-30 meters, 40 meters represents 30-40 meters, 50 meters represents 40-50 meters.

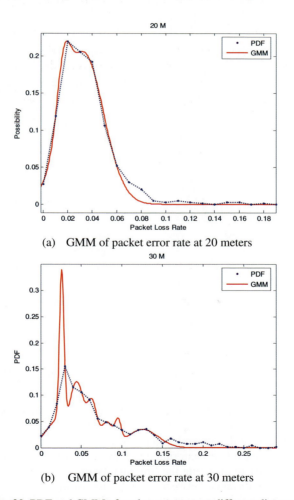

(a) GMM of packet error rate at 20 meters

(b) GMM of packet error rate at 30 meters

Fig. 20. PDF and GMM of packet error rate at different distances

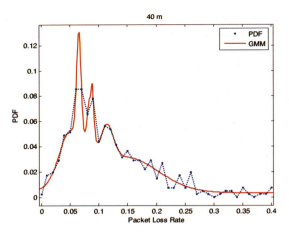

(c) GMM of packet error rate at 40 meters

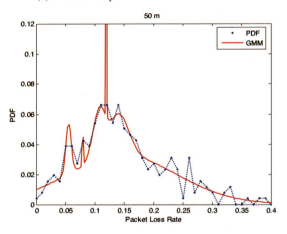

(d) GMM of packet error rate at 50 meters

Fig. 20. (*continued*)

According to $p(x)$, GMM $G_d(x) = \sum_{i=1}^{k} \omega_i N(x_t | \mu_i, \xi_i)$ can be get using EM algorithm.

Fig 20 shows probability distribution of packet error rate in VANET can be fitted by GMM, and the longer the distance, the larger the k value of GMM, as shown in Fig 20(c,d). It demonstrates that many factors affect packet error rates of VANET. In GMM, each Gaussian function represents a factor, and they are independent of each other in affecting the link quality of VANET and PDF of packet error rate. When the distance between vehicles increase, the factors affecting the link quality will become more complex, as a result, the value of k will increase.

Estimation of Packet Error Rate at Wireless Link of VANET

Unlike the statistical data sets, we abstracted another 10 sets of packet error rates, observed the first 45 beacon packets in one second, we obtained the packet error rate $x(t)$, according to GMM, we can obtained the Gaussian distribution $N_{MAP}(x_i | \mu_i, \xi_i)$ which $x(t)$ follows, and then estimate the packet error rate $x(t)$. The results are shown in Fig 21.

Fig 21 is estimation results of real-time packet error rates. Fig 21 illustrates the RPEE algorithm can estimate packet error rate well. Packet error rate which RPEE estimate is very close to the reference value, and the errors maintain at low level.

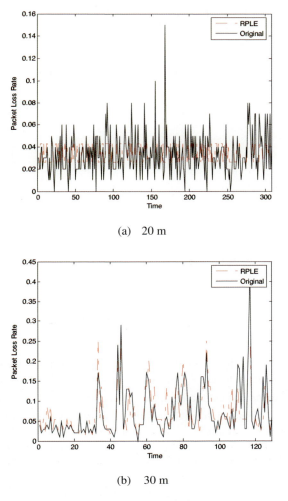

Fig. 21. Estimation results of real-time packet error rate

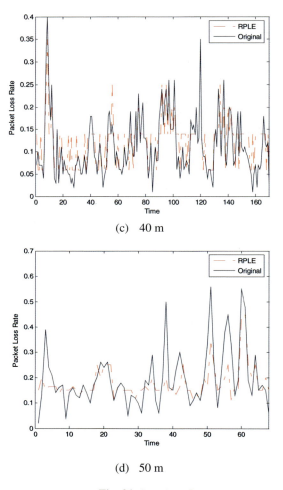

(c) 40 m

(d) 50 m

Fig. 21. (*continued*)

In order to quantify the estimation of the packet error rate , we define error on average ε, the relative mean $\bar{\varepsilon}$, mean square standard deviation σ_{MSE}. ε shows how the packet error rate estimated approach the real value . σ_{MSE} reflects discretion and stability .We normalized the results , so it shows the relative value.

$$\varepsilon = \frac{1}{N}\sum_{t=1}^{N}|\hat{x}(t) - x_R(t)| \qquad (3)$$

$$\bar{\varepsilon} = \frac{\frac{1}{N}\sum_{i=1}^{N}|\hat{x}(t) - x_R(t)|}{\frac{1}{N}\sum_{i=1}^{N}x_R(t)} \qquad (4)$$

$$\sigma_{MSE} = \sqrt{\frac{1}{N}\sum_{i=1}^{N}[\hat{x}(t)-x_R(t)-\varepsilon]^2} \quad (5)$$

To compare the performance of the algorithm, we compare RPEE with ETX (maximum transmission times) which is used widely in wireless networks. d_f and d_r represent transmission accuracy of the uplink and downlink. ETX uses 20 broadcast packets which contents 134 bytes to detect link per second, but the link of vehicle ad-hoc network changes rapidly, so 20 broadcast packets is not enough to get accurate in face, so we use 60 broadcast packets per second.

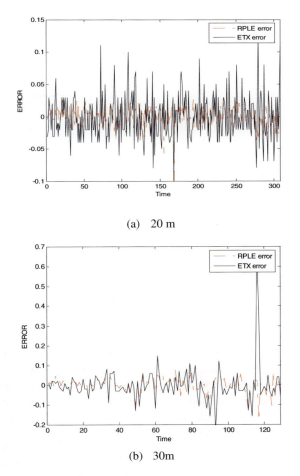

(a) 20 m

(b) 30m

Fig. 22. The packet error rate estimation error based on the maximum a posteriori

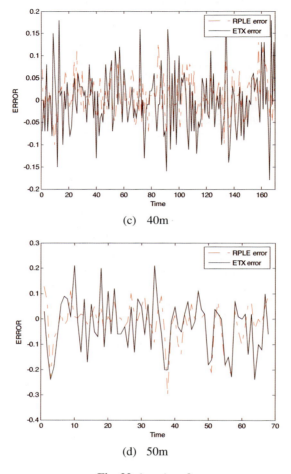

Fig. 22. (*continued*)

Table 6 shows parameters of the packet error rate error. As shown in the figure, RPEE is more accurate. In particular, when distance between nodes is 10 meters to 30 meters, absolute error ε of RPEE is -0.05 ~ 0.05, the relative error $\bar{\varepsilon}$ is under about 30%. With the ETX method, if the distance is 50 meters, the relative error $\bar{\varepsilon}$ will be 45%, and it goes as high as 79% if the distance is 20 meters. RPEE more stable and closer to the real packet error rate for its σ_{MSE} is lower and less discrete, at the same time, we can use fewer probe packets with this algorithm. However, when the node-to-node distance is too large, both of the two methods would have a larger bias, further observation and study will be taken.

Table 6. Parameters about the packet error rate estimation error based on MAP

	20 meters		30 meters		40 meters		50 meters	
	RPEE	ETX	RPEE	ETX	RPEE	ETX	RPEE	ETX
ε	0.0112	0.027	0.0269	0.0499	0.0366	0.0528	0.0595	0.0894
$\overline{\varepsilon}$	0.3307	0.7998	0.3736	0.6932	0.3286	0.4744	0.3036	0.4565
σ_{MSE}	0.0152	0.0338	0.0406	0.088	0.0453	0.0683	0.0861	0.1086

5 Conclusions

In this chapter, we have shown the analysis of the packet error characteristics of the VANET channel based on measurement in real urban road. It is found that the burstlengths and gaplengths statistics are almost similar in same scenarios and the correlation properties between gaps and bursts and it is shown that they are statistics dependence. Additionally we analyze the packet error rates of the same scenario have the same distribution if the distance is the same, no matter traffic are heavy or light. Following the measured statistics we have proposed PLM model (a four-state markov model) which is capable of meeting the statistics of burstlengths and gaplengths properly. We also propose RPEE (Real-time Packet Error Estimation) model to estimate packet error rates of VANET. It can be used to grasp the overall characteristics of the link, and support the data transmission control and routing algorithms in network protocols.

References

[1] Fan, L., Yu, W.: Routing in Vehicular Ad Hoc Networks: A Survey. Vehicular Technology Magazine 2(2), 12–22 (2007)
[2] Hartenstein, H., Laberteaux, K.P.: A Tutorial Survey on Vehicular Ad Hoc Networks. Communications Magazine 46(6), 164–171 (2008)
[3] http://pdos.csail.mit.edu/~rtm/papers/carnet00-abstract.html
[4] http://www.cartalk2000.net/
[5] http://cartel.csail.mit.edu/doku.php
[6] http://discolab.rutgers.edu/traffic/index.htm
[7] http://www.aqualab.cs.northwestern.edu/projects/C3.html
[8] http://www.winlab.rutgers.edu/pub/docs/focus/vehicular-networks.html
[9] Bouassida, M.S., Shawky, M.: On the Congestion Control within VANET. Wireless Days, 1–5 (November 24-27, 2008)
[10] IEEE WG, IEEE 802.11p/D2.01, Draft Amendement to Part 11: Wireless Medium Access Control (MAC) and Physical Layer (PHY) specifications: Wireless Access in Vehicular Environments (March 2007)

[11] Choi, N., Choi, S., Seok, Y., Kwon, T., Choi, Y.: A Solicitation-based IEEE 802.11p MAC Protocol for Roadside to Vehicular Networks. Mobile Networking for Vehicular Environments, 91–96 (May 2007)
[12] Perkins, C.E., Belding-Royer, E.M., Das, S.: Ad Hoc On Demand Distance Vector (AODV) Routing, IETF Request For Comments 3561 (2003)
[13] Johnson, D.B., Maltz, D.A., Hu, Y.-C.: The Dynamic Source Routing Protocol for Mobile Ad Hoc Networks (DSR), Internet Draft: <draftietf-manetdsr-10.txt>, July 19 (2004)
[14] Lochert, C., Hartenstein, H., Tian, J., Herrmann, D., Füßler, H., Mauve, M.: A Routing Strategy for Vehicular Ad Hoc Networks in City Environments. In: IEEE Intelligent Vehicles Symposium (IV 2003), Columbus, OH, USA, June 2003, vol. 1, pp. 156–16 (2003)
[15] Jerbi, M., Senouci, S.-M., Meraihi, R., Ghamri-Doudane, Y.: An Improved Vehicular Ad Hoc Routing Protocol for City Environments. In: ICC 2007, June 2007, pp. 3972–3979 (2007)
[16] Lochert, C., et al.: A routing strategy for vehicular ad hoc networks in city environments. In: IVS 2003, June 2003, pp. 156–161 (2003)
[17] Zhang, M., Wolff, R.: Routing Protocols for Vehicular Ad Hoc Networks in Rural Areas. Communications Magazine 46(11), 126–131 (2008)
[18] Schmitz, R., Leiggener, A., Festag, A., Eggert, L., Effelsberg, W.: Analysis of Path Characteristics and Transport Protocol Design in Vehicular Ad Hoc Networks. In: Proc. of the 63. IEEE Semiannual Vehicular Technology Conference (VTC-Spring), pp. 528–532 (2006)
[19] Sardar, B., Chand, P., Saha, D.: A novel version of Wireless TCP for Vehicular On-Board IP Networks. In: Proc. of Vehicular Technology Conference, 2006 Spring, pp. 876–880. IEEE Press, Grand Hyatt Melbourne (2006)
[20] Khorashadi, B., Chen, A., Ghosal, D.: Impact of Transmission Power on the Performance of UDP in Vehicular Ad Hoc Networks. In: ICC 2007, June 2007, pp. 3698–3703 (2007)
[21] Chen, A., Khorashadi, B., Ghosal, D., Chuah, C.-N.: Impact of Transmission Power on TCP Performance in Vehicular Ad Hoc Networks. In: WONS 2007, January 2007, pp. 65–71 (2007)
[22] Bouassida, M.S., Shawky, M.: On the Congestion Control within VANET. In: On the congestion control within VANET, WD 2008, November 2008, pp. 1–5 (2008)
[23] Torrent-Moreno, M., Santi, P., Hartenstein, H.: Fair Sharing of Bandwidth in VANETs. In: Proceedings of the second ACM International Workshop on Vehicular Ad Hoc Networks (VANET), pp. 49–58 (2005)
[24] Wischhof, L., Rohling, H.: Congestion control in vehicular ad hoc networks. In: Vehicular Electronics and Safety, October 2005, pp. 58–63 (2005)
[25] Sofra, N., Gkelias, A., Leung, K.K.: Link Residual-Time Estimation for VANET Cross-Layer Design. In: IWCLD 2009, pp. 1–5 (2009)
[26] Hsin-Mu, T., Wisitpongphan, N., Tonguz, O.K.: Link-quality aware ad hoc on-demand distance vector routing protocol. In: 1st Int. Symp. On Wireless Pervasive Computing, p. 6 (2006)
[27] Chang, R., Leu, S.: Long-lived path routing with received signal strength for ad hoc networks. In: 1st International Symposium on Wireless Pervasive Computing (January 2006)
[28] Kim, S.-C.: Analysis of Link Error in Reducing Broadcasting Redundancy of MANET AAPs. In: MSN 2008, December 2008, pp. 250–257 (2008)

[29] Wang B.-Z; Wang Y.-P; Wang W; Lou R.-Y. Inference of Wireless Link Performance in MANET, Convergence Information Technology, pp. 1481–1487 (November 2007)
[30] Babich, F., Lombardi, G.: A Markov model for the mobile propagation channel. IEEE Trans. on Veh. Technol. 49, 63–73 (2000)
[31] Gilbert, E.N.: Capacity of a burst-noise channel. The Bell System Technical Journal (39), 1253–1265 (September 1960)
[32] Elliott, E.O.: Estimates of Error Rates for Codes on Burst-Noise Channels. The Bell Systems Technical Journal 42, 1977–1997 (1963)
[33] Fritchman, B.D.: A Binary Channel Characterization Using Partitioned Markov Chains. IEEE Trans. Information Theory 13(2), 221–227 (1967)
[34] Nguyen, G.T., Noble, B.: A Trace-Based Approach for Modeling Wireless Channel Behavior. In: Proc. the 1996 Winter Simulation Conf., pp. 597–604 (1996)
[35] Chengxiang, W., Dayong, X.: A Study on Burst Error Statistics and Error Modeling for MB-OFDM UWB Systems. In: Ultra Wideband Systems, Technologies and Applications, pp. 211–248 (2006)
[36] http://en.wikipedia.org/wiki/Pearson%27s_chi-square_test
[37] Elliott, E.O.: Estimates of Error Rates for Codes on Burst-Noise Channels. The Bell Systems Technical Journal 42, 1977–1997 (1963)
[38] Karner, W., Nemethova, O., Rupp, M.: Link Error Prediction in Wireless Communication Systems with Quality Based Power Control. In: Proc. IEEE Int. Conf. Comm. (ICC), Glasgow, Scotland (June 2007)
[39] Karner, W., Rupp, M.: Measurement-Based Analysis and Modelling of UMTS DCH Error Characteristics for Static Scenarios. In: Proc. 8th Int. Symp. DSP and Comm. Systems (DSPCS), Sunshine Coast, Australia (December 2005)
[40] Subasingha, S., Murthi, M.N., Andersen, S.V.: On GMM Kalman predictive coding of LSFS for packet loss. Acoustics, Speech and Signal Processing, 4105–4108 (2009)
[41] Bilmes, J.A.: A gentle tutorial of the EM algorithm and its application to parameter estimation for Gaussian mixture and hidden Markov models[R]. ICSI TR-97-021, Department of Electrical Engineering and Computing Science, U.C. Berkeley, USA (1998)
[42] Shravan, R., Arunesh, M., Dheeraj, A., Sharad, S., Suman, B.: Diagnosing Wireless Packet Losses in 802.11:Separating Collision from Weak Signal. In: Infocomm 2008 (2008)
[43] Yun, J.-H., Seo, S.-W.: Collision Detection based on RE Energy Duration in IEEE 802.11 Wireless LAN. In: Comsware (2006)
[44] Miu, A., Balakrishnan, H., Koksal, C.E.: Improving loss resilience with multi-radio diversity in wireless networks. In: ACM MOBICOM (2005)
[45] Cheng, Y., Bellardo, J., Benkö, P., Snoeren, A., Voelker, G., Savage, S.: Jigsaw: solving the puzzle of enterprise 802.11 analysis. In: SIGCOMM 2006 (2006)

Author Index

Al-Ameen, Mahdi Nasrullah 43
Al-Shamma'a, A.I. 75

Boonyanant, Phakphoom 199
Boyle, David 1
Brenk, D. 125

Charoenkul, Sarot 199
Chinrungrueng, Jatuporn 199
Cionca, Victor 1
Cocco, Michele 25

Davidson, J. 221
Dumnin, Songphon 199

Essel, J. 125

Gaddam, A. 307
Gupta, G. Sen 307
Guru, Siddeswara Mayura 283

Heidrich, J. 125
Huang, Dongliang 101

Intarapanich, Apichart 199

Jiang, Hao 329

Kaewkamnerd, Saowaluck 199
Kittipiyakul, Somphong 199
Knight, C. 221

Lamont, Louise 177
Lee, Kang B. 243
Leung, Henry 101

Maheswararajah, Suhinthan 283
Manjunatha, P. 273
Markert, Juergen 25
Mason, A. 75
Mukhopadhyay, S.C. 307

Namerikawa, Toru 151
Newe, Thomas 1

Pongthornseri, Ronachai 199

Samphanyuth, Supat 199
Shaw, A. 75
Sieber, Arne 25
Song, Eugene Y. 243
Srividya, A. 273
Sunantachaikul, Udomporn 199

Takeda, Takashi 151

Verma, A.K. 273

Wagner, Matthias F. 25
Wang, Lin 329
Weigel, R. 125
Woegerer, Christian 25

Xu, Jun 329

Yang, Yang 329

Zhou, Yifeng 177

Printing: Ten Brink, Meppel, The Netherlands
Binding: Stürtz, Würzburg, Germany